Emergent Possibilities for Global Sustainability

It must be acknowledged that any solutions to anthropogenic global climate change (GCC) are interdependent and ultimately inseparable from both its causes and consequences. As a result, limited analyses must be abandoned in favour of intersectional theories and practices.

Emergent Possibilities for Global Sustainability is an interdisciplinary collection which addresses global climate change and sustainability by engaging with the issues of race, class and gender through an intersectional lens. The book challenges readers to foster new theoretical and practical linkages and to think beyond the traditional, and oftentimes reductionist, environmental science frame by examining issues within their turbulent political, cultural and personal landscapes. Through a variety of media and writing styles, this collection is unique in its presentation of a complex and integrated analysis of global climate change and its implications. Its companion book, *Systemic Crises of Global Climate Change*, addresses the social and ecological urgency surrounding climate change and the need to use intersectionality in both theory and practice.

This book is a valuable resource for academics, researchers and both undergraduate and post-graduate students in the areas of environmental studies, sustainability, gender studies and international studies, as well as those seeking a more intersectional analysis of GCC.

Phoebe Godfrey is an Assistant Professor-in-Residence at UCONN in sociology. She co-founded the non-profit CLiCK, in Willimantic, Connecticut, dedicated to a local sustainable food system.

Denise Torres is a doctoral candidate at the Graduate Center, City University of New York. The unifying theme of her work and publications is the authentic inclusion of silenced and marginalized groups in the systems that affect them.

Routledge Advances in Climate Change Research

A full list of titles in this series is availabe at: www.routledge.com/series/RACCR

Community Governance and Citizen-Driven Initiatives in Climate Change Mitigation
Edited by Jens Hoff and Quentin Gausset

The Two Degrees Dangerous Limit for Climate Change
Public understanding and decision making
Christopher Shaw

Ageing, Wellbeing and Climate Change in the Arctic
An interdisciplinary analysis
Edited by Päivi Naskali, Marjaana Seppänen, Shahnaj Begum

China Confronts Climate Change
A bottom-up perspective
Peter H. Koehn

Community Action and Climate Change
Jennifer Kent

Reimagining Climate Change
Edited by Paul Wapner and Hilal Elver

Climate Change and the Anthropos
Planet, people and places
Linda Connor

Systemic Crises of Global Climate Change
Intersections of race, class and gender
Edited by Phoebe Godfrey and Denise Torres

Urban Poverty and Climate Change
Life in the slums of Asia, Africa and Latin America
Edited by Manoj Roy, Sally Cawood, Michaela Hordijk and David Hulme

Strategies for Rapid Climate Mitigation
Wartime mobilisation as a model for action?
Laurence L. Delina

Emergent Possibilities for Global Sustainability

Intersections of race, class and gender

Edited by
Phoebe Godfrey and Denise Torres

LONDON AND NEW YORK

First published 2016 by Routledge

2 Park Square, Milton Park, Abingdon, Oxfordshire OX14 4RN
711 Third Avenue, New York, NY 10017

Routledge is an imprint of the Taylor & Francis Group, an informa business

First issued in paperback 2017

British Library Cataloguing-in-Publication Data
A catalogue record for this book is available from the British Library

Library of Congress Cataloging-in-Publication Data
Names: Godfrey, Phoebe (Phoebe Christina), editor. | Torres, Denise (Sociologist), editor.
Title: Emergent possibilities for global sustainability : intersections of race, class and gender / edited by Phoebe Godfrey and Denise Torres.
Description: Abingdon, Oxon ; New York, NY : Routledge, 2016.
Identifiers: LCCN 2016001739 | ISBN 9781138830059 (hb) | ISBN 9781315737478 (ebook)
Subjects: LCSH: Environmental sociology. | Sustainability—Social aspects. | Climatic changes—Social aspects.
Classification: LCC GE195 .E528 2016 | DDC 338.9/27—dc23LC record available at http://lccn.loc.gov/2016001739

ISBN: 978 1 138 83005 9 (hbk)
ISBN: 978-0-8153-6456-6 (pbk)

Typeset in Goudy
by FiSH Books Ltd, Enfield

This book is dedicated to:

Phoebe's sons, niece and nephews and to her students past, present and future and to the 'cultural creatives' of the world—may we all continue to do our part in the great transformations.

Denise's interns, students, and mentees—the universe gives us what we need.

'To achieve fair and sustainable futures, it will be essential to dissolve old hegemonic stereotypes like "womanhood" and "indigeneity" and nourish the seeds of systemic change with marginal wisdoms. With honesty, acuity, passion, and hope, an international groundswell of authors calls theory back to praxis through inspirational verse and the joys of movement building.'

Ariel Salleh, author of *Ecofeminism as Politics*

'In *Emergent Possibilities for Global Sustainability*, Godfrey and Torres creatively link intersectionality theory to the concept of *just sustainabilities* to provide both a framework and new tools to analyze the dynamics of the different social and material worlds as evidenced by the diverse voices of their contributors.'

Julian Agyeman, Professor of Urban and Environmental Policy and Planning, Tufts University, USA

'Informed by intersectional frameworks of race, class and gender, the editors assemble a disparate and wide-ranging constellation of perspectives to imagine innovative possibilities for the future. Only junior scholars with passion and commitment could pull off such a tour-de-force of dedication, imagination and old-fashioned hard work. One of-a-kind, this unique and provocative volume raises a new bar for environmental activism.'

Patricia Hill Collins, University of Maryland, College Park, USA

'Across a diversity of situated perspectives and locations, this volume makes a strong and necessary argument for the importance of seeing environmental justice advocacy and feminist intersectional politics as intertwined, not to be separated neither in theory nor in practice. The volume gathers together many passionate voices of activists-scholars-poets committed to struggles for globally sustainable and just feminist futures. A timely and urgently needed intervention in debates on global climate change!'

Nina Lykke, Professor of Gender Studies and co-director GEXcel Collegium for Advanced Transdisciplinary Gender Studies, Linkoeping University, Sweden

Contents

Illustrations

Figures

Tables

Boxes

Notes on contributors

Editors

Phoebe Godfrey, PhD, is an Assistant Professor-in-Residence at UCONN in sociology. She is an activist teacher, community organizer, gardener and artist. She co-founded the non-profit CLiCK, dedicated to a local sustainable food system. She lives with her wife and three cats in their creative house and garden in Willimantic, CT.

Denise Torres, LCSW, MPhil is a clinician, evaluator, and educator. The unifying theme of her work and publications is the authentic inclusion of silenced and marginalized groups in the systems that affect them. She is a New York native and doctoral candidate at the Graduate Center, City University of New York.

Contributors

Arwa Aburawa is a campaigner-turned-journalist who works for an environmental show called 'Earthrise' at Al Jazeera's English news channel.

Imna Arroyo's research focuses on African spirituality; and philosophical and aesthetic legacy in the African Diaspora. She has traveled to Ghana and Nigeria, West Africa; Salvador de Bahia, Brazil and has made subsequent trips to Cuba to continue her research. Her artwork has been reviewed and exhibited nationally and internationally.

Subhankar Banerjee is a photographer, writer, activist, and environmental humanities scholar. His photographs have been exhibited in more than fifty museum exhibitions, including the Rights of Nature: Art and Ecology in the Americas at the Nottingham Contemporary. He is the editor of Arctic Voices: Resistance at the Tipping Point, and received a 2012 Lannan Foundation Cultural Freedom Award.

April Karen Baptiste is currently an associate professor of environmental studies at Colgate University. Her research interests lie at the nexus of environmental psychology and environmental justice. Her current research explores world views of environmental problems in the Caribbean with a focus on fishers' perceptions of climate change.

Melica Bloom is an artist from Eastern Connecticut. Her focus is centered on large works of public art, encompassing history, culture, and color. She also works digitally to produce designs.

Dwayne Booth (Mr. Fish) has been a freelance artist for twenty-five years. His work has appeared in Harper's Magazine, the Los Angeles Times, the Village Voice and The Nation. His books include *Go Fish: How to Win Contempt and Influence People*, Akashic Books 2011, and *WARNING! Graphic Content*, Annenberg Press 2014.

Karen Bradley, Associate Professor and Director of Graduate Studies in Dance at the University of Maryland, College Park, is a Laban Certified Movement Analyst and author of *Rudolf Laban*, in Routledge's series on twentieth century performance practitioners. She has published on dance, neuroscience, movement education, arts policy, and leadership analysis.

Cynthia Bogard is professor and chair of sociology at Hofstra University, specializing in political sociology, sociology of the environment, poverty policy, and social problems. She has written on homelessness, civic engagement, the social construction of social problems, and poverty policy in the environmentally degraded nation of Haiti.

Vincci Cheng graduated from Colgate University, majoring in Environmental Studies and Psychology. Originally from the Hong Kong metropolis, she has always appreciated the natural side of the environment. She is particularly thankful to Professor Shangrila Joshi Wynn and the book editors for mentoring her with amazing inspirations in the chapter.

Candace Ducheneaux is a Lakota activist who lives on the Cheyenne River Lakota Reservation in South Dakota with her children and grandchildren. She is the founder and director of Mni, Grassroots for Water Justice, a nonprofit organization working on water restoration in the area. She has spoken at Tufts University and at UCONN.

Lindsy Floyd grew up in Salt Lake City, UT and has lived in New Mexico, California, Hawai'i, and Texas. With a Bachelor of Science in environmental studies and biology, and a Master of Science in environmental humanities, Ms. Floyd is passionate about education, community involvement, and climate activism.

Aleya Fraser is an educator, scientist and returning-generation farmer in Maryland. She co-manages a farm using agroecological practices steeped in social and ecological transformation. She obtained her BSc in physiology/neurobiology from UMCP and uses farming and education as a platform for social justice, inter- and intrapersonal transformation, and scientific discovery.

Sufia Giza Amenwahsu, an ancestrally inspired artist of Gullah-Geechee descent, is a poetess; recording artist; TV producer; documentary filmmaker; media literacy specialist and certified yoga instructor. She has an associate's degree in Spanish, a Bachelor's in bilingual education and a Master's in cross-cultural education.

Terran Giacomini is a PhD student in community development and adult education at the University of Toronto. Her research addresses food and energy sovereignty.

José G. González is a K-12 public educator in formal and informal education in the arts, education, conservation, and the environment. Jose contributes to *Green Chicano*, working on diverse environmental issues with an emphasis on the intersection of art, education, and conservation. Find him on Twitter @JoseBililngue, @Green_Chicano, and online at www.josegagonzalez.com

Paul K. Haeder, a journalist since 1977 and educator since a young man (a dive master at twenty), has traveled the world to gather stories and be with people. He has taught college English courses in Texas, New Mexico, Washington and Mexico. His work is published extensively. He has MAs in urban planning and English. A new book, *Reimagining Sanity: Voices Beyond the Echo Chamber*, comes out in April 2016.

Rachel Hallum-Montes holds a PhD in sociology and currently works as a healthcare systems researcher in San Mateo, California. Apart from her day-to-day work in public health, she also has over twenty years of experience in transnational environmental activism through her work with the Alliance for International Reforestation.

James Elias Hamue Torres is a New York City Outward Bound student with a passion for food, science, and literature. He recently co-founded the high school's science club and robotics team.

Marylee Hardenbergh is an artist-in-residence at the Center for Global Environmental Education at Hamline University in Minnesota. She is the artistic director of both Global Water Dances and Global Site Performance, and has received numerous awards for choreography including a Fulbright fellowship, National Endowment for the Arts fellowship, and several McKnight fellowships.

Irene Hardwicke Olivier Growing up in South Texas, living in Latin America, and finally moving to New York, I developed an intense longing to live in a wilder place. Now I live off the grid in the high desert of Oregon to be closer to nature in a more ancient, vulnerable way.

Daina Cheyenne Harvey is on the faculty of the College of the Holy Cross where he is an assistant professor in the Sociology and Anthropology Department. His work focuses on the urban and environmental conditions that result in both acute and chronic suffering and that produce eco-disparities in everyday life.

Tyler Hess is a young farmer, writer, traveler, and herbalist. Influenced by a suburban Catholic American upbringing, he now cultivates fruits, nuts, vegetables, medicines and community in the Driftless area of Wisconsin. He intends to continue facilitating land-based projects as leverage points for the times ahead.

Emily Hinshelwood is a freelance writer, performer and community arts facilitator. She runs a programme of arts and climate change projects in the UK. She writes poetry and plays that address the issues and she delivers poetry at sustainability conferences. Her current project 'Three Questions about Climate Change' is a walking/poetry project (www.emily-hinshelwood.co.uk/three-questions-about-climate-change/)

Jennifer L. Hirsch is an applied anthropologist fostering grassroots participation in sustainability action, using methods from anthropology, asset-based community development, participatory facilitation, and popular and museum education. She is the inaugural Director of the Center for Serve-Learn-Sustain at the Georgia Institute of Technology, equipping students to work collaboratively with community partners to create sustainable communities.

Madronna Holden taught anthropology, philosophy, folklore, ethnic studies, and women's studies at universities in the US and abroad and has published widely in these areas. She is currently retired but maintains a public intellectual dialogue on justice and environmental issues on her website, Our Earth/Ourselves (www.holdenma.wordpress.com), which has visitors from 180 countries.

Will Hooper is a permaculturist. Earth care, people care, fair share – starting with this ethical foundation, permaculture provided a framework to address the contradictions I saw between dominant cultural narratives and my own lived experiences. My journey provides an example of moving from hopeless defensiveness at the world to hopeful interaction with it.

Christopher Hrynkow is an assistant professor in the Department of Religion and Culture at St. Thomas More College, University of Saskatchewan, where he also contributes to the minors in Catholic Studies and Social Justice and the Common Good. Previously, Hrynkow taught at all three of the universities in Winnipeg.

Marc Hudson is a PhD candidate at the University of Manchester. He investigates the political, economic and technological strategies of the coal industry in response to climate change. An environmental activist for a long time, he writes at www.marchudson.net about how meetings and events could be more inclusive, less hierarchical and less patriarchal.

Lisa See Kim facilitates disruptive and creative thinking in informal spaces. With a background in art, design, and education, she is interested in using multimedia tools to empower communities. As youth commission director for

Mikva Challenge, she works with diverse Chicagoans to bring youth voice to the city's decision-making process.

Len Krimerman lives, works, dances, and dreams in rural Eastern Connecticut, and has helped build bridges between many varieties of grassroots democracy over the past five decades. In this, he has invariably been mentored by his amazing GEO colleagues, and by the imagination and support of his lifelong partner, Marian Vitali.

Vanessa Lamb of Uniondale, NY is a scientist and an artist. She graduated from UCONN in 2015 with a BSc in Chemistry. Currently, she works as a scientist at a pharmaceutical company.

Laura Levinson is a dancer and early childhood arts educator in Minneapolis, MN. She graduated from Macalester College with a BA in sociology; her thesis on intersections of race and gender in online portrayals of the dancing body was awarded first place in the fifty-first annual Midwest Sociological Society student paper competition.

Janet A. Lorenzen is an assistant professor at Willamette University in the Department of Sociology. Her work focuses on social and environmental change in the form of green lifestyles, environmental movement groups, and environmental policy making.

Ryan Lugalia-Hollon is director of Excel Beyond the Bell San Antonio. He previously served as an executive director for the YMCA of Metropolitan Chicago, worked as a program evaluator for Northwestern University, led major projects at The Field Museum and was a Justice fellow at Adler University. He earned his PhD in urban planning and policy.

Isis Rakia Mattei is an attorney, writer and entrepreneur who lives in New York with her husband and children, where she pursues broad interests including law, social and environmental activism, art, architecture, herbalism and the study of earth-based spiritual traditions.

Paul Mitchell has worked on climate change adaptation and sustainable development for over a decade. He has developed, implemented, monitored and evaluated projects and strategies across the Pacific, Asia, Africa and the Caribbean for government and non-government organizations. He currently researches the intersections of adaptation and sustainable development at RMIT University.

Dennis Patrick O'Hara began his career as a chiropractor (1979) and naturopathic doctor (1986). Since 2002, he has been a professor in the Faculty of Theology at the University of St. Michael's College in the University of Toronto, and the director of the Elliott Allen Institute for Theology and Ecology.

John E. Peck grew up on a farm in Minnesota. With a PhD in Environmental Studies from UW-Madison, he is executive director of Family Farm Defenders and also teaches at Madison College. He attended the 1992 UN Earth Summit and has participated in many global justice, food sovereignty, and climate justice events since then.

Ryanne Pilgeram is an assistant professor of sociology and director of the Certificate in Diversity and Stratification at the University of Idaho. She teaches a course on social inequalities and food justice. Her work has recently appeared in *Rural Sociology*, *Environmental Communication*, and *The Journal of Critical Thought and Praxis*.

Ryan Pleune is a teacher, mountain climber, bus driver and climate justice activist. He currently works as the service/adventure coordinator at Pacific Heritage Academy. Growing up in Colorado, his heart and soul are rooted in the Intermountain West but like all white people in the Americas his ancestors are immigrants.

Bandana Purkayastha, professor of sociology and Asian American studies, University of Connecticut, has published several books, articles and chapters on gender, class, race, religion, migration, transnationalism, human rights, violence and peace. She holds leadership positions in several national and international professional organizations. For more, see sociology.uconn/edu/Purkayastha.html

Cori Redstone is a social justice activist and multimedia artist. She grew up in the suburbs of Salt Lake City and has worked a farm and ranch in rural Southwest Colorado. Redstone is an expert in the aesthetics of social movements. She is based in Los Angeles, CA.

Sonalini Sapra is an assistant professor of political science and gender and women's studies at Saint Mary's College, Notre Dame. She was the Andrew Mellon Post-doctoral Teaching Fellow in International Studies at Kenyon College in 2009–2010. Her research and teaching interests are in gender and international relations, environmental politics, transnational feminisms, and feminist theory.

Joshua Sbicca is assistant professor of sociology at Colorado State University. His work focuses on the intersections between the contentious politics of food and agriculture, social movements, and social inequalities. He is also interested in coalition development and how social movements use food to resist and alter various power relationships.

Deric Shannon is a former line cook, now an assistant professor of sociology. He has written and edited books, book chapters, and journal articles on social movements, pedagogy, political economy, food justice, human rights, and political sociology. His most recent research focuses on people who live in Georgia's food deserts.

Phoebe Sheppard, LGSW, received her Master of Social Work, specializing in public health, from Morgan State University in 2015. She received her BA in sociology, specializing in social stratification, from the University of Maryland, College Park, in 2012. She currently works as a behavioral health clinician.

Tina Shirshac is an artist, gardener and practising reiki master and healing facilitator. She lives in Northeastern Connecticut with her wife Phoebe and their three cats. She has hopes that if we are grounded in the present moment we will be able to see that all things are connected.

Sarah Sommers has spent the last decade helping non-profits, foundations, and educational organizations develop tools to create change. Focusing on print and web design, Sarah builds upon her unique background in anthropology, graphic design, and mass communication in her work. Her clients span from coast to coast.

Milton Takei graduated from Kalani High School, Honolulu, Hawai'i, in 1966. He is an independent scholar with a Master's degree in political science, and he is a retired periodical indexer. He is the moderator of the Ecopolitics Discussion List. He lives in Eugene, OR, USA.

Shakara Tyler is a returning-generation farmer, educator and activist-scholar who engages in agroecology as a decolonizing practice. She is pursuing her doctorate at Michigan State University in community sustainability (CSUS) and explores decolonial pedagogies in the food justice and food sovereignty movements within the communal praxis of black agrarianism.

Laurens G. Van Sluytman's professional life is grounded in a personal commitment to critical thinking, inquiry and social justice. His scholarship focuses on increasing knowledge of the interactive forces that govern human potential, enforcing boundaries of human lives, and containing pathways to escape the bounds of oppression. He has published and presented widely.

Alexis Winter is an applied anthropologist who seeks to understand the role of culture in environmental conservation and collective climate action – in particular the ways that environmental knowledge and policy are produced, represented, shared, and implemented. She has worked with communities in and around Chicago, IL and Tampa, FL.

Nala Walla (MS, NTP) is a nutritional therapist, performing artist, and permaculture designer devoted to restoring vitality to "the Body" at multiple levels: personal, political, and planetary. She holds a Master's degree in integrative arts and ecology, plays with award-winning kindie-band The Harmonica Pocket, and maintains a private coaching practice in nutrition and somatics.

Karen Washington Since 1985, Karen has been a community activist, striving to make New York City a better place to live. As a farmer, community gardener and board member of various non-profit organizations, she has stood up and spoken out nationally for land preservation, food sovereignty and racial equity.

Foreword

The authors of this scholarly book that you are holding in your hands asked me, a fourth generation, white fisherwoman with a high school education, to write a forward to their book. You might ask why any professional writing a serious work on the state of environmental justice, sustainability, and equality on planet Earth in these dire times would lend their ears to a fisherwoman? Well, the better question might be "why not?", especially given the recent Paris climate talks where the indigenous world, the climate vulnerable, and the Earth were no better served than they were in the Copenhagen Accords, six years ago.

Yes, the rainforests burn, another two hundred species go extinct today, torrential rains and typhoons escalate, and Amazonia leaders and First Nations people tell stories of their culture's undoing and the terrible destruction to their land. On my personal home front, we, the gulf people, watch another thousand cuts being delivered to the Gulf of Mexico. These people and the land all suffer different fates but the outcomes are the same. The life of our planet is in crisis and we, the mostly unheard voices and the largely untaught, are on the front lines and fighting for our lives. We are the janitors, parents, fishermen, grocery clerks, musicians, feminists, teachers, farmers, dishwashers, artists, caregivers, laborers, immigrants, and students. We come from diverse backgrounds, but we have a similar goal: to defend this planet that is our only home.

Our strategies include the entire range of nonviolent direct action techniques, including sit-ins, blockades, protests, and demonstrations. Many are jailed and will continue to be. There are also actions committed to education and cultural work, crucial for a future that will demand strong local communities that embrace direct democracy, economies of support, universal human rights, and the rights of nature. As oppressive social and economic systems come down, cultures based on justice, sustainability, and a connection to the land must be built. The indigenous from many lands must be supported.

I am among the diverse. On the eve of my fortieth birthday in 1989, with my shrimp boat tied stern and bow to a run-down dock, I watched a half dozen federal trailers haul off twenty-five dead dolphins. Another hundred dolphins floated in the bay. My hometown, our fishing culture, the dolphins, the fish, and the bay were dying. I didn't realize we had so little time left.

It was a crying shame and an outrage but, like thousands of others who were on the frontlines of environmental destruction, I hadn't a clue what to do. Yes, what do you do when you don't have a clue? Well, do something! A phone call! A meeting!

These were the early days and I was pretty stupid. I did minor actions that were very much inside the box. The first reporter that wrote an article about my piddling action said I was an "environmentalist." I had to look the word up in the dictionary. An activist. Well, that's what the dictionary said and that's what the reporter said but lots of folks said different. I wasn't the activist type and not only wasn't I 'the type' but fishermen in general couldn't pull off that activist-type thing. Everybody knew fishermen would rather be on the water than running around being an activist. Too many meetings, plotting on spread sheets, and driving around in cars for them. Leave it to the experts.

I started my activism from scratch. Ground zero. I wasn't an expert. I wasn't a biologist. I didn't know the names of the chemical plants five miles down the road. I didn't know the name of the state environmental agency in Austin or what region Texas was located in within the Environmental Protection Agency. I didn't know what wastewater permits were or what chemicals were included. I had never heard of vinyl chloride or ethylene dichloride.

Yes, I started cold. Tomorrow or the next day or any day after that would get better. But nope, that wasn't what happened. In hindsight, I think I'd have skipped some of the dumb steps if I'd had someone like Howard Zinn wising me up, but he wasn't in Texas so I actually believed the politicians (local, state and federal) and agencies (local, state, and federal) who were elected, appointed, or hired to serve the people, tell the truth, and do the right thing. Well, that illusion died hard, probably because there had been no indication whatsoever, from elementary school on up to high school, that it might be a damn lie.

I didn't have time to give my poor innocence a decent burial because I was too busy stomping out environmental fires that started right under my nose with the corporate arsonists getting away scot-free. It was news to me if I was making any headway at all with my environmental activism because the vehicle under me was sputtering, recoiling two miles back for every mile forward, and refusing any shift other than low gear.

Illusions are funny things. They lay down quiet in their graves like they are dead as a hammer but a chance light flickering off their bones and there I go—again—believing the lie: politicians will lead you, agencies will protect you and your land, and corporations want to, and can, make a profit and "do good"!

Isn't it amazing how we distrust ourselves and our own cautionary tales? Amazing, too, how fast we are willing to toss aside what is the best part of ourselves. Eventually though, I realized (or maybe I just decided) that monkey wrenching can be done just as easily and more creatively. I was a natural at the thing. My first bright idea outside the box was to do a hunger strike. The idea just popped into my head. After I told a savvy environmentalist from Houston, and he told me I was crazy and that nobody in Texas did hunger strikes, I knew better than to tell anybody else about it—well, except the reporter who pretty much cemented every word from my mouth and put it in the newspaper. That hunger strike was so successful that I decided to do two more, but sinking the shrimp boat really did it.

After that boat stunt, I was approached by some well-meaning folks who said that my extreme and spontaneous actions were uncompromising and alienating the mainstream. I was harming the broader environmental movement. Maybe I ought to rein myself in some. Maybe I needed to do something a little more predictable. Act polite. Get some manners, girl! And get a plan with a budget! And never, never do something that the community-at-large—that is, the chamber of commerce's banquet halls, county commissioner's court rooms, city hall's cubby holes, economic development boardrooms, state capitol cafeteria eating areas, and intimate industry circles in the Texas chemical corridor—wasn't ready for!

Well, I blew that advice the day I scaled a seventy-foot chemical tower at Dow Chemical. Yes, you might say, but did it work? I don't know, but it sure made Dow Chemical nervous. I like nervous. It was almost as good as controversial. Environmentalist Paul Hawken (2009) said it best, "Do what needs to be done, and check to see if it was impossible only after you are done."

And so this is how my activist journey began. And has it ended? Well, I once had a lawyer and he said, "Tie a bow around your action and put it on a shelf! It's finished!" Oh, not so. My actions with the chemical plants morphed into concerns about injured workers and that's why I co-founded Texas Injured Workers. Then the oil refineries on the Gulf Coast and fossil fuel concerns morphed into the Iraq War and all those terribly ensuing issues of global conflict, destruction, weapons, and death. That's when I, along with a number of 'unreasonable women', co-founded Code Pink. In these twenty-five years of self-taught activism I have been on thirteen hunger strikes, jumped the White House fence, chained my neck to fences and trucks, and disrupted countless political jamborees. I've been arrested over forty times, and during my last foray into jail, I co-founded the Texas Jail Project.

So back to the question: "why not?" Because a book such as this that brings together the theory and practice of intersectionality can help us each both individually and collectively 'decide what needs to be done.' Then we can do it and then together we can decide what is possible. For certainly what we need now, more than anything, is a new understanding of what is possible, and the only way to get there is by co-creating it. You don't have to be an 'expert'—heck I wasn't—just start where you are, as you are, right now.

Diane Wilson
Social Justice Activist and an "Unreasonable Woman"
Seadrift, Texas
December 2015

Reference

Hawken, P. 2009. *Commencement: healing or stealing?* University of Portland Commencement Address. Accessed December 29, 2015, available at: www.up.edu/commencement/default.aspx?cid=9456

Preface

Maybe

There is a Taoist story of an old farmer who had worked his crops for many years. One day his horse ran away. Upon hearing the news, his neighbors came to visit. "Such bad luck," they said sympathetically. "Maybe," the farmer replied.

The next morning the horse returned, bringing with it three other wild horses. "How wonderful," the neighbors exclaimed. "Maybe," replied the old man. The following day, his son tried to ride one of the untamed horses, was thrown, and broke his leg. The neighbors again came to offer their sympathy on his misfortune. "Maybe," answered the farmer.

The day after, military officials came to the village to draft young men into the army. Seeing that the son's leg was broken, they passed him by. The neighbors congratulated the farmer on how well things had turned out. "Maybe," said the farmer.

Acknowledgements

This project has been three years in the making, and during that time many people have provided critical support. While it is impossible to capture all the individuals or the ways they have given meaning to this second volume, we want nevertheless to acknowledge key sources of inspiration and perspiration.

A joint "thank you" to:

> Malaena Taylor, our Assistant Editor, for your eagle eye, organizational skills, and professionalism;
> Chris Sneed, our supporting reviewer on numerous pieces;
> Louisa Earls, Margaret Farrelly, and Annabelle Harris at Routledge/Taylor Francis for your guidance and structure; and
> All our generous and patient contributors, both newcomers and those who have been with us since the inception.

From Phoebe:
I would also like to thank my wife, Tina Shirshac, for her unlimited support as an intellectual, creative, emotional, and spiritual companion. I'd also like to thank my loving mother for always enthusiastically believing in my endeavors.

From Denise:
Thank you to all my colleagues and collaborators who contributed to my personal and professional growth throughout the years, especially D. Berman, M. Garcia-Bigelow, S. Burghardt, S. Estrine, M. Fabricant, T. Mizrahi, A. Savage, and L. Van Sluytman. And a special thank you to my co-editor Phoebe for your vision, courage, and unfailing energy.

Introduction

Opportunities for renewal: intersectional praxis for just sustainabilities

Phoebe Godfrey and Denise Torres

> The master's tools can never dismantle the master's house.
>
> (Audre Lorde, 1979)

> Look and listen for the welfare of the whole people and have always in view not only the present but also the coming generations, even those whose faces are yet beneath the surface of the ground – the unborn of the future Nation.
>
> (The Constitution of the Iroquois Nations, c. 14th Century/2014, ¶ 28)

From Paris to theory to praxis and back

The much publicized 2015 Paris Climate Talks (COP 21) have resulted in 195 nations signing what has been referred to as a "landmark climate accord" (Davenport, 2015). However, a few key dissenting voices stand out from the rising sea of praises. Preeminent climate scientist James Hansen is quoted as saying "It's a fraud really, a fake … It's just bullshit for them to say 'We'll have a 2C warming target and then try to do a little better every five years.' It's just worthless words. There is no action, just promises" (Milman, 2015). Similarly, Bill McKibben, climate activist and founder of 350.org sees it as much too little too late, but adds the important recognition that any agreement at all "is testament to the mighty movement that activists around the world have built over the last five years" (McKibben, 2015). He goes on to remark that what is really needed is something much more severe in terms of cutting CO_2 and thereby he puts the onus on the continued growth of an even bigger, more engaged global movement.

A meeting held simultaneously in Paris was that of the International Rights of Nature Tribunal. There, indigenous leaders from around the world played key leadership roles in devising "Earth-driven, not market-driven solutions to climate change" (Global Alliance, 2015). Within this tribunal was the collective recognition of "the Universal Declaration for the Rights of Mother Earth and international human rights law," such that participants proposed "perpetrators of the crime of ecocide … be prosecuted before the International Criminal Court (ICC)" as has been done in Ecuador (Global Alliance, 2015).

It is into this worldwide differentiated and yet united call for action that we birth this international and interdisciplinary volume on the emergent possibilities for global sustainability. Aware that solutions are 'unsurprisingly interdependent' and ultimately inseparable from the diverse causes and consequences of anthropogenic Global Climate Change (GCC), this volume is a companion to and emerges from *Systemic Crises of Global Climate Change: Intersections of Race, Class and Gender* (Godrey and Torres, 2016; henceforth, *Systemic Crises*). As a crisis of extreme and increasing structural inequalities (Leopold, 2015) that perpetuate and exacerbate the status quo, GCC represents Audre Lourde's "master's house." Heeding her warning, we eschewed the master's tools and employed intersectionality theory, with its recognition of the dynamic and co-constructive intersections of race, class, gender, and other social variables, grounded in the scholarship of African American feminists and other marginalized people (Crenshaw, 1989; Collins, 2000) to begin to dismantle it.

Deconstructed and with its dysfunction and fundamentally violent internal systems exposed to the elements, both figuratively and literally, we begin reconstruction by linking intersectionality theory to the concept of *just sustainabilities*, that is itself a product of the linking between "concepts of sustainability and environmental justice and their practical actions" (Agyeman, Bullard, and Evans, 2003, p. 9). In the pursuit of social and environmental justice, just sustainabilities engages theoretically and in practice with issues of the environment, as well as the constructs of race, class, and gender (Taylor, 2000 in Agyeman *et al.*, 2003).

Sustainability is frequently depicted in Venn diagrams as the union or point of overlap of three intersecting circles representing the economic, social, and environmental spheres. At all unions, however, the colors blend, indicating the dynamic and transformative nature of interaction. Moreover, as Campbell (1996) notes in discussing the development triangle, there also reside social and economic tensions and conflicts. Our concern, then, is whether or not sustainability—or any of the other intersects—addresses these struggles and is in reality 'just' or merely "sustainababble" (Washington, 2015, p. 1) that further perpetuates the master's house.

It is therefore essential that any actions taken in the name of sustainability—whether they be by individuals, community groups, international NGOs or nations—be assessed in ways that recognize the intersectional complexity of social and ecological justice. Our explicit linkage of these concepts we believe helps to further reveal "how power works in diffuse and differentiated ways through the creation and deployment of overlapping identity categories" (Cho, Crenshaw, and McCall, 2013, p. 797) at the level of the individual and community as well as at the exo- and macro-levels in terms of the emerging category of change efforts in the difficult and relatively unknown territory of equitable solutions. The union of just sustainabilities and intersectionality theories creates a *means* to doing intersectionality and for evaluating its *ends*: Together they provide a malleable framework capable of linking the parts to the whole with the potential to theorize and put into practice the long-term goals of creating and sustaining a just and livable planet for all beings.

Furthermore, rather than making the environment an et cetera in an essentialized list of identities that are mentioned but never unpacked, we see the "'human/nature' or 'earth other axis'" (Lykke, 2009, p. 39) as a foundational component of the "rebuilt house," which, though historically hidden and ignored, needs to be recovered from the debris and privileged in our rebuilding efforts. As a framework, then, just sustainabilities combined with intersectionality provides new tools that are complex enough to analytically encompass the dynamism of both the social and material worlds so that we do not inadvertently reinforce what already exists. Indeed, as a multi-systems framework, intersectionality-informed just sustainabilities may be used as a personal tool, not just in some distant, hypothetical future, but now, within our individual lives, to evaluate how we interact with all the other lives on this planet, human and non-human.

Revisiting intersectionality

As part of a two-volume series, understanding *Emergent Possibilities* requires revisiting our intent in *Systemic Crises*. Mindful of our space constraints, we frequently draw directly from the latter. Although we may have reframed them to maintain clarity, our use of quotes also reflects our not wanting to muddy words after having worked hard to richly capture and communicate our meaning. Such 'recycling' also represents and demonstrates our belief in the inseparability of the fact that "inequality and injustice are the problems" and "equality and justice" are "the solutions" (Godfrey and Torres, 2016, p. 11).

In the introduction to *Systemic Crises* we proposed that intersectionality (Crenshaw, 1989; Collins, 2000) is ultimately the "holographic process" providing the layered analysis capable of generating a hologram that allows for the whole to be reproduced throughout all of the parts, with the theory itself acting as "the object or illumination beam focused onto the systemic ideologies and the corresponding structural inequalities and oppressive practices undergirding GCC" (Godfrey and Torres, 2016, p. 4). Insofar as intersectionality is "not committed to particular 'subjects nor to identities' but 'to marking and mapping the production and contingency of both' (Carbado, 2013, p. 815) and to recognizing the act of doing so from given perspectives" (Godfrey and Torres, 2016, p. 4), we carry this forward, aware if any just and viable solutions are to be theorized and ultimately practiced, they too must be scrutinized.

As a way of capturing point in time connections, intersectionality is "'heuristic in nature,' enabling users to gain insight into that which was previously obscured, one dimensional or nebulous" (May, 2015 p. 19). As before, our central concern remains postmodernisms' obfuscation of matter "to matter" (Barad, 2008, p. 120) and consequently for the solutions, regardless of how branded, to obfuscate the interdependence of the constructions and treatment of oppressed individuals, communities and groups by those with privilege in conjunction with the treatment of non-human nature. Alternately stated,

> in our use of intersectional theory and analysis we have sought to not only engage with the core social constructions of identity represented by race,

> ethnicity, social class, gender, sexuality and nationality, we also extend it to the imagined divide between the material and the social worlds.
>
> (Godfrey and Torres, 2016, p. 4)

Our inclusion of the "environments as a 'category' or 'phenomena' for consideration" was purposeful, a "push[ing of] 'the theoretical boundaries'" (Carbado, 2013, p. 841) as we added to Carbado's concepts of 'colorblind' and 'gender-blind intersectionality' the concept of "Nature-blind intersectionality, which includes the physical body, in that bodies as well as air and water are always present in any analysis even though they may only be highlighted when they are connected with negative human consequences" (Godfrey and Torres, 2016. p. 10). Such an extension is consistent with the fact that intersectionality "Is never done, nor exhausted by its prior articulations and movement; it's always already an analysis-in-progress" (Carbado, Crenshaw, Mays, and Tomlinson, 2013, p. 304). Here, we push the boundaries still further, "to engage an ever-widening range of experiences and structures of power" (Barad, 2008, p. 120) that include the practical and ideological opposition to just sustainabilities, as well as those movements and creative practices attempting to realize it.

Structurally, as an attempt to "engage intersectionality heuristically, throughout all aspects of the book," we embody matter by "intentionally link[ing] the ideological, hence the socially constructed worlds, with the material" through "elemental segmentation" (Godfrey and Torres, 2016, p. 7). Whether we recognize it or not, we are all simultaneously social and ecological beings, dependent on and ultimately returning to nature, or, as Pope Francis recently wrote, "we have forgotten that we ourselves are dust of the earth (cf. Gen. 2:7); our very bodies are made up of her elements, we breathe her air and we receive life and refreshment from her waters" (Pope Francis, 2015). Thus, in *Systemic Crises* we used the classical elements of Air, Earth, Fire, Water, and Chaos, with the last representing for us what Parenti (2012) has labeled as "catastrophic convergence" of GCC, failed states, inequality within and between nations, and global militarism/terrorism. Here, in *Emergent Possibilities*, we use Aether instead of Chaos to reflect the elasticity, movement, and dynamic interactional potential at the nexus of intersectionality and just sustainabilities.

Epistemologically, the structure also represents our commitment to "political intervention and intersectional praxis" as we bear witness to intersectionality's history and commitment to social justice practice through the inclusion of international and interdisciplinary "voices and perspectives not normally seen in either a social science volume nor in works addressing" sustainability as it

> is through a more holistic, embodied engagement with ourselves, each other, and our world that we will see more clearly what we are doing now that defies both the ideas of democratic social justice and a livable planet as well as how to better position us to imagine what we might do collectively and in solidarity to create authentic radical change.
>
> (Godfrey and Torres, 2016, p. 8)

We explicitly searched outside of the academy to include poets, activists, dancers, artists, and youth to begin to truly transform our system as well as "to confront and confound the taxonomic dichotomies typically utilized to grapple" with inclusion wherein the marginalization of the 'other' is reified through chapter "'ghettos' segregating people of color, the disabled/unwell, the poor or other 'vulnerable groups' into special sections" (Godfrey and Torres, 2016, pp. 7–8). Yet again, given this diversity, "there is an unevenness in terms of how each has chosen to apply intersectionality and how intricately the matrices of oppression and privilege have been unpacked and thickly described" so that our continued goal as editors has been to at the very least conceptually envelop all the pieces included into the material, through the elements as an indication of matter's presence, hence of ongoing yet dynamic intersection.

By continually emphasizing matter's presence we have addressed one of the prevailing themes in works on sustainability; that of "accepting reality" insofar as modern industrial society is "hugely unsustainable" (Washington, 2015, p. 193). The chrysalis metaphor extended in *Systemic Crises* is helpful here. We see sustainababble as capitalism's natural resistance to the awakening occurring across the globe (Ray and Anderson, 2000)—an immune response attempting to stave off its own dissolution—as a "'transformative, coalitional consciousness' (Keating, 2002, p. 6) among dispersed imaginal cells" (Godfrey and Torres, 2016, p. 7) overwhelms its defenses.

While we as editors "present the practices emerging from the goop of the current crisis, myriad unfolding butterflies" (Godfrey and Torres, 2016, p. 3), there are still imaginal cells cocooned and dormant, fearing what such a revolutionary process might entail. We suggest, as the aphorism frequently attributed to Buckminster Fuller does, that there is nothing in a caterpillar that tells you it's going to be a butterfly. Indeed, each of us can and must play a part in society's metamorphosis. As the contributions in *Emergent Possibilities* make clear, many have taken up the challenge to develop new models and tools—some by openly reclaiming those that have been ideologically and institutionally suppressed—to dismantle the master's house. Others interrogate how the rebuilding of our collective house has been framed. It is to this we now turn.

Old/new, strong/weak, and just sustainabilities

Given that what is 'real' and 'known' is socially constructed, the cosmologies, social structures and identities that dominate in any period represent the culturally specific embodiment of the evolving stories we tell (Berger and Luckman, 1966). For many writing about sustainability and how to achieve it, a key issue is the rewriting of Western culture's evolving story about who we think we are in relation to the Earth (Korton, 2012). This narrative has come to be conceptualized and lived as 'the earth was made for man,' in contrast to what can be categorized as a more ancient and indigenous perspective that 'man was made for and of the earth' (Quinn, 1998; Pope Francis, 2015). Washington (2015) correctly notes that "What we *mean* by sustainability today draws on a long history of people thinking (and feeling) about living in harmony and balance with Nature and recording this story as lore or law" [italics in original] (p. 6).

Labeling this "old sustainability," Washington (2015) observed that it also involved "reverence, sacredness, spirituality, respect, care, witness, reciprocity, custodianship, stewardship, beauty and *even* love" [italics in original], elements he laments "rarely appear in the new and sanitized 'sustainable development" (p. 8). As many of the contributors make clear, however, the *practice of being* in 'harmony' and 'balance' has never died: Ways of Knowing (WOK)—whether prefaced by 'indigenous', 'traditional', or 'native'—have resisted the predominant Western discourse and persisted despite attempts to overwrite and erase them. Indeed, we replace living with being to reaffirm that "things do not exist in of themselves but constantly gain form and meaning, hence 'mattering', in direct and dynamic intra-action with one another and that this process does not in fact have a designated beginning or end" (Godfrey and Torres, 2016, p. 2).

As with the plethora of Venn diagrams depicting it, there are numerous definitions of new-sustainability. The two most commonly referenced are that of the World Commission on Environment and Development (WECD, 1987) (known as the Brundtland Report), which reads "sustainable development is development that meets the needs of the present without compromising the ability of future generations to meet their own needs" and that of the World Conservation Strategy (IUCN, 1991), which focuses on the goal "to improve the quality of life while living within the carrying capacity of ecosystems" (Agyeman *et al.*, 2002, p. 80). While neither recognizes these concepts as 'old,' the absence of any acknowledgement to traditional WOK in Brundtland is noteworthy as it paraphrases what is known as the Seventh Generation philosophy first recorded by the Iroquois Confederacy, and referenced above. This principle's commodification by a corporation that markets as sustainable everything from paper towels and disposable wipes to landfill-gagging diapers underscores the need to ensure we are not co-opting and appropriating non-Western values and ideas in the name of 'green technologies and solutions' thereby creating "green illusions" (Zehner, 2012), "greenwashing" (Montevalli, 2011) practices, and/or promoting policies that are "sustainable failures" (Cable, 2012). Instead, as Washington (2015) warned, we "have an environmental crisis precisely because we (as a society) have forgotten the 'old sustainability' and the teaching and wisdom of the millennia" (p. 8). By remembering our collective indigenous histories we can begin to illuminate the inequalities and injustices that the Western capitalist story creates and uses to preserve itself, while further imagining what other worlds are possible.

In a seminal paper, Catton and Dunlap (1978) noted the "impasse" then occurring as the 'reality' story was being contested by activists and theorists and they castigated "the numerous competing theoretical perspectives" for their inherent anthropocentric worldview, calling it the "'Human Exceptionalism Paradigm' (HEP) [that] has become increasingly obstructive of sociological efforts to comprehend contemporary and future social experience" (p. 42). Among the many assumptions they contested was that of the difficulty of "perceiving the possibility of an era of contrived scarcity" and the "neglect of the ecosystem-dependence of human society" (p. 43). Yet, as observed by Agyeman (2008), while the New Environmental Paradigm (NEP) they offered encouraged environmental stewardship as well as a greater interest in biodiversity and intergenerational equity, NEP places too little focus on cultural diversity and intragenerational equity—for social justice for all in the here and now. Thus, NEP does too little to dismantle the oppressive and hierarchical structures and assumptions embedded within, and therefore supporting and maintaining, the Western story.

NEP serves as the basis for sustainability thinking and discourse (Agyeman, 2008). Yet, early on it was noted that sustainability as an

> idea has become hegemonic, an accepted meta-narrative, a given...We should therefore neither be surprised that no definition has been agreed upon, nor fear that this reveals a fundamental flaw in the concept. In the battle of big public ideas, sustainability has won: the task of the coming years is simply to work out the details, and to narrow the gap between its theory and practice.
>
> (Campbell, 1996, p. 27)

Unfortunately, the tensions at the gap between theory and practice persist, as evidenced by the two most commonly cited definitions' failure to even reference "justice and equity" (Agyeman *et al.*, 2002, p. 80).

Just sustainabilities, defined as "the need to ensure a better quality of life for all, now and into the future, in a just and equitable manner, whilst living within the limits of supporting ecosystems" (Agyeman, 2003, p. 2) addresses this failure. Such a position comes under what is recognized as "a strong sustainability approach" in that to achieve it would "require a more dramatic transformation of society's values and practices" (Jacques, 2015 p. 41) in contrast to "weak sustainability," which "argues we can continue to consume ecological goods and services at a growing rate, whereas strong sustainability insists on limits to consuming ecosystems" (p.41). In so doing, it can help "mind the gap" not just between individual words and behaviors (Kolmuss and Agyeman, 2002), it also holds the promise of closing the gap between the hegemonic rhetoric of new-sustainability and the prescience of old-sustainably by re-centering social and ecological justice as both an inter- and intragenerational imperative (Agyeman, 2008).

This re-centering is critical as sustainability has mostly discounted the environmental justice (EJ) movement. For example, in a review of web sites for the largest cities in the US, over 40 per cent had "sustainability projects" listed "but only five of these dealt with environmental justice" (Warner, 2002, p. 37). Perhaps because at its core EJ is a movement for equity, equality, and democracy that grew out of the larger US Civil Rights Movement. Recognizing this, we must state emphatically, as Bullard (1994) did earlier, that:

> People of color have *always resisted* actions by government and private industry that threaten the quality of life in their communities. Until

> recently, this resistance was largely ignored by policymakers. This activism took place before the first Earth Day in 1970; however, many of these struggles went unnoticed or were defined as merely part of the "modern" environmental movement.
>
> (p. 3) [italics added]

Hence, while this resistance is not 'new,' it is both worthy of repeating and essential to do so.

Such a diffuse emergence, however, makes it difficult to say with any certitude the exact date of EJ's onset. Still, those at the forefront of "the convergence of the environmental and social justice movements into a new movement" (Bullard, 1994, p. xx) mark the First National People of Color Environmental Leadership Summit held on October 27, 1991 as a watershed moment as activists, scholars, and citizens from across the country convened to mobilize and drafted principles for action (Chavis in Bullard, 1994, p. xi; Grossman in Bullard, 1994; Lazarus, 2000; Principles of Environmental Justice, 1996). The summit itself was the result of various earlier civil disobedience actions and reports from 'sacrifice zones' (cf. Bullard, 1994), the most significant being the *Toxic Wastes and Race in the United States*, released by the Commission for Racial Justice of the United Church of Christ (UCC) (Chavis Jr. and Lee 1987).

The UCC report was a response to a number of top-down attempts to quiet growing claims and grievances, including a 1983 Government Accountability Office's (GAO) report, *Siting of Hazardous Waste Landfills and Their Correlation to the Racial and Economic Status of Surrounding Communities*. In contrast to that report, the UCC documented nationally the "clear patterns which show that communities with greater minority percentages of the population are more likely to be the sites of such facilities. The possibility that these patterns resulted by chance is virtually impossible" (p. xv) regardless of income. Notably, in the UCC was an explicit definition of institutional racism that helped former National Association of the Advancement of Colored People (NAACP) director and report co-author, Rev. Dr. Benjamin Chavis Jr., coin the term "environmental racism" (cf. Chavis in Bullard, 1994; Lazarus, 2000).

Owing to this history, EJ advocacy focuses on the *fundamental rights* of humans and non-humans to clean air, land, water, and food; self-determination; informed consent; not to have to choose between unsafe or no work; full compensation for injustices; and civic participation in decision-making (Principles, 1996). Unfortunately, such ideals have yet to be reached creating a continued "equity deficit" (Agyeman, 2013 p. 4) and resultant disparity in the distribution and siting of environmentally degrading projects. Just sustainabilities' exhortation that sustainability "cannot be simply a 'green, or 'environmental' concern" (Agyeman, Bullard, and Evans, 2002 p. 78) helps reframe EJ struggles in policy and planning discourses to promote community participatory and procedural equity, well-being, and capital (physical, human, social, cultural and economic) (Agyemen, 2008).

While we hope such reframing succeeds in unseating the NEP and its empty promises, for us a real concern is that just sustainabilities will, instead, be emptied of its promise. Indeed, that the singular form has taken root suggests universalization has begun and with it diminishing attention to the unique constitution of community actors (i.e., species, built and natural environments, text and Internet, etc.) and the multiplicity of scales (e.g., block, neighborhood, town, county, state, region, etc.) necessary to mobilizing, extending, and replicating the work (Bickerstaff and Agyeman, 2009). To avoid this, in our selection of contributions we privileged the efforts and actors, recognizing that although many do not engage directly with the concept, they do advocate for social and ecological justice using an intersectional lens and thereby support a strong sustainability approach.

Achieving just sustainabilities requires maintaining a nuanced and complex analysis of all the variables and levels involved simultaneously—a significant challenge, especially as "the domination of sex, race, and class and the domination of nature are mutually reinforcing" (King, 1989, p. 20). It is for this reason the extension of intersectionality to just sustainabilities is necessary; it helps illuminate the transformative and mutually constitutive ways in which this reinforcement actually transpires structurally and individually, in time and in physical space, including within the body. Intersectionality's heuristic ability enhances our 'seeing' how these social signifiers intersect and co-construct in relation to each other and the environment. For example, while Agyeman and colleagues note "the rich can ensure that their children breathe clean air" (2003, p. 1), for us air itself becomes an intersection whose quality—or lack thereof—adds to the creation of how given humans understand themselves and are understood. As we stated in *Systemic Crises* "the use of intersectionality and the intra-actional ideal we sought addresses the tension between the materiality of air by unpacking the 'baggage of concretion' that engenders disparities" by "captur[ing] the interference patterns we experience in a manner that problematizes statuses and their relationship to each other" in the holographic process (Godfrey and Torres, 2016, p. 8). Indeed, for us, if the concept of just sustainabilities is to maintain its meaning then it must seep out to all our social engagements.

Awakening and reclaiming who we are

In his book title, Paul Hawken (2007) communicates his view that the multiple movements described represent a "blessed unrest … restoring grace, justice and beauty to the world": *Emergent Possibilities* is our contribution. We use the Western tools of numbers, theories, and proofs and reclaim those 'old' yet timeless, essential expressions of our socially constructed identities that have been contained, othered, and marginalized by the Western story—heart, spirit, and love. Thus, we sought to include "every manner of art" to assist our readers to "establish an original and felt relationship with the universe" (Swimme, 1990, p. 19). We hope that through this relationship all can "reweave new stories that acknowledge and value the biological and cultural diversity that sustains all life" (Diamond and Orenstein, 1990, p. xi) and that we have helped to "dispel the notion that poetry and politics, spirituality and activism, scholarship and vision are to remain forever divided, either from each other or within the same person" (Diamond and Orenstein, 1990, p. vii). In fact, while in our first volume the five sections had introductions written by us, here, since the emphasis on highlighting a diversity of voices is even more salient, we have decided to allow the pieces to speak for themselves. While some may deride us for having gone 'native,' we suggest that in fact it represents a 'doubled consciousness' we share contemplatively so that "notions of the relation of humans to nature and other taken-for-granted assumptions about reality could be transformed" (Tresch, 2001, p. 316).

Indeed, for us, the answer is, and perhaps always was, *both/and*, as put forth by intersectionality, not *either/or*. We carefully chose the title to this volume in that nothing is definite and, like the Zen story in our preface, the verdict and the possibilities for sustainability—in particular those that are just—'may be' and 'may not be.' Nevertheless, any 'emergent possibilities' for just sustainabilities are increased if intersectionality as a theory and as a practice is more widely engaged, seeing through the created hologram of ourselves in others and the 'other' in ourselves, across all of matter. Arriving at such an empathic intersection—and averting the looming reality of ecological and social collapse—requires remembering who we are and reweaving our story, to affirm, as the 1st Principle of the Environmental Justice (1996) movement does, "the sacredness of Mother Earth, ecological unity and the interdependence of all species, and the right to be free from ecological destruction."

References

Agyeman, J. (2013) *Introducing Just Sustainabilities: Policy, Planning and Practice*, New York: Zed Books.

Agyeman, J.(2008) Toward a 'just' sustainability? *Continuum – Journal of Media and Cultural Studies*, 22(6), 751–757.

Agyeman, J., Bullard, R. D. and Evans, B. (2002) Exploring the nexus: bringing together sustainability, environmental justice and equity, *Space and Polity*, 6(1), 77–90.

Agyeman, J., Bullard, R. D. and Evans, B. (2003) *Just Sustainabilities: Development in an Unequal World*, Cambridge, MA: MIT.

Barad, K. (2008) 'Posthumanist performativity: toward an understanding of how matter comes to matter' in S. Alaimo and S. Hekman (eds), *Material Feminisms*, Bloomington, IN: Indiana University Press, 120–156.

Berger, P. and Luckmann, T. (1967) *The Social Construction of Reality: A Treatise in the Sociology of Knowledge*, New York: Anchor.

Bickerstaff, K. and Agyeman, J. (2009) Assembling justice spaces: the scalar politics of environmental justice in north-east England, *Antipode*, 41(4), 781–806.

Bullard, R.D. (1994) *Unequal Protection: Environmental Justice and Communities of Color*, San Francisco, CA: Sierra Club Books.

Bullard, R.D. (1999) Dismantling environmental racism in the USA, *Local Environment*, 4, 5–19.

Bullard, R.D. (2000) *Dumping in Dixie: Race, Class and Environmental Quality*. Boulder, CO: Westview Press.

Cable, C. (2012) *Sustainable Failures: Environmental Policy and Democracy in a Petro-dependent World*, Philadelphia, PA: Temple University Press.

Campbell, S. (1996) Green cities, growing cities, just cities? Urban planning and the contradictions of sustainable development, *Journal of the American Planning Association*, 62(3), 296–312.

Carbado, D.W. (2013) Colorblind intersectionality, *Signs: Journal of women in culture and Society* 38(4) 811–845.

Carbado, D.W., Crenshaw, K., Mays, V.M., and Tomlinson, B. (2013) 'Intersectionality: mapping the movements of a theory', *Du Bois Review*, Fall 10(2), 303–312.

Catton, W.R. Jr. and Dunlap, R.E. (1978) Environmental sociology: A new paradigm, *The American Sociologist*, 13, 41–49.

Chavis, Jr., B.F. and Lee, C. (1987) *Toxic Wastes and Race in the United States: a National Report on the Racial and Socio-economic Characteristics of Communities with Hazardous Waste Sites*, New York City: Commission for Racial Justice United Church of Christ/Public Access Data, Inc.

Cho, S., Crenshaw, K., and McCall, L. (2013) Toward a field of intersectionality studies: theory, application and praxis, *Signs: Journal of Culture and Society*, 38(4), 785–810.

Collins, P.H. (2000) *Black Feminist Thought: Knowledge, Consciousness, and the Politics of Empowerment* (2nd edn), New York: Routledge.

Constitution of the Iroquois Nations (2014) The Great Binding Law-Gayansahagowa, prepared by Gerald Murphy, The Cleveland Free Net, National Public Telecomputing Network: Accessed December 18, 2015 www.indigenouspeople.net/iroqcon.htm

Crenshaw, K. (1989) Demarginalizing the intersection of race and sex: a black feminist critique of antidiscrimination doctrine, feminist theory, and antiracist politics, *University of Chicago Legal Forum*, 139.

Davenport, C. (2015) *New York Times*, December 12, 2015, accessed December 19, 2015. Available online at www.nytimes.com/2015/12/13/world/europe/climate-change-accord-paris.html

Diamond, I, and Orenstein, G.F. (eds) (1990) *Reweaving the World: The Emergence of Ecofeminsm*, San Francisco, CA: Sierra Club Books.

General Accountability Office [GAO] (1983) *Siting of Hazardous Waste Landfills and their Correlation to the Racial and Economic Status of Surrounding Communities*. Washington, DC: Author.

Global Alliance (2015) Earth-driven, not market-driven solutions to climate change, December 9, 2015, accessed December 19, 2015. Available online at www.therightsofnature.org/ron-events/tribunal-offers-earth-driven-not-market-driven-solutions-to-climate-change/

Godfrey, P.C. and Torres, D. (eds) (2016) *Systemic Crises of Global Climate Change: Intersections of Race, Class and Gender*, London: Routledge.

Hawken, P. (2007) *Blessed Unrest: How the Largest Movement in History is Restoring Grace, Justice and Beauty to the World*, New York: Penguin Books.

Jacques, P. (2015) *Sustainability: The Basics*. London: Routledge.

Keating, A. (2002) Charting pathways, marking thresholds…a warning, an introduction. In Anzaldúa, G.E. and Keating, A. (eds), *This Bridge We Call Home: Radical Visions for Transformation*, New York: Routledge (6–20).

King, Y. (1989) *The Ecology of feminism and the feminism of ecology*. In Judith Plant (ed), *Healing the Wounds: The Promise of Ecofeminism*. Philadelphia and Santa Cruz. New Society Publisher.

Kollmuss, A. and Agyeman, J. (2002) Mind the gap: why do people act environmentally and what are the barriers to pro-environmental behavior? *Environmental Education Research*, 8(3), 239–260.

Korton, D. (2012) *The Great Turning: From Empire to Earth Community*, New York: Kumarian Press.

Lorde, Audre. (1984) The master's tools will never dismantle the master's house. *Sister Outsider: Essays and Speeches*, ed, Berkeley, CA: Crossing Press, pp. 110–114.

Lazarus, R.J. (2000) Environmental racism! That's what it is, *University of Illinois Law Review*, 2000, 255–274.

Leopold, L. (2015) *Runaway Inequality: An Activists Guide to Economic Justice*, New York: The Labor Institute Press.

Lykke, N. (2009) Non-innocent intersections of feminism and environmentalism, *Women, gender and research*, 18(3–4), 36–44.

McKibben, B. (2015) Op-ed, *New York Times*, December 13, 2015, accessed December 19 2015. Available online at www.nytimes.com/2015/12/14/opinion/falling-short-on-climate-in-paris.html

May, V. M. (2015) Pursuing Intersectionality, Unsettling Dominant Imaginaries. New York: Routledge.

Montevalli, J. (2011) The history of greenwashing: how dirty towels impacted the green movement, *Daily Finance*, Feb. 12, accessed December 27 2015. Available online at www.dailyfinance.com/2011/02/12/the-history-of-greenwashing-how-dirty-towels-impacted-the-green/

Parenti, C. (2012) *Tropic of Chaos: Climate Change and the New Geography of Violence*, New York: Nation Books.

Pope Francis (2015) Encyclical letter, On care for our common home, paragraph 2, accessed December 20, 2015. Available online at w2.vatican.va/content/francesco/en/encyclicals/documents/papa-francesco_20150524_enciclica-laudato-si.html

Principles of Environmental Justice (1996) Preamble, accessed December 21, 2015. Available online at www.ejnet.org/ej/principles.html

Quinn, D. (1992) *Ishmael: An Adventure of Mind and Spirit*, New York: Bantam Books.

Milman, O. (2015) James Hansen, father of climate change awareness, calls Paris talks 'a fraud', *The Guardian*, December 12, 2015, accessed December 20, 2015. Available online at www.theguardian.com/environment/2015/dec/12/james-hansen-climate-change-paris-talks-fraud

Ray, P.H. and Anderson, S.R. (2000) *The Cultural Creatives: How 50 Million People are Changing the World*, New York: Harmony Books.

Swimme, B. (1990) How to heal a lobotomy. In Diamond, I. and Orenstein G.F., eds., *Reweaving the World: The Emergence of Ecofeminsm*, San Francisco: Sierra Club Books, 4–14.

Taylor, D.E. (2000) The rise of the environmental justice paradigm: injustice framing and the social construction of environmental discourses, *American Behavioral Scientist*, 43(4), 508–80.

Tresch, J. (2001) Ongoing native: Thomas Kuhn and anthropological method, *Philosophy of the Social Sciences*, 31(3), 303–322.

Warren, K., ed. (1997) *Ecofeminism: Women, Culture, Nature*, Bloomington, IN: Indiana University Press.

Warner, K. (2002) Linking local sustainability initiatives with environmental justice, *Local Environment*, 7(1), 35–47.

Washington, H. (2015) *Demystifying Sustainability; Toward Real Solutions*, London: Routledge.

Zehner, O. (2012) *Green Illusions: The Dirty Secrets of Clean Energy*, Lincoln, NE: University of Nebraska Press.

Part I
Air

Figure I.1 Air

1 You probably still have doubts

Anonymous

It's a fallacious argument.
I will prove that there is no such thing.
Global warming???
Yeah it's hot today. It's supposed be,
It's summer Duh!!

There you have it.
The world is FLAT,
There was NEVER a moon landing
And there was only one shooter in the depository
Everything that you "believe" is nothing more than propaganda.
Neo-commie leftist hoo-ha.
Have you ever seen the world spin?
That's why I home school my kids,
This notion that the world is a ball spinning in space is ludicrous.
Schools have globes. Lies. All lies.
The world is flat
And heaven is right above the clouds where God watches
over us in a big Chair with Jesus at his right and St. Peter at the gates (even
commies like yourself).

And in His wisdom he gave us good people to dispel this ridiculous idea that carbon emissions are
harming this sweet land.
Why, every time I see a smokestack all I see is industry. I see jobs.

Don't talk to me about hurricanes and ocean levels rising and the polar caps melting.
Tsunami?
Well that's just nature. Nature doing her thing
Just as the Good Lord designed it.

You probably still have doubts.
Simply put, if the Good Lord did not want this we wouldn't have it.

You must be a pinko commie atheist
I will pray that The Lord forgives you for your evil ways and questioning His creation.
And don't talk about planting any more trees.
Please,
I don't want to have to explain how it was His design that we chop the trees raze the forests burn what
we need…
God wants us to be INDUSTRIOUS. It's in the bible

There is no such thing as Global Warming.
Now if we are done, I am going to watch some television.

You know? I will never understand how the people get into that box.
Praise Jesus. It's a miracle.

2 The Virgin and the seed

Phoebe Godfrey

Figure 2.1 Virgin. Oil and mixed media on canvas 78" × 52" × 62"

This triptych altarpiece speaks to the more than 2000-year-old story of the Christian 'savior' and the 'mother figure' and brings to light our current environmental crisis and what I believe is the need to return to Earth-based spirituality. On the right are symbols of patriarchal religious domination, as in the bishop's crook, and a stained glass window depicting extinct birds and animals. There is a clock face at the top of the window representing the idea that the Christian savior is the alpha and omega—the beginning and end of time. In the middle is the Virgin, aka Pachamama, who has been used, along with her infant son, as the feminine face of the dominating force but who has now left the institution, giving up the male child who has been used to lead us away from the true sources of salvation—our connections with ourselves, hence nature. In her arms is the seed, sprouting so strongly that it is coming out of the canvas. On the left are the four elements: Earth, Air, Water and Fire—all inviting us to return to their healing powers. This panel includes bark, small tree branches and roots that are freely growing out and off the canvas. Above the Virgin is the Moon, the ancient symbol of the many goddesses who have reigned in the night sky. The power expressed in this piece is about the indomitable life force within us all and that asks and needs to be reclaimed.

3 Womanism and agroecology

An intersectional praxis seed keeping as acts of political warfare

Shakara Tyler and Aleya Fraser

> Earth is my home – though for centuries white people have tried to convince me I have no right to exist, except in the dirtiest, darkest corners of the globe. So let me tell you: I intend to protect my home.
>
> (Alice Walker, 1983, p. 342)

> Women of color warriors are constant warriors who dig in bare earth to feed the hungry child, who pray for health the bedside of the sick when there is no medicine, who fashion a toy to make a poor child smile, who take to the streets demanding freedom, freedom, freedom against armed police. Every act of survival by a woman of color is an act of resistance to the holocaust and the war. No soldier fights harder than a woman warrior for she fights for total change, for a new order in a world in which can finally rest and love.
>
> (Barbara Omolade, 1994, p. 220)

***

Today we are barren fields completely dry after a fall harvest.
Yesterday we were our mothers' spring garden, full of life and color.
Vivid as your imagination and as free-form as the kinks growing out of my head.
Alas, time is relative and today is yesterday and tomorrow is today and today is never-ending as I sit here in this barren field full of art
reflecting on how we got here.

My children always ask me…
do you hate all of the Washichus?
I tell them that my heart does not have the capacity for hate
and neither does their own.
I do wish they would be banished from Aya
and sent back to their own time space continuum
which they called Earth.
They came here from a parallel universe on a parallel planet
60,000 years ahead of the time that Aya was when they arrived.
They came looking for refuge from the ultraviolet rays

making their own world uninhabitable due to a myriad of human and natural causes. They came with the sole purpose of farming the land and the peoples of our Earth
in the same unsustainable ways
that turned all but underground caverns on their planet uninhabitable.
They came like an unchecked virus set loose on our precious ecosystem, infecting every part with their desperation.

The plants speak to us but do not speak to the Washichus
so they appointed us as stewards of the land.
It is written in our DNA to care for the land as an extension of our being
much like rings of growth on a tree
every year of our maturity grants us secret understanding of a new life classification widening our understanding of our own beings to include all life
and relationally all matter.
If it were not for the blatant disregard of the strangers
to our circles of compassion
we would have loved being stewards of the land
but only in a harmonious system.
The Wasichus instead look at the land as property to be owned conquered
and raped to death.
Almost no one on their place called Earth could actually hear the plants speak
feel what they hear and see what they say.
Electromagnetic radiation, pollution and negative energy
clouded the plant to human pathways
which we here on Aya depend on for survival.
We were twins in every sense
except for the way in which we cared for the land.

Fast forward to today…
as I sit here staring at this barren field
just fifty years after we were stolen from the land
across the sea that was already turning uninhabitable
by the Washichus due to the hot dry conditions and large amount of UV radiation.
The other side of our sphere has turned cooler, cloudier and rainier in climate
so we were brought here where the aliens can actually survive and have already driven those before us to extinction.
It was here on the other side that the aliens learned our secret.
The fact that melanated folks with feminine energy (women and third genders) hear plants the loudest. We cannot ignore them, we are one with them.

This naturally made us (the backbones as they call us) very valuable as they sought and continue to seek ways to dominate our life sources, our gods, our knowledge, our wisdom, our plants.

We have resisted in many ways
from not sharing all of our knowledge to seed keeping.
However they still have found ways to steal our secrets and use them to harm us, the Earth and ironically themselves.

A revolution is brewing.
For decades we have been secretly rebuilding our seed bank of beautiful, rare and important seeds to our spiritual and physical livelihoods.
Varieties that will be especially useful in the barren climate of our homeland
which has been turned into a desert by the Washichus abuse.
We are spiritually connected to that land and must return.
Our mitochondrial pull back to our original land is so deep that it is now all the plants speak to us about.
Sorghum speaks the loudest.
She says that we must take her there to repair the soils
and feed us and our animals
until we can begin growing her brothers and sisters.
She is drought and heat-tolerant because its roots run deep
and strong branching like the veins on the back of my hand.
Her grains can be used for animal feed
popped like popcorn
boiled like rice
ground into flour
and fermented into beer
which my mother used to trade in the marketplace.
Beautiful grains with beautiful individual stories.
The stalks can be used to make molasses
broom sticks
building material.
Her stalks and leaves converting sunlight into energy
in efficient ways that my grandmother has never witnessed in any other plants
and she has known millions.
Sorghum, in the same voice of every woman I know said
she will provide us all that we need to fight our oppressors and I trust her.
I trust her seeds.

***

Introduction

Womanism, as coined by Alice Walker in *In Search of our Mother's Garden* employs a critical cultural lens to validate the special vantage point of black women from many intersecting paradigms – black lesbian, black queer, black poor women, and the very junctions of these identities. These multiple vantage points are unearthed to better recognize and understand the interlocking systems of oppression while embodying humanism for collective liberation. Womanism breathed life into intersectionality, as a theory and practice. As conceptualized by Walker, womanism contextualizes black women's resistance to various faces of oppression as acts of loving nature and nurturing the gardens from which black people's resilience grew. Black women's resistance can also be described akin to warfare with black and brown women as "constant warriors who dig in bare earth to feed the hungry child, who pray for health at the bedside of the sick when there is no medicine" (Omolade, 1994, p. 220). Black people's resilience during slavery and post-slavery eras was agroecologically cultivated by ancestral knowledge of sustaining the land and all things that flow from the land including seeds.

Agroecology is a dual process of ecological agricultural production and organizing and building community self-determination that builds upon ancestral and cultural knowledge (WhyHunger, 2015). Women and our knowledge, values, vision, and leadership are critical to agroecology as a practice, science, and social movement (Anderson *et al.*, 2015). This narrative theorizes womanism and agroecology as an intersectional praxis building on the theory of intersectionality. The intent is to provide a way of Afro-futuristically visioning black women's role as agroecological warriors keeping seeds to creatively plant and cultivate an alternative world in which the "imperialist white supremacists capitalists patriarchy" (hooks, 2013, p. 4) becomes transparent and we can practice different ways of being (Alexander, 2000). We are black women immersed in an intersectional matrix where we can stand in our purpose as womanists, returning generation farmers, educators, warriors, artists and scholars. These intertwined identities are the roots of our resistance, healing, and empowerment that construct our intersectional activism across both wombs of womanism and agroecology.

While no other group in the United States (US) has had their identity socialized out of existence as have black women (hooks, 1981), US black women have also continued to produce social thought designed to oppose this invisibility and oppression (Collins, 2009). This narrative intends to be an exhibition of our social thought where – like the black literary tradition – "escape for the body and freedom for the soul went together" (Walker, 1983, p. 5). Opening this narrative with a visionary fiction story represents the wedding of our bodily and soul freedom. While we acknowledge the "matrix of domination" that refers to how intersecting oppressions are organized (Collins, 2009), we focus on the regenerative notions of intersectionality to resist oppression. By dwelling not on the oppressive reality constructed by patriarchal domains, but dwelling on the liberation visions of black women in agroecology through an Afro-futuristic seed story, we unveil a new sun Octavia Butler speaks so confidently of in her visionary work (Butler, 2012).

Visionary fiction offers new models for creating new possible worlds (Ferguson is the Future, 2015). According to Octavia Butler, "There is nothing new under the sun, but there are new suns" (in Canavan, 2014). Seed keeping stories as acts of political warfare are an incubation of an alternative agroecological world rooted in our womanist imaginations – imaginations where we reclaim our mothers, grandmothers, great grandmothers, and great greats. We preface the theoretical exploration of womanism and agroecology with an Afro-futuristic narrative to plot a different course forward by discussing the historical, present and future manifestations of an agroecological reality centering the experiences of black women, the *herstories*. It becomes imaginatively emblematic of the larger freedom struggle for humanity to return to the rightful communion with mother earth. For "[i]n a world of possibility for us all, our personal visions help lay the groundwork for political action" (Lorde, 2003, p. 27) and we theorize womanist agroecological practice as a creative incubating space to grow new visions of the worlds we want to exist.

This is not about acknowledging and valuing women's role as a 'debt' that agroecology has yet to pay (Lopes and Jomalinis, 2011) because we are not waiting for the debt to be paid. That would mean waiting to be freed, and essentially, we are already free. We know our power and we believe in our capacity to use our intersecting oppression as inspiration for creativity to transcend it. We honor the complexity of our identities and find solace in a framework that explicitly speaks to the "audacious, courageous, and willful behavior" (Walker, 1983, p. xi) of black women claiming intersectionalilty as seeds of power all the while knowing we may "not have racial, sexual, heterosexual, or class privilege to rely upon" (Combahee River Collective, 1978, p. 7). This dual contradiction of the visible oppression and invisible privilege is the source of our greatest strength, it is the strength to find the erotic in our lives and use it creatively because our erotic knowledge empowers us (Lorde, 1984).

We know from Paulo Freire when the oppressed discover themselves to be the "hosts" of the oppressor they will contribute to the "midwifery of their liberating pedagogy" (Hederman, 1982, p. 58). Rather than being hosts of the oppressor we are the hosts of our liberation. We are not "exquisite butterflies trapped in an evil honey" (Walker, 1983, p. 232); we are not the "*mules* of the world" (Walker, 1983, p. 232, her emphasis); nor are we using the master's tools to dismantle the master's house (Lorde, 2003). We are the "women who literally covered the holes in our walls with sunflowers" (Walker, 1983, p. 242). We are the women preserving ourselves as acts of political warfare (Lorde, 1988). We are the women boldly saying to any intruder on our personal or ecological sovereignty that "I intend to protect my home" (Walker, 1983, p. 342). We are sisters actively crafting meaning out of our circumstances beyond our relative disadvantages (Harris Perry, 2011, p. 46). Like the nineteenth-century black female, we are not relying on any group to provide us with a blueprint for change rather we are the makers of blueprints (hooks, 1981). This narrative – comprised of a theoretical exploration and a visionary herstory – is our blueprint.

Intersectionality

The intersection of womanism and agroecology is complimented by intersectionality as a theory and practice. Theoretically, intersectionality most often refers to particular forms of intersecting oppressions such as race, gender, sexuality and nation recognizing the many branches of oppression that work together to produce injustices (Collins, 2009). The elements that constitute the global, Eurocentric, capitalist model of power do not stand separately from each other; intersectionality reveal what is not seen when race and gender categories are conceptualized separately (Lugones, 2007). Isolating these categories that have become influential identity markers homogenizes and forces a cherry picking of the dominant in the group (Crenshaw, 1995). This "logic of categorical separation distorts what exists at the intersection, such as violence against women of color" (Lugones, 2007, p. 193). Intersectionality has become the praxis to counter this erasure of compounding categorical oppressions.

Womanism is critical rhetoric to use within the intersectionality context for two reasons. Rather than the first wave feminism of the twentieth century that countenanced no one else's gender oppression (Lugones, 2007) other than white woman's and the second wave feminist efforts to transnationally (Mendez, 2015) include 'others' from which black feminism grew (Combahee River Collective, 1978), we choose to employ a womanist framework that emphasizes the collective liberation of humanity. It becomes a necessary point of reference in this context because Alice Walker coined the term in direct resistance to the lack of space women of color, and more specifically black women, possessed in the non-intersectional first and second waves of the U.S. feminism movements (James *et al.*, 2009), which essentially pleaded for white women to become honorary men within the system that oppressed them. Additionally, like bell hooks (1981), we choose to "re-appropriate the term 'feminism'" (p. 195), to illuminate that to be 'feminist' in any authentic sense of the term is to want for all peoples' liberation from sexist role patterns, domination, and oppression.

In essence, this is the crux of womanism and the discernment between feminism and womanism detracts from the multilayered narrative we choose to create. In agreement with Patricia Hill Collins (1998), the controversy over naming is a political distraction that diverts attention from the dire need of black women's liberation. Though we root this narrative in Alice Walker's universalist womanism, we do not intend to generalize across all black women and we purposefully speak from our intersectional position as black women living and loving "womanishly" (Walker, 1983) within the ancestral practice of agroecology. Our black 'womanish' identities agroecologically communing with land presents another layer of intersectionality. In *Building Houses out of Chicken Legs*, Psyche Williams-Forson theorizes black women, food, and power by adding another dimension to the usual race, class, gender intersection: food. She argues producing one's own knowledge of self and community persists at this tetralogy and food has become an important cultural mediator preserving cultural customs and rituals. Building on this notion, we add relationships with the earth to the intersectional mix because an intimate relationship with food as a cultural practice is undeniably tied to an intimate relationship with the earth.

There lies an intersectionality rooted in an agroecological consciousness with humans and land communing for joint liberation. Carolyn Merchant in Shades of Darkness; Race and Environmental History states, "Slavery and soil degradation are interlinked systems of exploitation, and deep-seated connections exist between the enslavement of human bodies and the enslavement of the land" (Merchant, 2003, p. 380). As the ecology of the soil is broken, the ecology of the people – who are so deeply tied to the land – also becomes broken. The role of women is critical in this intersectional enslavement as the 'ecofeminist' principle illuminates the 'material resourcing' – or less euphemistically coined, the violence – of women and nature structurally interconnected in the capitalist patriarchal system (Mies and Shiva, 2014). The violence against women of color and the violence of land degradation constructs a dual intersectionality of women of color who work the land. Thus, the integrity of the land becomes contingent upon the integrity women possess in deeply connecting to the land because as we know from the 'ecofeminism' framework, both women and nature are producers of life (Shiva, 1989). This is where agroecology meets feminism, and in our black women's lens: womanism. As womanist, we expound upon this connection to focus exclusively on the role of black women in agroecology, as an intersectional praxis.

Womanism and agroecology: an intersectional praxis

Womanism is a nuanced framework with a variety of orientations. Alice Walkers's universalist lens of centering the liberation of all people while appreciating and preferring women's culture and emotional flexibility (Walker, 1983) differs from Clenora Hudson-Weems's nationalist lens that theorizes an ideology created and designed for all women of African descent and sits separately from white and black feminisms (Hudson-Weems, 1994). Africana Womanism centers the 'unique' experiences, struggles, needs, and desires of Africana women and differs from African feminism (Hudson-Weems, 1994). Within the multiplicities of these womanist perspectives, Alice Walker's universalist perspective is the most salient in conceiving the merged praxis of womanism and agroecology given the explicit attention to human liberation that comes with black women's liberation. As bell hooks says, "the only way out of domination is love" (Yancey and hooks, 2015, p. 1) and womanism and agroecology, as we know it, are erotic acts that lovingly confront systems of domination. In *In Search of our Mother's Garden* Walker excavates the powerful experiences black women possess to what we consider to be agroecological systems. She says, "[f]or these grandmothers and mothers of ours were not Saints, but Artists, driven to a numb and bleeding madness by the springs of creativity in them for which was no release. They were Creators" (Walker, 1983, p. 233). What they were creating was an alternative ecology to the physical and psychological enslavement, displacement, and exploitation of human bodies and the land. Acts of creation and creativity were acts of resistance against the dual enslavement of the body and land, as previously mentioned.

A consistent theme in the people's agroeoclogical discourse is women leading creative processes of knowledge building in void of dominant institutional support. Speaking from the global social movement of agroecology, a woman from the Agroecological Movement of Latin America stated: "When state or government is lacking, women have been creative" (Anderson *et al.*, 2015). "It is the nature of the oppressed to rise against the oppression" (Walker, 1983, p. 275) and the robust ecology of the people is rooted in the robust ecology of (re)communing with the land – which counteracts the oppression. By standing in solidarity with the land and everything that flows from it, what becomes tangible is a womanist agroecological praxis "committed to survival and wholeness of entire people, male and female" (Walker, 1983, p. xi). As Walker states in *Revolutionary Petunias* (1983): I "*raise* them because you just put them in any kind of soil and they bloom their heads off … like black people tend to do" (p. 268, italics added). In other words, the feminine creative energy has facilitated the resiliency of black peoples' rooting in the most unfertile soils but continue to thrive – "as if by magic" (Walker, 1983, p. 241).

Historically, black women developed a unique set of perspectives by way of the gardens they grew as slaves and then as freedwomen (Glave, 2003). "They gardened within a gendered and racial milieu that gave the application of these simple instruments of skill a complex social potency" (Glave, 2003, p. 396). Fannie Lou Hamer's Freedom Cooperative is an example of this space led by a black woman farm activist in the Mississippi Delta in the late 1960s. Fannie Lou Hamer's experience of starvation as a political weapon allowed her to identify this structural obstacle to collective black progress and develop Freedom Farms as a locale of resistance against the oppressive white establishment (White, 2015). Developing the Freedom Farm Cooperative as a political institution of communal well-being was a womanist act that ensured housing, health care, employment, education, and access to healthy food were available when purposefully withheld by the white power structure of rural Mississippi. The womanist trait of being "[t]raditionally capable" (Walker, 1983, p. xi) is reflected in Hamer's intersectional being as the child of sharecroppers, farmworker that was eventually kicked off the plantation and a community organizer who sang spirituals as a calming mechanism during contentious moments (White, 2015). Within this framework, Hamer's resilient spirit that centered her organizing around food, freedom, and land is the epitome of the intersectional praxis of womanism and agroecology.

In her womanist prose, Walker often centers the rooted resilience of black women in the Post-Emancipation era. These are the "women who literally covered the holes in our walls with sunflowers" (Walker, 1983, p. 242), planted things that grew "as if by magic" (Walker, 1983, p. 241), and creatively colored childhood poverty with the flower gardens of sunflowers, petunias, roses, dahlias, and other blooms. At the heart of it all, these creative creations were agroecological acts, which births what we consider womanist agroecology. Like the times her mother worked in her flowers was her only moment of radiance, agroecology meets womanism when black women are fulfilled by creatively cultivating the earth. Thus, agroecology is a womanist act where she "is involved in work her soul must have" (Walker, 1983, p. 241). This is 'the ecological approach' that centers self (and societal) transformation happening intimately at the level of the planet (Gumbs, 2010) because of the inseparable intersection of the two. Womanist agroecology illuminates the power of artistry where an ecological approach means artistically moving energy and inspiration through (Gumbs, 2010) and also being an artist like Alice Walker's mother who ordered "the universe in the image of her personal conception of Beauty" (Walker, 1983, p. 241) through simple acts of earth work.

The intersectional praxis of womanism and agroecology is rooted in the "creative spark, the seed of the flower" our mothers and grandmothers "never hoped to see" (Walker, 1983, p. 240). In their honor, we plant the seeds they left behind as conveying seedkeeping as womanist agroecological acts of political warfare. Remembering the soul stirring words of Audre Lorde (1988): "Caring for myself is not self-indulgence, it is self-preservation, and that is an act of political warfare" (p. 132), our freedom as womanist agroecologists is tied to our self-preservationist acts of keeping seeds.

Seedkeeping as an act of political warfare

To resist detrimental effects of climate change, seed keeping is a critical act of survival and resistance to retain biodiversity and is a critical act of resistance against the capitalist agricultural system. According to Vandana Shiva in *Seed Sovereignty, Food Security*, "Women are showing the way to having both bread and freedom" (2015, p. 13) and are in the vanguard of defending seed freedom and sovereignty (Shiva, 2015). As such, the agroecology paradigm, and more particularly seed keeping is heavily influenced by women's knowledge and practice. Black women – as many women around the world – have had a paramount role in seed keeping. As previously mentioned via Williams-Forson, food, and seeds as a preface to food play significant roles in creative knowledge production for self and community.

Black women were the creative sources of gardening and/or farming in their communities from slavery to the early twentieth century (Glave, 2003). They manipulated and interpreted garden spaces for sustenance, comfort, joy, and sometimes profit (Glave, 2003). Perhaps this is why Parham says: "Pungent and composite, the smell of farmyards is the fragrance of the woman" (p. 55). Though the emphasis here is on black women in the gardens, it is important to acknowledge black women were known to do backbreaking work in the US South (Lugones, 2007) or to complete the same agricultural tasks as men (hooks, 1981) in the enslaved plantation fields and on sharecropping farms and beyond. And, one of the critical roles of black women in the agricultural matrix was seed keeping. More than likely, these enslaved women used generationally 'kept' seeds to grow vegetables like okra, watermelon, white sweet potatoes, and many more crops (Glave, 2003) that reminded them of their homeland somewhere in the ecologically rich womb of Africa. The seeds represented more than food production. They also symbolized the cultural stories and secrets that came with the seeds and were passed on from generation to generation.

Our black female ancestors were pivotal in diffusing the African dietary preference in the Americas. For example, before embarking on the horrors of the Middle Passage, African woman hid rice and other grain crops in their hair and their children's hair, "bestowing a gift of lifesaving food from Africa" (Carney and Rosomoff), 2009, p. 76). Black women led these seed keeping acts throughout the slavery and post-slavery eras through tending home gardens, cooking African indigenous crops in the home, and passing on seeds and their stories from generation to generation as acts of struggle, survival, and self-determination. This is an agroecological act of political warfare – that also engulfs sociocultural warfare – aiming to counteract the raping of land and people perpetuated by the slave trade and what became capitalist agriculture.

Black women guarded seeds with their Afro and dreadlock hairstyles (Carney and Rosomoff, 2009). They were seed keepers that guarded seeds by entrusting their children to the life giving sources of cultural heritage (Carney and Rosomoff, 2009). And, they were seed keepers that inspired us to continue the tradition because the preservation of families, communities and ourselves undeniably depends upon it. Like the story of Alice Walker's mother who planted a flower garden in the 1930s and 1940s, "[o]ther stories are waiting in the hands, arms, shoulders, and backs" (Glave, 2003, p. 407) of black women who womanize agroecology through seed keeping and other acts of creative resistance. These are "womanish" practices that represent the "audacious, courageous, and *willful*" (Walker, 1983, p. xi) actions where seed keeping represented acts of self-preservation.

Seeds are the springs from which everything flows, and seeding the discourses to (re)conceive womanism within the fertile womb of agroecology (and vice versa) will lead to an abundant harvest for both wombs. Womanism and agroecology as an intersectional praxis essentially reflects Barbara Omolade's (1994, p. 220) words in *The Rising Song of African American Women:* "Every act of survival by a woman of color is an act of resistance to the holocaust and the war. No soldier fights harder than a woman warrior for she fights for total change, for a new order in a world in which can finally rest and love." Seed keeping as acts of political warfare are the womanist love songs that militantly serenades the earthly destruction back to its rightful place of peace and wholeness. Keeping seeds and the stories that come with them are among many of the creative acts that will make the earthly transformation we all seek become immanently tangible within our struggle.

References

Alexander, M. J. (2005) *Pedagogies of Crossing: Meditations on Feminism, Sexual Politics, Memory, and The Sacred*, Durham and London, NC: Duke University Press.

Anderson, C., Pimbery, M., and Kiss, C. (2015) Building, defending and strengthening agroecology: a global struggle for food sovereignty, *AgriCultures Network*, available online at www.agriculturesnetwork.org/library/253979.

Butler, O.E. (2012) *Parable of the Sower*, New York: Open Road Media.

Canavan, G. (2014) 'There's Nothing New / Under The Sun, / But There Are New Suns': recovering Octavia E. Butler's lost parables, *Los Angeles Review of Books*, available online at www.lareviewofbooks.org/essay/theres-nothing-new-sun-new-suns-recovering-octavia-e-butlers-lost-parables.

Carney, J.A. and Rosomoff, R.N. (2011) *In the Shadow of Slavery: Africa's Botanical Legacy in the Atlantic World*, Berkeley, Los Angeles, CA, London: University of California Press.

Collins, P.H. (1998) *Fighting Words: Black Women and the Search for Justice, Vol. 7*. Minneapolis, MN: University of Minnesota Press.

Collins, P.H. (2009) *Black Feminist Thought*, New York: Routledge.

Combahee River Collective (2009) the Combahee River Collective statement, 1978. In James, S.M., Foster, F.S., and Guy-Sheftall, B. (eds), *Still Brave: The Evolution of Black Women's Studies*, Feminist Press, 3–11.

Crenshaw, K. (1995) Mapping the margins: intersectional, identity politics and violence against women of color, *Critical Race Theory: The Writings that Formed the Movement*, New York: New York University, 363–377.

Ferguson is the future: speculative arts + social justice (2015) *Black to the Future: An Imagination Incubator* available online at https://blacktothefuture.princeton.edu/.

Glave, D.D. (2010) *Rooted in the Earth: Reclaiming the African American Environmental Heritage*, Chicago, IL: Chicago Review Press.

Gumbs, A.P. (2010) This is what it sounds like (an ecological approach), *Scholar and Feminist Online*, Issue 8.3, Summer 2010, available online at www.sfonline.barnard.edu/polyphonic/gumbs_01.htm

Harris-Perry, M.V. (2011) *Sister Citizen: Shame, Stereotypes, and Black Women in America*, New Haven, CT: Yale University Press.

hooks, b. (1981) *Ain't I a Woman: Black Women and Feminism*, Boston, MA: South End Press.

hooks, b. (2013) Racism: naming what hurts, in hooks, b. (ed), *Writing Beyond Race: Living Theory and Practice*, New York: Routledge, 10–25.

Hudson-Weems, C. (1994) *Africana Womanism: Reclaiming Ourselves*, New York: Bedford Publishers.

James, S.M., Foster, F.S., and Guy-Sheftall, B. (eds) (2009) *Still Brave: The Evolution of Black Women's Studies*, New York: Feminist Press.

Lewis, R. and Mills, S. (2003) *Feminist Postcolonial Theory: A Reader*, London: Routledge.

Lopes, A. P. and Jomalinis, E. (2011) *Feminist Perspectives Towards Transforming Economic Power. Agroecology: Exploring Opportunities for Women's Empowerment Based on Experiences in Brazil*, available online at www.issuu.com/awid/docs/fpttec_agroecology_eng

Lorde, A. (1984a) Age, race, class, and sex, *Sister Outsider*, Trumansburg, NY: Crossing Press, 114–123.

Lorde, A. (1984b) The uses of the erotic: the erotic as power, *Sister Outsider*, 53-59.

Lorde, A. (1988) *A Burst of Light: Essays*, Ann Arbour, MI: Firebrand Books.

Lorde, A. (2003) The master's tools will never dismantle the master's house, in *Feminist Postcolonial Theory: A Reader*. New York: Routledge, 25–27.

Lugones, M. (2007) Heterosexualism and the colonial/modern gender system, *Hypatia*, 22(1), 186–219.

Merchant, C. (2003) Shades of darkness: race and environmental history, *Environmental History*, 8(3), 380–394.

Mendez, X. (2015) Notes toward a decolonial feminist methodology: revisiting the race/gender matrix, *Trans-Scripts*, 5, 41–59.

Mies, M. and Shiva, V. (1993) *Ecofeminism*, New York: Zed Books.

Nash, J.C. (2008) Re-thinking intersectionality, *Feminist Review*, 89(1), 1–15.

Omolade, B. (1994) *The Rising Song of African American Women* New York: Routledge.

Parham, M. (2009) *Haunting and Displacement in African American Literature and Culture*, New York: Routledge.

Shiva, V. (1989) *Staying Alive: Women, Development, and Ecology in India*, New York: Zed Books.

Shiva, V. (2015) *Seed Sovereignty, Food Security*, New Delhi: Women Unlimited.

White, M. (2015) 'A Pig and a Garden': Fannie Lou Hamer as organic intellectual and the freedom farms cooperative. In *Freedom Farmers: Agricultural Resistance and the Black Freedom Movement*, unpublished manuscript.

WhyHunger (2015) *Agroecology: Putting Food Sovereignty into Action*, New York [online] available at whyhunger.org/uploads/fileAssets/6ca854_4622aa.pdf

Yancy, G. and hooks, b. (2015) bell hooks: Buddhism, the beats and loving blackness, *NYT Opinion Pages* available online at http://nyti.ms/1XYuz9N

4 An economy of hope

The surprising rise of a "Grassroots Democratic Economy" (GDE)

Len Krimerman

Introducing the GDE

The notion of a GDE may seem both obscure and far-fetched. Don't the Big Banks and the multi-national corporations control our economy, as well as that of the entire planet?

Perhaps so, but I will argue that the GDE is a steadily emerging reality, and despite being a relatively small and recent one, it is gaining significant strength, sustainability, and impact throughout all sections of this country.

Let's start with a rough definition. What I mean by a GDE is *an economy shaped by those who labor and consume within it, and not by those who possess privileged amounts of wealth*. In short: where the 99 percent who struggle to make ends meet are in charge, rather than the 1 percent, whose average income far exceeds $1 million. Note, the 1 percent are not excluded or banished, but they do not control the goals, products, services, or institutions of economic life.

A closer look

But what, more precisely, does the GDE look like in this country? Let's approach this question by briefly examining its historical development.

Worker controlled cooperatives had been developed before in the United States (US), e.g., by the Knights of Labor in the late nineteenth century, but this century's worker co-op movement did not get started – as a movement – until the 1970s. It was during that decade that the first technical assistance organization aimed exclusively at supporting worker co-ops emerged: the Industrial Cooperative Association in Boston.

At first, and through the late 1990s, growth was meager, and most often one small co-op at a time. But things began to change with the new century, and most noticeably during the Great Recession years. Today, small, halting, and gradual growth is being steadily replaced by worker cooperative expansion in many directions, especially in new sections of the country and economic sectors.

Instead of isolated co-op enterprises on the east and west coasts, we now see a network of regional cooperative development hubs throughout almost all parts of the country, as well as the formation of a national organization, the United

States Federation of Worker Cooperatives (USFWC). These new organizations function as incubators for the growth of multiple cooperatives, and as catalysts for collaboration among those enterprises.

Institutional barriers that restricted GDE's growth are now breaking down. For example, government agencies such as the Small Business Administration (SBA) and the US Department of Agriculture now recognize worker cooperatives as genuine and sustainable economic alternatives. And many labor unions now support the development of worker cooperatives by their members, rather than being suspicious of or hostile towards them.

Critique

If the GDE is now developing at an unusual pace and in many unexpected ways, a conclusion that one might be tempted to draw is that joining or starting a worker cooperative is both a worthy and a prudentially wise decision. Such a decision, ultimately, might be warranted, but at this point, it could well strike us as too good, or too simple, to be true.

More specifically, two questions can be raised about the GDE. The first concerns its *sustainability*; the second focuses on its *intersectionality*, or its capacity to assist in building an economy that works for all groups in our society, especially those now oppressed or marginalized.

As for 'sustainability,' let's start by distinguishing two forms of it – which I'll call 'internal' and 'external.' The first of these focuses on whether the GDA's workplace democracy community can sustain itself, whether *its own recent growth can be expected to keep expanding in the future*. External sustainability is quite different: it turns on whether this new economy can meet the needs of the current generation *without undermining or threatening the needs of future generations*. At this point, it is difficult to predict with full certainty the future growth of the GDE, or its workplace democracy dimension. As many others have noted, we might be experiencing a paradigm shift or just a bubble, bound to soon burst. Still, I think there are good grounds for modest optimism.

Over four decades ago, a movement in many ways similar to GDE came to life in the US, and elsewhere. It challenged the conventional and dominant way learners were treated and learning organized. Calling for detachment from both the public and private systems of (mis-)education, it offered a wide spectrum of grassroots democratic 'free schools,' 'learning communities,' 'experimental colleges'; by some accounts well over a thousand of these formed in the late 1960s and early 1970s. Most of these were started by groups of parents and their children, and most kept their expenses minimal, so as to reach out to low-income families. *But the movement was not sustainable*; it died as quickly as it arose, and by 1975 it became hard to find even a hundred of them, and those remaining tended to be extremely expensive.

Will something like this be the fate of the now-emerging GDE? I don't think so. One way to see why is to examine how the current GDE differs from the Free School movement.

First of all, free schools were all on their own, isolated, unable to draw support from one another. This is far from the case within today's GDE: as we have already seen, it has formed large-scale regional associations (such as the Eastern Conference for Workplace Democracy and the Western Worker Cooperative Conference), as well as widely dispersed regional development hubs (in Austin, the Pioneer Valley in Massachusetts, Philadelphia, the San Francisco Bay Area, Madison, Portland, Oregon, New York City, western North Carolina, and several others). And, of course, there is the national organization, the USFWC, now in its twelfth year, along with its recently formed national affiliates, Democracy at Work Network (DAWN) and the Democracy at Work Institute (DAWI). All of these provide their members with welcoming environments and experienced mentoring through which they can learn and collaborate together, share lessons and resources, and begin to develop a sustainable culture of cooperation.

Each has its own vision and constituency, but all would endorse something like this Vision statement from the Philadelphia Area Cooperative Alliance:

> We envision a vibrant and growing network of cooperative enterprises that operate in all sectors of the economy to build a better world – providing essential services, strengthening democratic organizations, creating quality jobs, building community wealth, reducing poverty, strengthening our local economy, protecting the environment and increasing community engagement.
>
> (See www.philadelphia.coop)

Second, free schools often were not very diverse; on the whole, most served middle class white families. The opposite holds for the GDE: marginalized groups are well represented, in terms not only of membership, but also of leadership. In addition, workshops given at conferences within the GDE are increasingly held in both Spanish and English, and occasionally, in Portuguese. And workshops on developing sensitivity to diverse cultures have become common in these conferences (See www.philadelphia.coop/about-us/).

Finally, with the marginal exception of teachers, free schools did not in general generate much, if any, income for themselves. On the contrary, parents had to pay for places in them, though far less than for private schools. But workers in a cooperative receive wages for being part of that enterprise, and share in the overall profits or surplus the enterprise produces. Indeed, of all progressive social movements, worker cooperatives are unique in offering their members living wage compensation and workplace benefits along with the opportunity to make the world a better place for all. This should certainly be counted towards their own internal sustainability (think, burn-out).

Aside from this comparison with free schools, there is an important additional reason for thinking that the GDE is here to stay: *it is pragmatic, rather than ideological.* Years ago, when I helped found a chicken-processing cooperative in Willimantic, CT, I was chastised by numerous comrades as a 'traitor' for accepting funds from a government agency and a Catholic charity. Today, the USFWC, to support its new Democracy at Work Institute (DAWI) affiliate, has accepted quite substantial funding from the USDA and from that very same Catholic charity – so long as the money can be used to meet its mission. But at the same time, the Federation *also supports local or regional groups who reject such external sources of funding* as undermining the self-directed capabilities of cooperative practitioners. This pragmatic position creates a 'big tent' or large network of support, in which opposing strategies are tolerated and even combined. And it helps keep everyone distanced from ideological disputes.

As for 'future generations' or external sustainability, worker co-ops have a number of built-in incentives to recognize its importance. Unlike owners of capital-driven corporations or businesses, worker-owners typically live within or near, rather than far away from, their workplaces. If their cooperative enterprises are polluting their community's wells or rivers, or emitting dioxins into the air it breaths, those worker owners and their families will reap the dreadful outcomes along with their neighbors.

In addition, by residing close to the businesses they own, cooperative members will reduce their own carbon footprints, and they will also remain far more accessible and accountable to the communities where they reside and to the longer term needs and goals of those communities. For example, should the current recession continue in certain parts of this country, worker owned enterprises would be far more likely to seek and find ways to keep those enterprises local and running, rather than moving them, as so many corporations have, to other parts of the country or overseas.

What, then of intersectionality?

There are a number of cross-sector initiatives afoot now within the overall GDE, e.g., the Next System Project (see http://thenextsystem.org/) and the New Economy Coalition (neweconomy.net/), that embody economic intersectionality. Worker cooperatives and cooperative development organizations have joined these organizations and are collaborating with them and their separate organizational members.

A definite shift over the past decade towards intersectionality has taken place. Here's a story that illustrates this shift:

In 2006, to begin a USFWC workshop in New York City, I suggested that we examine a disturbing question: "If worker cooperatives are such a good thing, why are there so few of them?" My own answer to this question was that, unintentionally, worker co-ops were themselves part of this problem, in that their vision and missions remained narrow and self-regarding. That is, they were behaving as if cooperatively owned and managed enterprises – *by themselves* – could turn around our exploitive, anti-democratic, soul- and planet-crushing undemocratic economy.

Cooperative or self-managed enterprises may be organized democratically and may become allied with one another; they may even make charitable donations or provide discounted prices to neighboring non-profits. But, I argued, they typically have little incentive or support to go beyond their self-imposed walls or to merge cooperative loyalty with any wider or more transformative forms of solidarity. They will thus feel no responsibility to collaborate, coalition, partner, or organize in mutually beneficial ways with labor, immigrant, social justice, participatory budget, time bank/community currency, restorative justice groups or constituencies other than those within the what I called 'the cooperative family.'

At that 2006 USFWC gathering, this notion of cross-organizational collaboration outside the co-op family got a very mixed reception. In contrast, it was very well received a month later at the Canadian Worker Cooperative Federation conference in Edmonton (see www.geo.coop/node/57). Many were the objections; e.g. "It asks too much of us." "It violates the need for co-ops to be politically neutral"; "How can we tell which other groups or organizations to collaborate with?"; "It's too confrontational."

But that was then. Today a very different story is emerging. At the 2014 USFWC conference in Chicago, you could find this workshop description:

Building grassroots coalitions around worker ownership

> A critical element in the growth of large-scale worker cooperative movements around the world is their alliances with other grassroots movements, running the gamut from organized labor to communities of faith to movements for economic, racial and language justice.
>
> This session looks at possibilities for building such coalitions in the U.S., looking at models for organizing and already existing coalitions. Exploring how faith-labor-immigrants' rights, and racial justice, coalitions have begun to connect with worker cooperatives, we will discuss how to build coalitions to increase our scale and power.
>
> (See http://conference.coop/schedule/building-grassroots-coalitions-around-worker-ownership-desarrollo-de-coaliciones-de-base-en-torno-la-propiedad-de-los-trabajadores/)

Moreover, the 2016 national conference of the USFWC will have *the theme of collaborating with other social justice movements*.

In short, there is now enthusiasm for, rather than resistance to, cross-organizational collaboration, for 'moving movements together.' Elements within the GDE no longer see themselves as the sole or primary initiative to bring about transformative change. This, for me, is a definite sign of a mature and inclusive movement and a further reason to see this new economy as prioritizing both external sustainability and cross-sector alliances.

Concluding thoughts

So much for a critique of the GDE, and some responses to that critique. What conclusion(s) can now be reached?

Let's start with what I'm not claiming. My point is not that a full-blown GED has come to life, or even that the road ahead will now be smooth or without extremely difficult challenges. My claim is only that this emerging economy has made an excellent start, in several crucial directions, and has the capacity and willingness to keep itself alive, well, and growing.

Specifically, it has created local, regional, and national associations to overcome isolation and spark collaboration, as well as practitioner-directed support networks to assist start-ups and established enterprises. It is diverse and culture-sensitive in many ways, based on decades of preparation and hands-on experience. It provides a living wage income for many of those who are most engaged, is pragmatic, open to conflicting approaches, and non-ideological; it is willing to build cross-organizational coalitions with a variety of grassroots and social justice movements.

Nor have I characterized this emerging GDE as 'problem-free' or without flaws or gaps. As I don't need to argue, we all have clay feet, and our share of moderate if not major imperfections. For example, my friends and comrades in the GDE are exuberant and love to dance, but I am still waiting for the first rousing song to be written by and for this movement.

More seriously, the issue of cooperatives and other democratic workplaces being 'growth aversive,' keeping employment and the number, as well as size, of enterprises relatively small, has not yet been addressed in any depth. (The complaint, occasionally voiced by labor unionists, that worker co-operatives substitute 'enterprise' for 'class' consciousness raises a very similar concern.)

Way back in the 1990s, Tim Huet of the San Francisco Bay Area's Arizmendi Association of Cooperatives raised this issue in an article published by GEO. Here's a small part of what he wrote:

> … worker cooperatives tend to stop growing once members feel their business has reached a healthy and productive size. Unlike capitalist cancers which grow for their own sake and destroy their host environment, cooperatives aim for homeostasis, a healthy balance. Unfortunately this pro-social characteristic of cooperatives can be a fatal weakness in economic competition with capitalist businesses. The cooperative grocer, bookstore, etc. enjoying its homeostasis can be devoured by a malignant capitalist growth (chain, mega-wholesaler, etc.)
>
> (See http://geo.coop/archives/huet.htm)

In 2006, Tim presented a version of his 'growth aversive' thesis at that same USFWC Conference just mentioned above. Covering that Conference for GEO, I wrote a critique of this presentation, but other than that, the issue Tim raised – certainly a crucial one for the future of any GDE – seems to have been ignored. For my reply to Tim, see www.geo.coop/node/57.

But in any case, I have not attempted to provide a full-scale assessment of the current GDE; that's a task for a much longer research project. Rather, my aim has been to showcase and appreciate the substantial distance already traveled by the tenacious shapers of this grassroots democratic work, and to offer evidence that they are providing the resources, opportunities, lessons, and human capital to build and sustain a great deal more grassroots democratic growth.

Some have contended that it will take 'several generations' to develop a well-rooted GDE. GEO's Michael Johnson, for example, claims that:

> [t]his shared vision of abundance and solidarity at the core of these emerging local and regional dynamics will probably take about a generation to root firmly, if it does. In three or four generations these changes could possibly become a substantial part of the political economics in the US. Possibly.... So we are talking about a process that will take a long time. A long, long time beyond our lifetimes.
>
> (See http://geo.coop/story/another-world-emerging-well-maybe)

His somewhat discouraging speculation might be true or it might not, but believing it will only tend to make it self-fulfilling. So also is the apparently opposed view of Kevin Zeese and Margaret Flowers, echoing Tom Paine: "We are at a critical convergence in history, where ... we have it in our power to begin the world over again" (see http://truth-out.org/news/item/14076).

My own view differs from both of these. I see the future of GDE as fundamentally unpredictable: who would have predicted, even ten years ago, where we now stand? Nor do I know how to assess claims about 'critical historical convergences.' It is better, I'd suggest, to set aside speculative predictions, and instead *celebrate what we have already done and commit ourselves to doing ever more of it*. The point is to stay on the path, and enjoy the journey for its own sake, rather than attempting to identify if, when, or how some pre-fixed destination, or worldwide transformation, will be reached. As Václav Havel, the Czech playwright and dissident, put it, "Hope is definitely not the same thing as optimism. It is not the conviction that something will turn out well, but the certainty that something makes sense, regardless of how it turns out." (Havel, V. 1993. 'Never Hope Against Hope', *Esquire*, October, p. 68).

I have that sort of hope and certainty about detaching ourselves from capitalism by continuing to support and build today's GDE.

But what about the immense power and brutal militaristic force of the dominant capitalist regime? Yes, detachment from it to a GDE will likely not be sufficient to overthrow that evil regime; for that we will also need protest, resistance, opposition – on a scale even larger than the recent marches against climate change. Still, neither will resistance by itself suffice; something new, different, and much better will need to be available and welcoming. Resistance and reconstruction, indignant rage and constructive creativity, need to be seen as allies, rather than enemies. Neither alone will get us far; both are essential.

And both can be and have been combined. A recent article by Judy Rebick, entitled "Long Live Occupy: Occupy Three Years Later," reveals that this sort of collaboration has already begun. The author quotes Toronto Occupy organizer Sakura Saunders:

> I believe that Occupy, Idle No More and Quebec's student movement showed us that there are hundreds of thousands of people, if not millions of people, who want dramatic change but are on the sidelines waiting for a movement that inspires them to hope, and this is what I hold onto in continuing to organize past these moments of heightened activity.

Saunders then concludes:

> In my view, Occupy as a movement continues in the particular forms of activism we are seeing emerging around the world. Whether Occupy Gezi in Turkey or Occupy Hong Kong, all these actions have similar qualities. While they include big marches, their focus is on building alternative communities, however short-lived, that are providing the necessities of life, political debate and cultural activities, as well as protests. They are all organized collectively with relatively flat structures, participatory democracy like the general assemblies and a valuing of whatever skills and ability people bring into the community.
>
> (See http://interoccupy.net/blog/long-live-occupy-occupy-three-years-later/)

Moreover, stories of alliances between resistance and reconstruction/detachment are far from uncommon. Here's just one example:

> Begun in the spring of 2013, WORCs (Worker-Owned Rockaway Cooperatives) is an initiative to rebuild after Sandy in a way that addresses both the storm's impact and the long-term systemic issues in the neighborhood. The program's goal is to equip Far Rockaway residents with the skills and financing to launch small, worker-owned businesses that fill a need in their community. More than 40 residents have already joined the weekly, three-month training program –people who now work for others as babysitters, house cleaners, exterminators and upholsterers but dream of being their own bosses and keeping wealth inside their neighborhood. So far, the group has developed plans to launch five worker-owned businesses: a construction company, a bakery, a health food store, an entertainment collective and a restaurant. They are now in the process of turning these plans into realities.
>
> (See www.theworkingworld.org/us/341-2/worker-owned-rockaway-cooperatives)

WORCs is a joint initiative between Occupy Sandy and The Working World, an incubator of worker co-ops in both the US and elsewhere. Occupy Sandy organizers approached members of the Working World, who they knew through Occupy Wall Street, and together they came up with the idea.

Let's make these collaborative initiatives less and less uncommon!

5 Intersectionality, ecology, food

Conflict theory's missing lens

Deric Shannon

Scholars of food and ecology quite often – and rightfully so – point out various systemic inequalities in food supply chains, industrial agriculture, and the distribution of environmental harms (see e.g. Alkon and Agyeman, 2011; Carolan, 2013; DuPuis, Harrison, and Goodman, 2011; White and Heckenberg, 2014; Gottlieb and Joshi, 2012; Patel, 2007; Shiva, 2000). Environmental justice scholars in particular are quick to show how marginalized groups are disproportionately affected by environmental harm, paying close attention to how people of color, colonized and indigenous groups, poor and working class people, and women effectively bear the brunt of capital's tendency toward growth-at-any-cost and, importantly, how those various relations of inequality overlap, interact, and produce differentiated and stratified social experiences. It has become obvious in my own work on food movements (Shannon, 2011) and food justice (Shannon, 2014a and 2014b) that intersectionality is an excellent starting point for showing how structured inequalities are built into our food system.

In my own discipline (an ironically fitting word to describe our academic corralling into walled-off areas of expertise and expectations), sociology, the theory typically employed for looking at structured inequalities is referred to as *conflict theory*. Conflict theorists "focu[s] on the distribution of resources, power, and inequality" in societies and the theory's proponents argue that "social structures and organizations" tend to reflect those inequalities – indeed, "some people have more resources (i.e. power and influence), and use those resources to maintain their positions of power in society" (Boundless, 2015). Given those theoretical commitments, conflict theory provides important conceptual tools for utilizing an intersectional analysis and applying it to food supply chains and environmental harms.

Somewhere, in our collective re-telling of history, the story of conflict theory came to be told something like this: There was a period in Europe where the Enlightenment, various social revolutions, and the Industrial Revolution collided as social forces, when liberal, capitalist democracies began to replace the old feudal system of political and economic relationships. In the process of these upheavals, people began to look for explanations for these large-scale social changes. Utilizing the positivist ideas of Auguste Comte combined with the scientific method that was emerging at that particular historical juncture and geographical region, the discipline of sociology was born.

So the story goes, a major thinker, a fiery radical named Karl Marx (see especially, 1867) arose to challenge the view that society was a stable, functioning unit. Rather, society, according to Marx, was not typified by harmony, stability, and function. Instead, the emerging capitalist system was a *class society* – one that was marked by inequality and conflict and stratified by *class*. According to Marx, it was this conflict, a great struggle between social classes – those who must rent themselves in order to earn a living, the *proletariat*, and those who made a living by virtue of owning land, capital, or the means of producing commodities and services, the *bourgeoisie* – that was the defining aspect of social life. Thus, to Marx, all of human history should be looked at through the lens of class struggle, and it was class struggle that produced new institutional arrangements and would eventually lead to a classless social order, what Marx referred to alternatively as *socialism* or *communism*.

During the tumultuous period of the 1960s and 70s, new conflict theories were created to understand structured inequalities not reducible to Marx's theory of class. This set the stage for a new kind of analysis that was materializing – what became known as *intersectionality*. Perhaps the first major statement about interlocking systems of oppression was the *Combahee River Collective* Statement (Combahee, 1977). The Combahee River Collective (CRC) was a group of Black feminists and revolutionary anti-capitalists who argued that, through organizing, study, reflection, discussion, and then more organizing (what some radicals refer to as *praxis*), their political position came to be "actively committed to struggling against racial, sexual, heterosexual, and class oppression" and that they viewed their "particular task" to be "the development of integrated analysis and practice based upon the fact that the major systems of oppression are interlocking" (Combahee, 1977). The term intersectionality, however, was first coined by the critical legal scholar, Kimberlé Crenshaw (1989) and has since served as a foundation among social scientists for theorizing and studying multiple, overlapping, and interlocking forms of oppression, exploitation, and structural advantage and disadvantage.

Thus, the narrative of conflict theory is often told (though this is necessarily a simplified version given spatial restraints). Out of the ashes of feudal Europe came positivism, sociology, and a theory of class conflict driven by Marx. As history unfolded and our collective understanding began to account for multiple forms of oppression, we eventually settled into intersectional analyses that theorize the simultaneity of oppressions, how they intersect and interact, and importantly, how they constitute each of us as social, viable beings – each of us both resisting these structural inequalities at times, and likewise socially reproducing them at others.

Mind the historical gap

But this narrative leaves much to be desired (perhaps all historical narratives do). For one, this history leaves out non-Europeans who were producing social science in very different historical and cultural contexts (i.e. the Indian economist, Chanakya, the Arab historical sociologist, Ibn Khaldun, or early African American women social scientists like Ida B. Wells). This centering of European and male knowledge production could be the focal point for a book length work on histories of the social sciences (it already has many times over, perhaps making the observation banal at this point).

But for social science generally, and the sociology of food and ecology in particular, I want to draw on a perspective that has all but been summarily ignored in my discipline – that of anarchism. People, and their varied standpoints, can be disciplined and written out of histories or into histories, as authors try to capture the lives of ideas, either consciously or unconsciously. But so, too, can *ideas* be marginalized themselves. I want to suggest in this chapter that there are some historical and contemporary insights that we might take by including anarchism in our understanding of the history of social science and the sociology of food and ecology. Along the way, I hope to make a few comments about the utility of intersectionality as a starting point and how we might make use of anarchist ideas to broaden its scope.

Before I begin in earnest, please allow me a few caveats. First, anarchism can be seen as both a sociological perspective, and in some cases, a critique of sociology or, at the least, in tension with the sociological enterprise (for an excellent exposition of this, see Williams, 2013). Therefore, suggesting the use of an anarchist lens for intersectional approaches to the sociology of food and ecology doesn't come without its own contradictions, particularly since anarchists tend to be critical of hierarchy writ large – including those embedded in the academic production of knowledge and the disciplinary power of the disciplines, of which sociology is not abstracted.

Secondly, after years of being at the forefront of the alter-globalization movement, a wide array of environmental movements and food movements, and being the primary inspiration and backbone for Occupy Wall Street (Bray, 2013), anarchism as a theory has been making increasing inroads into public discussions (though often as a foil for why Occupy failed, or as Marxism's more idealistic and utopian lesser-known cousin). Perhaps this provides a social space favorable for writing a piece such as this, suggesting that we might benefit from including anarchist ideas in conflict theories. However, I don't want to overstate engagements with anarchist ideas – they remain few and far between, and even less so good faith encounters, most typically written by people antagonistic to anarchist ideas (or radical ideas generally).

I outline below some historical and contemporary examples of anarchist people, ideas, and events that we can productively draw from in discussions of intersectionality, food, and ecology. Conflict theory, by ignoring anarchism, has been missing a viable lens. *Anarchism* is a diverse set of ideas and practices centered on a critique of hierarchy, and the various conflicts those hierarchies produce. The close interconnections between anarchism and intersectionality have already been probed theoretically (see especially, Rogue and Volcano, 2012) and below I will outline some ways that we might put this anarchist conflict theory lens back into focus by both unearthing history *and* our present moment.

A missing past and present

The possible anarchist historical contributions to our understandings of inequality, food, and ecology within conflict theory would be difficult to understate. While our ostensible "father" of conflict theory, Karl Marx, focused on his industrial working class – the *proletariat* – as the historical revolutionary subject, anarchists tended to pay special attention to farm and agricultural laborers in addition to the industrial working class, perhaps best articulated by theorists of *anarchist communism*, arguably most obvious in Peter Kropotkin's (1889) exposition that sought to recognize fields, or agricultural work, alongside factories and workshops in his post-revolutionary vision. This largely stood in contrast with Marx, who assumed that the rise of urbanization and increased mechanization of farm labor confirmed his position that it was industrial workers who would be the agents of social transformation.

To be clear, Marx *did* write about food and agriculture – the human necessity to eat, and thus to grow food, was not lost on him nor to Marxist theorists generally (if one considers Mao a 'Marxist', this is doubly true since Mao saw a central revolutionary role for the Chinese peasantry). Marx (1867), for his part, saw the separation between town and country and the development of capitalist agriculture centrally connected with environmental harm, particularly soil degradation. And Marx's friend and co-author, Engels (1993, orig. 1845), wrote explicitly on food inequality in his investigations into working class life in England, railing against the system that often traded hunger and deprivation in exchange for grisly working conditions. Expanding on Marx, writers within the tradition of Marxist conflict theory have productively built upon those themes (for two seminal examples, see Mann and Dickinson, 1928; McMichael, 2009).

But because these works rely on Marxism for explanatory power, they tend to look at complex questions around inequality and domination with a lens that privileges production and investigations into commodity chains – that is, they tend to be economistic. This is not to say that they have no value. Scholars of food, ecology, *and* intersectionality should absolutely be putting them to use, particularly given the sheer amount of liberal work within the social sciences that treats institutional arrangements like capitalism and the state as eternal and without alternatives. Marxist theory can be an important corrective to that (so can anarchism, but more on that later).

Historically, however, while Marxism is often utilized by conflict theorists, anarchism has largely been ignored within sociology and food scholarship, as a result of a combination of factors. For example, the ascendency of the USSR and (one particular interpretation of) Marxism as the radical theory attached to its development led many to rely on Marxism as *the* radical alternative to capitalism, ignoring anarchism. Secondly, anarchism, as a critique of hierarchical power relationships generally, cannot ignore the power imbalances embedded in academic life. This makes for, at the least, an uncomfortable theory for academics and other professional-managerial class people, as it puts their social position under critical scrutiny. But we miss a great deal by ignoring anarchism's rich tradition on food, ecology, farming, and how that tradition can be combined with an intersectional lens.

Perhaps one of the greatest historical examples of anarchist theorists we might benefit from engaging with is the Russian prince-turned-revolutionary, Peter Kropotkin (1892), who titled what arguably became his most well-known work *The Conquest of Bread.* In this book, Kropotkin argues forcefully that securing food must be the first, and primary, task of any revolutionary movement. This book was followed by his visionary tome outlining what he thought, in that historical moment, were the possibilities of organizing agricultural and industrial production. In *Fields, Factories, and Workshops,* Kropotkin (1898, emphasis from the original) turns Malthus on his head, excoriating him for "assert[ing], in reply to Godwin, that no equality is possible; that the poverty of the many is not due to institutions, but is a natural *law.*" Kropotkin is on to something important here: no egalitarian alternative to capitalism is possible without first recognizing that capitalism and its concomitant poverty are not inevitable, that another world is possible! For people who use sustainability as a framework for understanding society, with its three pillars of environment, society, and economy, this might serve as a beginning of a much more holistic theory of where we might go – a view rooted in social transformation, rather than (or, along with) reform.

Other anarchist revolutionaries followed suit in their demands for food as a basic right rather than a grouping of commodities to be sold for a profit on the market. Emma Goldman, the great writer, labor organizer, and orator, was infamous for advising workers in the United States to ask for work and bread, and if they were given neither, to *take* bread. The Mexican revolutionary, Ricardo Flores Magón, emphasized food in his conception of human freedom when he famously declared, "We are free, truly free, when we don't need to rent our arms to anybody in order to be able to lift a piece of bread to our mouths."

Many of these ideas – of food as a right instead of a privilege – were put into practice during the Spanish Civil War, likely one of the liveliest historical experiments in anarchism, as large parts of Andalusia and Aragón were organized by majority-anarchist populations and methods. Perhaps displaying a "proto-intersectionality," the *Mujeres Libres,* or Free Women of Spain, argued against "many anarchists [who] treated the issue of women's subordination as, at best, secondary to the emancipation of workers, a problem that would be resolved 'on the morrow of the revolution'" (Ackelsberg, 2005, p. 38). To the *Mujeres Libres,* a meaningful revolution would be "dedicated to the liberation of women from their 'triple enslavement to ignorance, as women, and as producers'" (Ackelsberg, 2005, p. 21). Thus, productivist Marxist (and, to be fair, at the time some anarchist) thinking was discarded in favor of a theory of domination that was much more holistic, that saw the intersecting relations of inequality that working class women faced in revolutionary Spain (and this was in the early 1900s!). Recovering this anarchist history allows us to note that even early radicals "insisted that power has its own logic and will not be abolished through attention to economic relations alone" (Ackelsberg, 2005, p. 37).

De Santillán's (1936) study of the political economy of anti-state socialism paid careful and close attention to agriculture. This study of anti-capitalist economics was a lens into the Spanish Civil War practices in which the *Mujeres Libres* were rooted. One of the primary divisions among the anarchists of the time, as noted by Ackelsberg (2005), was that between the *collectivists*, who were primarily in the cities, and the largely agrarian *libertarian communists*, who thought that access to the social product should be freely organized, decommodifying all of daily life. Interestingly, those closest to food production, soil, and the complex ecology of human/land interactions (i.e. the peasantry) tended to put into practice what would today look like a utopian arrangement – one where people produced and shared freely instead of assuming that the motive to profit was inherent human nature. This shows the limits of neo-classical economists who assume that humans are naturally rational, utility-maximizing machines, which is the guiding thesis, for example, of the philosophy of Ayn Rand's (1961) objectivism in her book *The Virtue of Selfishness*.

Anarchist ideas connecting humanity with our environs – or *ecology* – were likewise important during the Spanish Civil War. Naturist anarchists began engaging in a wide variety of practices that they thought would lead to a reconnection to the land, other animals, and our own animality. Particularly among relatively poor Southern peasants, nudism, vegetarianism, free love, and trips to unmanaged land became a methodology for *connecting*, put forward, for one, by the *Sol y Vida* (Sun and Life) group and other clusters of naturists. As Ortega (2003) notes:

> We must be aware that the naturist ideas expressed in [the publications of *Sol y Vida*] matched the desires that the libertarian [i.e. anarchist] youth had of breaking up with the conventions of the bourgeoisie of the time. That is what a young worker explained in a letter to 'Iniciales' He writes it under the odd pseudonym of 'silvestre del campo', (wild man in the country). 'I find great pleasure in being naked in the woods, bathed in light and air, two natural elements we cannot do without. By shunning the humble garment of an exploited person, (garments which, in my opinion, are the result of all the laws devised to make our lives bitter), we feel there no others left but just the natural laws. Clothes mean slavery for some and tyranny for others. Only the naked man who rebels against all norms, stands for anarchism, devoid of the prejudices of outfit imposed by our money-oriented society.

Of course, these ideas were not limited to revolutionary Spain. Clark and Martin's (2013) recent edition of the writings of the French anarchist communist, Elisée Reclus, shows us a portrait of a French anarchism deeply committed to living in communion with both the earth *and* other animals. Reclus declared, "I also include animals in my feeling of socialist solidarity" and insisted that a coherent anarchism would be a movement dedicated to humans living in harmony with their natural environment (Clark and Martin, 2013, p. 32). These naturist ideas traveled as far afield as Central America, where Shaffer (2005) notes naturist anarchism as an important third pole of anarchist theories in early Twentieth-Century Cuba, alongside anarcho-syndicalism and anarcho-communism.

None of this is limited to the past. Even into the late Twentieth-Century, the early 2000s, and up to the present, anarchist commitments to the freedom to food, healthy ecologies, and their intersections with white supremacy, capitalism, hetero-patriarchy, and other relations of domination are central to the philosophy and movement. Indeed, how can people enjoy the fruits of living in a world free of institutionalized domination without clean air, drinkable water, full bellies, and a livable climate? Within existing movements, anarchism had a heavy influence on the alter-globalization movement, was absolutely central to Occupy Wall Street (see especially Bray, 2013), has a substantial role in Food Not Bombs (see e.g. Shannon, 2011; Parson, 2014), and beyond. Theoretically, it inspired the early Bookchin (1993) and his "social ecology" as well as what might be considered contemporary versions of some forms of naturist anarchism, like anarcho-primitivism and anti-civilization forms of anarchism (see e.g. Landstreicher, 2007; Killjoy, 2010; Zerzan, 2009).

Food, ecology: intersectionality and beyond

Nearly four decades ago, anarcha-feminist Peggy Kornegger (1979) described her education like this:

> I had never heard of the word "anarchism" – at all. The closest I came to it was knowing that anarchy meant "chaos". As for socialism and communism, my history classes somehow conveyed the message that there was no difference between them and fascism, a word that brought to mind Hitler, concentration camps, and all kinds of horrible things which never happened in a free country like ours. I was subtly being taught to swallow the bland pablum of traditional American politics: moderation, compromise, fence-straddling, Chuck Percy as wonder boy. I learned the lesson well: it took me years to recognize the bias and distortion which had shaped my entire "education". The "his-story" of mankind (white) had meant just that; as a woman I was relegated to a vicarious existence. As an anarchist I had no existence at all. A whole chunk of the past (and thus possibilities for the future) had been kept from me. Only recently did I discover that many of my disconnected political impulses and inclinations shared a common framework – that is, the anarchist or libertarian tradition of thought. It was like suddenly seeing red after years of colourblind grays.

Unfortunately, in the proceeding three and a half decades little has changed. In sociology, it is common for textbooks by 'experts' to outline the alter-globalization movement, the women's movement, the environmental movement, or, more specifically, the Battle of Seattle or even Occupy Wall Street, without even giving anarchism so much as a whisper, for similar reasons as I outlined above.

This is frustrating as a sociologist who primarily studies food, ecology, and political economy, in large part because of how much we are missing by ignoring the anarchist tradition. Conflict theory's missing lens also gives us a number of useful tools for intersectional analysis. I tried to briefly outline some of the ways that anarchist ideas, people, and movements have been engaged in the larger social world around food and ecology. Beyond that, since intersectionality is a framework for viewing life by recognizing multiple and overlapping sites of domination and structured inequality, and anarchism is a philosophy and movement targeting all forms of hierarchy, they seem well-suited to work in tandem.

Certainly, anarchists have engaged contemporarily with a wide variety of forms of domination and perspectives that attempt to address those inequalities. Anarchists have engaged with feminist criticisms of patriarchy (see especially Dark Star Collective, 2012), queer theory and sexuality (see especially Heckert and Clemson, 2011), racism and critical race theory (see especially Revolutionary Anti-Authoritarians of Color, 2002), coloniality and postcolonial theory (see especially Ramnath, 2011). I made the case above that anarchists have made a number of interesting contributions to understandings of food and ecology. And, of course, some have written on the possible connections between anarchism and intersectionality (see especially Rogue and Volcano, 2012). But I have largely left the question of what scholars might *do* with this recovered lens. The topic could be the subject of a number of books, but I make two brief suggestions below.

First and foremost, many readers will note that I have avoided the term 'sustainability' in this chapter, often preferring instead 'ecology.' This is because, without specific ethical commitments – particularly those arising from radical politics – 'sustainability' is so much populist mush. Everyone from huge multinational corporations, to repressive states, to liberals, soft progressives, and even a few radicals claim they are for this thing called 'sustainability.' If it has such wide appeal among so many social actors, can the term have much meaning? Perhaps if we operationalize it with an eye toward social transformation it could.

Sustainability, after all, gives us a shared language to talk about environmental harm, social inequality, and economic equity (the three pillars of sustainability). It gives us a view of evaluating actions – whether they satisfy the needs of existing people without compromising the needs of people to come. Anarchism, and transformative movements generally, could intervene in the public discussion of sustainability and, again, as a movement against all structured forms of inequality, anarchism is well placed to make *intersectional* arguments for a holistic and meaningful sustainability.

Secondly, because intersectionality is rooted in Combahee's expansion of identity politics, anarchist ideas are well-suited to move us beyond intersectionality, or perhaps to expanding it. While liberals insist on the progressive possibilities of the state, for example, anarchists argue that power centralized into the economy *or* the state enacts relations of domination. This follows critical legal studies in food, where "progressive" legal processes enacted to protect workers have, in some cases, had the latent effect of destroying traditional farming practices of Hmong refugees, to name one example, exacerbating both racism and colonial practices of othering (Minkoff-Zern *et al.*, 2011). This also creates space for recognizing and analyzing other forms of disciplinary power that might not be reducible to identity, but instead are embedded in notions of normalcy.

The case for scholars of food, ecology, and intersectionality to put anarchist histories and ideas to use nearly makes itself. Unfortunately, in my own field of study, many scholars are ignorant of this history as well as its contemporary applications. I have made the argument in this chapter that we might try to re-encounter conflict theory's missing lens, not as a set of ideas that we all must agree with, but as a beautiful idea, ethic, and a movement that we should, at the least, honestly engage with. The possibilities are only limited by how much we allow our imaginations to be.

Bibliography

Ackelsberg, M. A. (2005) *The Free Women of Spain: Anarchism and the Struggle for the Emancipation of Women*, Oakland, CA: AK Press.

Alkon, A. H. and Agyeman, J. (eds) (2011) *Cultivating Food Justice: Race, Class, and Sustainability*, Cambridge, MA: The MIT Press.

Bookchin, M. (1993) What is social ecology? Anarchy Archives, available online at http://dwardmac.pitzer.edu/Anarchist_Archives/bookchin/socecol.html (accessed 5 July 2015).

Boundless (2015) The conflict perspective, *Boundless Sociology*, available online at www.boundless.com/sociology/textbooks/boundless-sociology-textbook/sociology-1/theoretical-perspectives-in-sociology-24/the-conflict-perspective-156-974/ (accessed 5 May 2015).

Bray, M. (2013) *Translating Anarchy: The Anarchism of Occupy Wall Street*, Winchester, UK: Zero Books.

Carolan, M. (2012) *The Sociology of Food and Agriculture*, New York: Routledge.

Carolan, M. (2013) *Reclaiming Food Security*, New York: Routledge.

Clark, J. P. and Martin, C. (eds) (2013) *Anarchy, Geography, Modernity: Selected Writings of Elisée Reclus*, Oakland, CA: PM Press.

Combahee River Collective (1977) *The Combahee River Collective statement*, available online at www.circuitous.org/scraps/combahee.html (accessed June 12, 2015).

Crenshaw, K. (1989) Demarginalizing the intersection of race and sex: a black feminist critique of antidiscrimination doctrine, feminist theory and antiracist politics, *University of Chicago Legal Forum*, 1989, 139–67.

Dark Star Collective (2012) *Quiet Rumours: An Anarcha-Feminist Reader*, Oakland, CA: AK Press.

De Santillan, D. A. (1936) After the revolution, *LibCom*, available online at www.libcom.org/book/export/html/33181 (accessed 14 May 2015).

Dupuis, E. M., Harrison, J. L. and Goodman, D. (2011) Just food? In Alkon, A. H. and Agyeman, J. (eds.) *Cultivating Food Justice: Race, Class, and Sustainability*, Cambridge, MA: The MIT Press.

Engles, F. (1845, reprint 1993) *The Condition of the Working Class in England*, Oxford: Oxford University Press.

Gottlieb, R. and Joshi, A. (2012) *Food Justice*, Cambridge, MA: The MIT Press.

Killjoy, M. (2010) Anarchism vs civilization, The Anarchist Library, available online at www.theanarchistlibrary.org/library/margaret-killjoy-anarchism-versus-civilization (accessed 3 July 2015).

Kornegger, P. (1979) Anarchism: the feminist connection. In Ehrlich, H. (ed) *Reinventing Anarchy: What are Anarchists Thinking These Days*, Boston, MA: Routledge and Kegan Paul.

Kropotkin, P. (1892) *The Conquest of Bread*, Anarchy Archives, available online at http://dwardmac.pitzer.edu/Anarchist_Archives/kropotkin/conquest/toc.html (accessed 28 June 2015).

Kropotkin, P. (1898) *Fields, Factories, and Workshops: Or Industry Combined with Agriculture*, Anarchy Archives, available online at http://dwardmac.pitzer.edu/Anarchist_Archives/kropotkin/fields.html (accessed 28 June 2015).

Landstreicher, W. (2007) A critique, not a program: for a non-primitivist anti-civilization critique, The Anarchist Library, available online at: http://theanarchistlibrary.org/library/wolfi-landstreicher-a-critique-not-a-program-for-a-non-primitivist-anti-civilization-critique (accessed 3 July 2015).

Mann, S. and Dickinson, J. (1978) Obstacles to the development of capitalist agriculture, *Journal of Peasant Studies* 5, 466–81.

Malatesta, E. (2005) *At the Café: Conversations on Anarchism*, London: Freedom Press.

Marx, K. (1867). *Capital, Volume 1: A Critique of Political Economy*, Marxist.org, available online at: www.marxists.org/archive/marx/works/1867-c1/ (accessed 30 June 2015).

McMichael, P. (2009) A Food Regime Genealogy, *Journal of Peasant Studies* 36(1), 139–169.

Minkoff-Zern, L., Peluso, N., Sowerwine, J. and Getz, C. (2011) Race and regulation: Asian immigrants in California agriculture. In Alkon, A. H. and Agyeman, J. (eds) *Cultivating Food Justice: Race, Class, and Sustainability*, Cambridge, MA: The MIT Press.

Ortega, C. (2003) Anarchism – nudism and naturism, *Revista ADN*, available online at www.naturismo.org/adn/ediciones/2003/invierno/7e.html (accessed 5 July 2015).

Parson, S. (2014) Breaking bread, sharing soup, and smashing the state: food not bombs and anarchist critiques of the neoliberal charity state, *Theory in Action*, 7(4).

Ramnath, M. (2011) *Decolonizing Anarchism*, Oakland, CA: AK Press.

Patel, R. (2007) *Stuffed and Starved: The Hidden Battle for the World Food System*, Brooklyn, NY: Melville House Publishing.

Rand. A. (1961) *The Virtue of Selfishness* New York: Signet Books.

Revolutionary Anti-Authoritarians of Color (2012) An anarchist introduction to critical race theory, The Anarchist Library, available online at www.theanarchistlibrary.org/library/revolutionary-anti-authoritarians-of-color-an-anarchist-introduction-to-critical-race-theory (accessed 6 July 2015).

Rogue, J. and Volcano, A. (2012) Insurrection at the intersections: feminism, intersectionality, and anarchism. In Dark Star Collective (eds), *Quiet Rumours: An Anarcha-Feminist Reader*, 3rd edn, Oakland, CA: AK Press.

Shannon, D. (2011) Food not bombs: the right to eat. In Armaline, W. T., Silfen Glasberg, D. and Purkayastha, B. (eds), *In Our Own Backyard: Human Rights in the US*, Philadelphia, PA: UPenn Press.

Shannon, D. (ed.) (2014a) Food justice and sustainability, *Theory in Action*, 7(4).

Shannon, D. (2014b) Operationalizing food justice and sustainability, *Theory in Action*, 7(4), 1–11.

Shiva, V. (2000) *Stolen Harvest: The Hijacking of the Global Food Supply*, Cambridge, MA: South End Press.

White, R. and Heckenberg, D. (2014) *Green Criminology*, New York: Routledge.

Williams, D. M. (2013) A society in revolt or under analysis? Investigating the dialogue between nineteenth-century anarchists and sociologists, *Critical Sociology*, 40(3), May, 469–492.

Zerzan, J. (2009) Future primitive, The Anarchist Library, available online at www.theanarchistlibrary.org/library/john-zerzan-future-primitive (accessed 5 July 2015).

6 Global warming as North-South conflict

The role of unconscious racism[1]

Milton Takei

Nicholas Stern, former chief economist at the World Bank, said on US National Public Radio in December, 2009 that greenhouse gas emissions must level off before 2020 (Stern, 2009). Stern wrote a report on global warming for the government of Great Britain. The planet's atmosphere is complex, so scientists can at best give only an estimate for when the build-up of greenhouse gasses will reach a point of no return. But clearly, time is short.

In this chapter, I treat global warming as an international relations problem, as an aspect of conflict between the poorer countries and the richer countries, that is, between the countries of the Global South and those of the North. In international negotiations surrounding the global warming crisis, many countries of the Global South are demanding that the Global North countries change their lifestyles. I argue that due to racism many people in richer countries are unwilling to make concessions to people of color in the Global South. I use Pierre Bourdieu's (1988, pp. 8, 134) concept of *habitus* (unconscious predispositions) as a theoretical framework to explain why racism, as well as its intersection with social class and sex/gender, might lead some people to act or not act on the global warming crisis, without necessarily causing them to behave in a given way.

Habitus and racism

The common assumption is that people do things rationally, that they consider their options and consciously make decisions to act or not to act. Pierre Bourdieu's (1988, pp. vii–viii) concept of habitus (unconscious predispositions) represents a challenge to the model of a conscious, rational person acting in the world. Although society imposes restrictions on people, they acquire proclivities in all aspects of life that often cause them to act without being consciously aware of their motivations (Postone, Li Puma, and Calhoun, 1993, p. 4).

Racist behavior can be one example of unconscious predispositions, and such mental processes can affect international relations. Psychologists have been able to demonstrate the existence of unconscious racial bias, beginning in early childhood (Baron and Banaji, 2006). The fact that people have a certain predisposition does not automatically mean they will behave in a certain way (Bourdieu, 1988, pp. 133–134). Throughout US history, a minority of the population resisted acting according to the dominant racial understandings (Holt, 2000, p. 124).

Specifically, racism helps to explain risk and risk aversion, especially as it relates to the environment. Dan M. Kahan and colleagues argue that group membership affects people's risk perceptions (Kahan et. al., 2007, pp. 267–270, 492–493). This is supported by a quantitative study that asked people in the US to rate the public health risk of twenty-five hazards; women and people of color perceived a greater risk from global warming than white males, perhaps because white males have more power and control, and feel less vulnerable (Flynn, Slovic, and Mertz, 1994, pp. 1101–1108). When examined through the lens of class, however, Linda Kalof and colleagues note that 'white men who are economically disadvantaged and exposed to high levels of environmental risk may remain inattentive to their circumstances because of strong socialization within the dominant ideology' (Kalof, Dietz, Guagacino, and Stein, 2002, p. 124). Their theory seems to indicate that lower class white males feel they have ample power and control, or at any rate they wish to present a confident face to the world. Jeffrey Lustig (2004, p. 56) says that white workers feel that they have something in common with powerful people and thus have trouble grasping the tenuous nature of their economic achievements.

Despite a continued white domination, racism has become less overt since the passage of civil rights legislation in the US. Instead, John R. Dovidio and Samuel L. Gaertner (1988) point to a hidden 'aversive racism,' which white people might not perceive or might deny exists. Racism can operate through the unconscious associations people make between political issues and people of color. Thus, whites in general and all who have been socialized into racist behavior in the US are less openly racist, but racism still exists. People of color, regardless of social class, recognize this when they have regular contact with white people and are the victims of everyday microaggressions (Sue, *et al.*, 2007). I argue that international relations is still to some extent affected by the unconscious predispositions of white elites and ordinary people toward people of color regardless of social class. Some people in the Global North use overpopulation discourses to put the blame for the global warming crisis on people of color.

International economic relations – population and power

White elites and many whites in general see international economic relations in racial terms (Winant, 2004, pp. 71, 141); thus, people in the US racialize global warming because of the important role of the Global South. China today is a rising Asian power intent on economic development. If the Global South continues to follow the same path as the richer countries did in pursuing economic development goals, the global warming crisis will lead to planet-wide disasters (Baer, 2002, p. 393). Both India and China place a high priority on their economic development goals (Rajan, 1997, pp. 25–37; Metzger and Meyers, 1998, pp. 23, 26). The United Nations Framework Convention on Climate Change of 1992 enshrines the right of development in international law

(Moellendorf, 2009). Simply put, people in China want cars, too (*The Economist*, 4 June 2005, pp. 24–26). In the future, China and India could attempt to continue on their current economic development path, despite a decline in global production of oil; they would need to rely on coal, which would last beyond the probable year when greenhouse gas emissions need to level off worldwide (Li, 2007; Harris, 2007). Apart from the gains in wealth that economic development brings, both rich and poor citizens of 'underdeveloped' countries have a lower status than people in richer countries (Dhaouadi, 1988). Governments of the Global South are seeking to escape a degrading condition.

Attacking the causes of the global warming crisis requires a reduction in consumption, not population. The particular racialized predispositions mean some environmentalists are concerned that the rise of China threatens US global dominance; they increasingly blame consumers in China for global warming (Zizer and Sze, 2007, pp. 392–393). This is particularly important in terms of China and India because of their large populations (Roberts and Parks, 2007, pp. 48, 152). Population numbers are rising fastest among the world's poorest people, many of whom do not have enough to eat. However, such people are responsible for very little in the way of greenhouse gas emissions; many countries with the lowest population growth have showed the greatest increases in emissions (Satterthwaite, 2009, pp. 547, 550, 554).

One quantitative study failed to find that world population growth affected carbon dioxide concentrations in the atmosphere during the years 1960 and 2010 (Tapia-Granados, Ionides, and Carpinterio, 2012, pp. 58–60). Blaming overpopulation for global warming represents an attempt to shift the responsibility to the Global South (Sawyer and Agarwal, 2000, p. 92) and is also a form of racism. For example, in the 1970s, women of color in the US were the victims of sterilization abuse; in the 1990s, population control proponents pushed Norplant (a long-acting contraceptive with bad side effects) on African-American women, and women in the Global South (Roberts 1997, pp. 90–95, 104–112, 122–127, 139–142).

Promoting social justice in international economic relations

The narratives around global warming and fears of overpopulation miss an essential point: The best way to slow down population growth is through measures seeking to promote social justice (Hartmann, 1995, pp. 289–303). When infant mortality declines, population growth slows; if parents think that some of their children might die before reaching adulthood, they might have extra children to insure enough of them will survive, and they might end up with more children than they wanted (Herr and Smith, 1968, pp. 105–106). A high infant mortality rate is a sign of a great deal of absolute poverty. Reducing infant mortality involves preventing malnourishment of mothers, improving the education and socioeconomic status of women, and improved health care (Bale, Stoll, and Lucas, 2003, pp. 6, 9, 21–23, 39–40). Fidel Castro's government in Cuba reduced fertility through measures that redistributed wealth, raised the status of women, and improved health care; better education also played a role (Diaz-Briquets and Perez, 1982, pp. 518–520, 523–524). At any rate, the world would not be able to change population dynamics fast enough, given the short amount of time remaining for countries to get greenhouse gas emissions to decline.

Convincing people to give up privilege will be a challenge. Psychologically, people hate losses more than they appreciate the same amount of gain (Kahneman and Tversky, 1982). People in the Global South may not understand how much the richer countries would be sacrificing, and people in the US, Europe, and Japan may not realize how badly the Global South wants to escape its poverty. In the US, even the poor might have privilege they need to surrender. In both rich and poor countries, more affluent people contribute to global warming by driving automobiles, living in excessively large housing units, eating too much meat, and consuming goods transported from the far corners of the globe (Takei, 2012, pp. 141–142). A part of the greenhouse gas emissions of China and other poor countries derives from exports to richer countries (Schroeder, 2011, pp. 176–177).

According to the theory of the Jevons paradox, reduced demand for energy leads in the end to a rebound in consumption because prices decline (Foster, Clark, and York, 2010). However, prices of energy could remain high despite lower demand because (1) OPEC countries reduce oil production in order to support prices (2) governments increase the price of energy, say through a carbon tax (3) A decrease in global oil production (peak oil) keeps prices high.

Less consumption will cause some workers to lose jobs. Sociologist Michael Dreiling suggests a version of the late labor leader Tony Mazzocchi's idea of 'paying workers not to work' (Leopold 2007, pp. 413–416, 442). If government measures fighting global warming displace workers, they could get the equivalent of social security disability checks. Since most people would probably prefer to work (the payments would be small), the benefit amount could be taxed away if they earn over a certain amount per year. A report of the New Economies Foundation finds that global economic growth is not compatible with keeping a global rise in temperature to two degrees centigrade (*The Guardian Weekly*, 29 January 2010, p. 18). However, Peter A. Victor (2012, pp. 210–211) has created a scenario in which Canada's GDP decreases so that it has a smaller ecological footprint, but has less unemployment and poverty in 2035 than in 2005. The key factor seems to be a redistribution of wealth.

In terms of its attitude toward international economic negotiations, people in the Global North should be conscious of their racial predispositions and how these impact countries' willingness to accept and proffer concessions. As the largest countries of the Global South, India and China are leaders in the struggle against global economic inequality. The question remains if China's increased consumption of coal and rising automobile use will make greenhouse gas limits possible (Higgins, 2010). At the December, 2011 Durban global warming conference, China and the rest of the Global South agreed that they would be willing, for the first time, to have their greenhouse gas emissions cut under a legally binding accord (Harvey and Vidal, 2011). Because India is poorer than China, it would likely have less stringent limitations than China under a new agreement.

The US Congress must act if the US is to make effective commitments; *The Economist*, 7 June, 2014, shows the weakness of executive branch regulations, which at a 5 percent cut in emissions are seen as '…tiny compared with the overall cuts needed to reign in climate change' (pp. 12–13).

Conclusion

Shankar Vedantam (2010, pp. 6, 7) calls the mental processes that produce unconscious bias 'the hidden brain.' Vedantam (2010, pp. 188–191, 199–208) describes a psychological study that found that people make unconscious associations between concepts, and argues that voters in the US associate welfare, crime, illegal drugs, and immigration with people of color. He could add global warming to the list, because of the importance of India, China and the rest of the Global South in negotiations over the crisis. Habitus is another form of unconscious bias. If people can make their decision making conscious when responding to global warming negotiations, perhaps they can overcome their unconscious bias.

I think that people in the Global South might have a difficult time understanding what a great sacrifice reduced consumption would be for affluent people. I also think that people in the richer countries might have a difficult time appreciating the importance the Global South attaches to its right to development. The fate of planet Earth rests on how well people can comprehend the thinking of those on the other side of the North-South divide, as well as their own national and regional class, racial and gender divides.

Note

1 This chapter is an updated version of Milton Takei's article, 'Racism and Global Warming: The Need for the Richer Countries to Make Concessions to China and India,' *Race, Gender & Class*, vol. 19, nos. 1 & 2, Spring, 2012.

References

Baer, P (2002) 'Equity, greenhouse gas emissions, and global common resources'. In *Climate Change Policy: A Survey*, S.H. Schneider, A Rosencranz and J Niles (eds), Washington, DC: Island Press, pp. 393–408.

Bale, JR; Stoll, BJ and Lucas, AO (eds) (2003) *Improving Birth Outcomes: Meeting the Challenge in the Developing World*, Washington, DC: National Academies Press.

Baron, AS and Banaji, RJ (2006) 'The Development of implicit attitudes: evidence of race evaluations from ages 6 and 10 and adulthood', *Psychological Science*, vol. 17, no. 1, pp. 53–58.

Bourdieu, P (1988) *Practical reason: On the Theory of Action*, Cambridge: Polity.

Dhaouadi, M (1988) 'An operational analysis of the phenomenon of the other underdevelopment in Arab World and the Third World', *International Sociology*, vol. 3, no. 3, pp. 216–234.

Diaz-Briquets, S and Perez, LL (1982) 'Fertility decline in Cuba: a socioeconomic interpretation', *Population and Development Review*, vol. 9, no. 3, pp. 513–537.

Dovidio, JF and Gaertner, SL (1998) 'On the nature of contemporary prejudice: the causes, consequences, and challenges of aversive racism' in *Confronting Racism: the Problem and the Response*, JL Eberhardt and ST Fiske (eds), Thousand Oaks, CA: Sage, pp. 3–32.

Foster, JB; Clark, B and York, R (2010) 'Capitalism and the curse of energy efficiency: the return of the Jevons paradox', *Monthly Review*, vol. 62, no. 6, pp. 1–12.

Harris, J (2007) 'Coal rush puts US on path to disaster', *The Guardian Weekly*, September 7, p. 7.

Harvey, F and Vidal, J (2011) 'Global pact in sight on carbon emissions', *The Guardian Weekly*, December 16, pp. 4–5.

Hartmann, B (1995) *Reproductive Rights and Wrongs: The Global Politics of Population Control*, Boston, MA: South End Press.

Herr, DM and Smith, DO (1968) 'Mortality level, desired family size, and population increase', *Demography*, vol. 5, no. 1, pp. 104–121.

Higgins, A (2010) 'China goes solar for green profits', *The Guardian Weekly*, May 28, p. 30–31.

Holt, TC (2000) *The Problem of Race in the Twenty-first Century*, Cambridge, MA: Harvard University Press.

Kahan, DM; Braman, D; Gastil, J; Slovic, P and Mertz, CK (2007) 'Culture and identity-protective cognition: explaining the white male effect in risk perception', *Journal of Empirical Legal Studies*, vol. 4, no. 3. pp. 465–505.

Kahneman, D and Tversky, A (1982) 'The psychology of preferences', *Scientific American*, vol. 246, no. 1 pp. 160–173.

Kalof, L; Dietz, T; Guagacino, G and Stein, PC (2002) 'Race, gender and environmentalism: the atypical values and beliefs of white men', *Race, Gender & Class*, vol. 9, no. 2, pp. 112–130.

Leopold, L (2007) *The Man Who Hated Work and Loved Labor*, White River Junction, VT: Chelsea Green.

Li, M (2007) 'Peak oil, the rise of China and India, and the global energy crisis', *Journal of Contemporary Asia*, vol. 37, no. 4, pp. 449–471.

Lin, C-P (1994) 'The new PLA', *Asian Wall Street Journal*, February 28, 1994.

Lustig, JR (2004) 'The tangled knot of race and class in America' in *What's Class Got to do With It?: American Society in the Twenty-first Century*, M Zweig (ed), Ithaca, NY: Cornell University Press, pp. 45–60.

Metzger, TA and Myers, RH (1998) 'Chinese nationalism and American policy', *Orbis*, vol. 42, no. 1, pp. 21–36.

Moellendorf, D (2009) 'Treaty norms and climate change mitigation', *Ethics & International Affairs*, vol. 23, no. 3, pp. 247–265.

Postone, M; Li Puma, E and Calhoun, C. (1993) 'Introduction: Bourdieu and social theory'. In *Bourdieu: Critical Perspectives*, M Postone, E LiPuma and C Calhoun (eds), Chicago, IL: University of Chicago Press, pp. 1–13.

Rajan, MG (1997) *Global Environmental Politics: India and the North-South Politics of Global Environmental Issues*, Delhi: Oxford University Press.

Roberts, D (1997) *Killing the Black Body: Race, Reproduction, and the Meaning of Liberty*, New York: Pantheon.

Roberts, JT and Parks, BC (2007) *A Climate of Injustice: Global Inequality, North-South Politics,and Climate Policy*, Cambridge, MA: MIT Press.

Samuels, RJ (1994) *'Rich Nation, Strong Army': National Security and the Technological Transformation of Japan*, Ithaca, NY: Cornell University Press.

Satterthwaite, D (2009) 'The implications of population growth and urbanization for climate Change', *Environment & Urbanization*, vol. 21, no. 2, pp. 545–567.

Sawyer, S and Agrawal, A (2000) 'Environmental orientalisms', *Cultural Critique* no. 45, pp. 71–108.

Schroeder, P (2011) 'Sustainable consumption and production in global value chains' in *China's Responsibility for Climate Change: Ethics, Fairness and Environmental Policy*, PG Harris (ed), Bristol, Policy Press, pp. 169–193.

Sue, DW; Capodilupo, CM; Torino, GC; Bucceri, JM; Holder, AMB; Nadal, KL and Esquilini, M (2007) 'Racial micro-aggressions in everyday life: implications for clinical practice', *American Psychologist*, vol. 62, no. 4, pp. 271–286.

Stern, N (2009) '"Global deal" author's blueprint for climate action'. Interview with Guy Raz, National Public Radio, December 12. Available online at www.npr.org/templates/story/story.php?storyId=121380462

Takei, M (2012) 'Racism and global warming: the need for the richer countries to make concessions to India and China', *Race, Gender & Class*, vol. 19, no. 1 and 2, pp. 131–19.

Tapia-Granados, JA; Edward L; Ionides, EL and Carpintero, O (2012) 'Climate change and the world economy: short-term determinants of atmospheric CO_2', *Environmental Science & Policy*, vol. 21, August, pp. 50–62.

Vedantam, S (2010) *The Hidden Brain: How Our Unconscious Minds Elect Presidents,Ccontrol Markets, Wage Wars, and Save Our Lives*, New York: Spiegel and Grau.

Victor, PA (2012) 'Growth, de-growth and climate change: a scenario analysis', *Ecological Economics* vol. 84, December, pp. 206–212.

Wang, Z (2012) *Never Forget National Humiliation: Historical Memory in Chinese Politics and Foreign Relations*, New York: Columbia University Press.

Winant, H (2004) *The New Politics of Race: Globalism, Difference, Justice*, Minneapolis, MN: University of Minnesota Press.

Zizer, M and Sze, J (2007) 'Climate change, environmental aesthetics, and global justice cultural studies,' *Discourse* vol. 29, no. 2 and 3, pp. 384–410.

7 The Air Around Me

Tyler Hess

the air around me contains heavy metals,
elements and compounds of indescribable shapes
slyly slicing into us with legally untraceable harm.
they end up in my lungs, sir – in my brain, sir

i've never heard of … heavy metals, you say?
HA cheap electricity in exchange for a few coughs? man up!
¿you wanna shut off the lights, leave the rivers without dikes?
¿the mines to stop and the markets to flop?
¿my industry to cease, just cuz of your cough and metals increase?
huh, boy?

actually, yes, sir.
the dozers gotta die so the forests can live
do we bulldoze the boreal, overrun the last orangutan?
for whom, for what?
we thus dedicate our lives to burning it all to squat.
amazon desertification, ocean life elimination
agricultural degradation, entire arctic evaporation,
all approaching *our* collective planet, sir

let's shift the focus to whom my words influence the most
how do <u>I</u> act with the privilege I've been granted
Noam says with more privilege, there's more opportunity
& with more opportunity, there's more responsibility
thus, what am I responsible for doing?

what should i grow? how do i live?
what should I build? how do i give?
pause.

let's remember, there is no doomsday
rather, i am one whose daily doom,
has merely been delayed & reframed
i am being poisoned by TV, not TB – for now.

shedding the screens & abandoning the Adderall
es mas fácil que
being black in Brooklyn,
growing grains in Gaza,
mining metals in Mongolia,
marching in Manama, Moscow, Mali, or Misrata

so sirs, madams, comrades
of all lands and ideological jams
how are we to locally and globally,
internally and continually
obstruct the omnipresent omnicide?

this is what's in my brain, sir

****

i breathe and transform,
the anger and hate, for
it is the air that gifts life,
not the CEOs

springs not coal seams water my soul
soil nourishes body, fire warms heart
returning to the source, seeing with new eyes,

we reconnect to our new, & ancient, story
moving from competition, control, and censorship
to connection, communication, & compassion

we are the bridge generation,
initiating the Great Turning
for the epochal evolution underway,
is in each moment

yet as chaos continues,
Oliver's question remains
what is it you plan to do
with your one wild & precious life?

I love to plant trees, harvest elders, & give hugs.
will you care for the dying, cultures and peoples
or help to disassemble the prisons & barracks
salvaging stones & souls
to help life repair and rebuild, our humble homes,
with Iraqis and Iroquois, Kurds & Klansmen

There is a seat for all at the table
We pray for peace, but keep knives nearby
As Che said, be guided by revolutionary love.

Let us replant our forests, families and fortitude;
reseed the pastures, the clouds, the minds;
rehydrate our land, our biology, our psychology
for we are still, like the water's sill
seeping spirit into soil and soul, deep down
where so much goes unnoticed, ignored
Let us cease being bored-ed up from the magic & mystery

for now, I lean back, sit still, and peer through
the fruit trees, the lenses, the perspectives
to awe at the patterns, between the pixels, that paint
a new story for this moment, for this land, this canvas
with its own water & colors, that sprout a masterpiece
when welcomed back, within a new design of mind
and we find, that things flow better, this way,

this water, has ways of slowing, spreading, sinking,
vitality and life, into soil, into you
let us see the beauty everyday
and source our lives from its presence

Figure 7.1 Mr. Fish, "Save the trees!"

8 Community schools as tools for climate change adaptation in impoverished nations

The example of Haiti

Cynthia Bogard

There is a sea view from the hill where I live now in my crowded neighborhood of Bel Air with my daughters and their children in a tin-roofed shanty we built. The sea, no longer provides for us. The large conch shellfish – we call them *konk* – are all but gone from the local area, and *wouj pwason* – the red snapper, a common fish not too long ago – has become harder to find and it is so expensive only the high-up people and the *blan* – foreigners – can afford to eat it. I have heard fishermen say that the reef is dying. It is brown, not colorful as it once was. It is nearly empty of fish.

When I was young, a jungle still grew on some of the steep mountains surrounding the city. Today, the trees are no more. The big beautiful trees were cut and sold to the Americans and Europeans. These days even young trees have little chance because people cut them to make charcoal, the only fuel most of us can afford. No one seems to be planting new trees, at least not in the city. When I was growing up in my village we had many mango trees – and plantain and *zaboka* – avocado trees. We ate their fruit for free. Not here.

Before my husband died of malaria, I did not live on this hot and dusty hill. We were farmers and lived in a small village. Even then the soil seemed worn out. It didn't produce like it had in granpapa's day. But also, US rice was so cheap it was difficult to sell enough of our rice to pay for school for the children. I left and came here with my two teenage daughters. When the rains come a river runs through the middle of our dirt floor and we have to place our cooking pots to catch the leaks from the roof.

We spend our days, my daughters and I, trying to make enough *lajan* – money – to feed the children. One daughter got a job with the Red Cross after the earthquake but it was just for a few months. So now we buy necessities like shampoo and soap from the wholesalers and then try to sell them at a small profit. In the quest for *lajan*, we have become a city of sidewalk peddlers. Some – though not my daughters – they are good, God-fearing mothers – have taken to selling drugs or their bodies and some of the young men have formed gangs that steal or charge protection money.

The children too, work at small jobs, running errands or selling gum, snacks and other small items. Some – but not my grandkids – have even become beggars. Other children go to work for the big people as servants; they become *restaveks*. Some rich people who have restaveks treat them well and even send them to school if they do their household duties well but others are treated like slaves and even beaten. Not so many of our children here in Bel Air go to school – it costs too much. By sixth grade, most kids in the neighborhood have dropped out. My eldest grandson – he's twelve years old – can read and write some because he went to school through the fourth grade. We don't have the funds for that now.

The streets of my neighborhood are littered with garbage. Dogs, chickens and pigs search among the plastic bags of waste for food. There is no running water. The better off buy it from a truck and keep it in a cistern below ground or, if they have a proper roof, they collect rainwater, which drains from the sloped cement roofs into the underground cisterns. People like me buy buckets of drinking and washing water at two separate prices when we have the money and carry the heavy plastic pails back to our shanty. There is power for only one dim street light in the neighborhood. Electricity mostly goes to the big people who live in guarded compounds and use diesel generators to fill in during the blackouts. We use oil lamps or candles after dark when we have the money for them or just go to bed early.

Sometimes people organize marches to complain about the conditions but things don't seem to get better. We vote in elections, hoping that some politician might bring jobs or services to the neighborhood. And some work at organizing the neighbors to provide security or services to Bel-Air but it's hard to make change when so much needs doing and the people are so poor. We work hard to survive around here. Not much time for anything else.

Introduction

The living conditions depicted by the fictional, but factual Haitian grandmother, compiled from research observations and comments from local residents, are not from some imagined future. It is a description of life in early twenty-first century Port au Prince, Haiti, in a section of the neighborhood of Bel-Air, which once was home to Haiti's intellectual and artistic community. It is not much different from life in many impoverished neighborhoods in crowded cities around the world, where millions live in abject poverty in self-built shantytowns. Global climate change (GCC) didn't produce these sad and squalid sub-cities but climate instability is likely to make living conditions in them even more desperate.

Specific socio-historical processes – some of them quite intentional – have produced these urban communities where few would choose to live if they had viable alternatives. Among the most important is colonialism, with its racialized hierarchies and class antagonisms aimed at natural resource extraction. In the post-colonial era, more modern globalized capitalist relations continue extractive economic practices (Acemoglu and Robinson, 2012).

Unlike citizens in high-consumption nations such as the United States (US), Haitians contribute very little to GCC, although they are likely to suffer very significantly even from small or occasional changes in climate. Indeed, in 2012, Haiti was ranked the most vulnerable country in the world to the ravages of climate change (Slagle and Rubinstein, 2012). Beleaguered by natural disasters such as hurricanes and earthquakes, Haiti is also plagued by extreme inequality, an extractive economy, and an ineffective national government, all residue of its colonial past and post-colonial present. Of particular relevance for Haiti is its racialized and rigid class structure, initiated when Haiti was a French slave colony (Dupuy, 2004). After slavery ended, Haiti continued to be dominated by political forces outside its boundaries, particularly, in the last century, by the US (Farmer, 2005).

In nations beset by histories of colonial exploitation and corrupt current governments, sustainable environmental policies might best be pursued at the village or neighborhood level. For example, community schools that are flexible and forward-looking could serve as the basis for developing and modeling new modes of learning about and relating to the environment. Ideally, community schools are places where critical history can be learned, oppressive race, class, and gender structures can be explored, and new models of increased equality and climate change adaption can be practiced. These institutions potentially could serve as incubators for civil rights, democracy, and civic empowerment – necessary ingredients to live sustainably.

Haiti's proud and tortured past

Haiti comprises the western third of the Caribbean island of Hispaniola, which it shares with the Dominican Republic. When Christopher Columbus arrived in 1492, Hispaniola was a lush, forest-covered, mountainous island with an Arawak-speaking, indigenous Taíno population of at least 100,000. By the mid-1500s, the Taíno population was decimated by European diseases and enslavement by the Spaniards (Zinn, 1995, p. 4). Meanwhile, the Spaniards, and later the French, cut timber from the jungle and converted the land to agriculture. For labor, European colonizers increasingly relied on slaves captured in Africa, who had resistance to European diseases.

Spanish interest in the portion of Hispaniola that now constitutes Haiti waned in the 1600s so that in 1660 the French slave colony of Saint Domingue was created. Described by the colonial powers of the day as the 'Pearl of the Antilles' because it exported huge quantities of agricultural products, by the mid-1700s the prosperity of the slave colony was harnessed to world demand for sugar and other slave-produced exports. This arrangement "condemned slave societies to underdevelopment, eventual stagnation and political disaster" (Fox-Genovese and Genovese, 1983, p. 37) that still haunts Haiti today.

By the late 1700s, more than 40,000 slaves were brought to Saint Domingue a year. The numerous insurrections against the *grand blancs*, or French-origin, Creole plantation owners, were put down harshly, although some slaves found freedom and organized communities in Haiti's mountainous terrain. By 1789 (the year of the French Revolution) the African slave population of the colony was 450,000 while white plantation owners and merchants numbered only 40,000 (Saint-Mery, 1797, 1: 28, p. 111, as cited in Dupuy, 1985). A free Euro-African (*gens de couleur*) community also eventually developed due to sexual relations between white masters and slave women. This mixed-race population grew to about 28,000 by the end of the 1700s. Mixed race people were defined as free under French law and though not equal to whites, they could own property, including slaves, and run businesses. "These people were intermediate in color, status, and power and socially situated between the *grand blancs* (large planters) on the one hand, and the vast mass of enslaved Africans (the majority African-born) on the other" (Mintz 1995, p. 75). By 1789, mixed race free people (also known as *affranchis*) owned about one third of the plantations and a quarter of the slaves. "As the 18th century wore on," writes anthropologist and longtime observer of Haiti, Paul Farmer, "Haiti came to be defined by rigid social categories, each with its own linguistic and religious practices, each with its own hatreds" (2006, p. 58).

The French Revolution inspired Saint Domingue's slaves: in 1791, a rebellion in the north spread to the entire colony and started what was to become the Haitian revolutionary period. Most *gens de couleur* aligned themselves with whites to keep the slavery and plantation system intact, emphasizing how class interests in Haiti trumped racial solidarity (James, 2001[1938]). Later, Napoleon Bonaparte sent a force to restore slavery, but after years of fighting, the army of former slaves, including mountain runaways, managed to defeat the French army and most whites fled the island (James, 2001 [1938]). The victorious former slaves declared independence, renamed the island Haiti, the Taíno word for 'land of mountains' in 1804, and thus became the only republic to be founded by slave revolutionaries, a history of which Haitians are justifiably proud.

Unsurprisingly, Haiti's revolution was viewed as a threat by the European and American powers, which were still engaged globally in slavery and colonization. Consequently, Haiti became economically isolated and the several decades after the revolution were internally tumultuous. In particular, the mixed-race minority, along with the few families directly descended from revolutionary leaders, and the large population of former slaves and their descendants grew into opposing classes. The former retained control of most of Haiti's resources and continued the exclusionary, exploitative, and extractive economy. Civil unrest returned in 1911, with six presidents killed or forced into exile in just four years. The US, urged by US bankers to which Haiti was heavily indebted, then sent Marines to the island and occupied it until 1934, during which more timber was cut for export to the US. Later, Haitian government officials and business elites (most *affranchis* descendants), brokered deals with foreign powers, particularly the US, to continue to extract natural resources, thus continuing the pattern of enriching foreigners and Haiti's tiny ruling class. This trend of corrupt enrichment of the Haitian elite while the poor received little attention continued during the Duvalier years when first the father (François, known as 'Papa Doc') and then his son (Jean-Claude or 'Baby Doc') ruled Haiti for thirty years

(1957–1986), using brutally repressive means to silence proponents of civil reform. The Duvalier reign of terror, including imprisonment, rape, torture, assassination, and constant intimidation, was implemented by the infamous *Tonton Macoute* – the Duvaliers' private paramilitary organization (Abbott, 2011). But international forces also continued to shape Haiti.

As repressive as they were, the Duvalier regimes did not end the tariffs on rice imports to the country and, as a result, Haiti was able to grow sufficient rice to feed its population until the early 1980s, when the US pressured the nation to open its market to US agricultural products (Chavla, 2010). Wealthy nations and global entities often promise agricultural aid to poor nations in return for lowering import barriers to staple crops such as rice and corn. The result is dependence on global agribusiness and its purveyors, such as the US, and heavy debt for the loans needed to buy food staples from wealthy nations and seeds and fertilizers from global agribusiness companies (Escobar, 2011[1994]). Ironically, in 1990 Haiti eventually lowered the rice tariff from 30% to 3% under President Jean Bertrand Aristide, one of the few Haitian leaders to advocate for the poor. After that, local rice production plummeted as cheaper US-produced and subsidized rice (and other products, especially corn) flooded the market, to the profit of US farmers and corporations. Before the tariff on rice importation fell, Haitians imported about 19% of their food; as of 2010, 51 percent of all food and 80 percent of rice was imported (Katz, 2010).

The drastic tariff reduction on foreign grain was the final blow for many farms. Without a market for their grain, farmers moved to the capital city in search of ways to make a living. As formerly rural Haitians flooded Port au Prince, the city grew from 200,000 to nearly three million living in the metropolis – a third of Haiti's total population. What modest water, sanitation, and electricity infrastructure there was in the capital was quickly overwhelmed. New arrivals found few jobs – half of residents of Port au Prince lacked formal employment even before the 2010 earthquake (World Bank [WB] 2006). Needing to cook, and with the only accessible and affordable fuel being charcoal, tremendous tree-cutting occurred around the capital and beyond, which loosened the soil on the steep hillsides that dominate the landscape, causing mudslides and massive erosion (Than, 2010). These new destitute city residents had no other recourse but to build shanties on rented land.

Like Haiti, many formerly colonized regions of the globe continue to suffer from the entrenched class and race-based hierarchies that marked colonial regimes. Common to colonial states is the extraction of natural resources and conversion of forested terrain and subsistence farming by natives to agricultural production for export, using an imported or local population as slave or near-slave laborers. These social processes – wherein a small, often foreign elite create systems of extreme inequality for the purpose of extracting from a region or nation as much wealth as possible to profit themselves – define what Acemoglu and Robinson term an "extraction economy" (2012). Such economic systems leave a legacy of social and environmental destruction that impedes self-sufficiency and will increase the impact of climate change.

A glimpse of present-day Haiti

Haiti's population is young and impoverished, with a median age of 21.4 and an average pre-earthquake annual per capita income reported to be between $611 (UNDATA, 2013) and $1,200 so that eighty percent of Haitians live below the poverty line (CIA World Factbook [WF], 2013). Haiti was already the most impoverished nation in the Western Hemisphere before it suffered a massive earthquake in 2010, with most of its population living in shanty housing built from found materials.

Farming in Haiti is difficult due to incursions of foreign food imports and the introduction of seeds that require chemical fertilizers to grow (and the money to purchase them). Additionally, with unprecedented deforestation, 98 percent of its original forest gone and Haiti's steep slopes experience massive soil erosion (Katz, 2013). The continued use of charcoal for cooking by 90 percent of Haitians has ensured trees will continue to be cut unsustainably (UN Brundtland Commission, 1987; Than, 2010).

The coral reefs surrounding Haiti, the basis for aquatic life in the Caribbean ecosystem, have been smothered by soil washing onto them from the denuded mountains (Slagle and Rubinstein, 2012). Haitian waters are polluted with raw sewage and farm fertilizer runoff and are overfished with poisonous methods such as arsenic, because there is no enforced marine policy in Haiti. The director of Reef Health, a non-profit dedicated to monitoring coral reefs, called Haiti's "the worst case of overfishing … anywhere in the world" (McDonald, 2011, p.A9). Increasing water temperatures and acidification of the oceans due to rises in global CO2 levels have also hastened reef destruction and local fish stock depletion (Slagle and Rubinstein, 2012).

Haiti's ever-growing population also puts severe pressure on a mountainous land with mostly thin tropical soils ill-equipped to sustain 10 million people. The majority of Haitian women have little access to education, healthcare, or contraception. Beyond general impoverishment, Haiti is crippled with a severe estrangement between its tiny, wealthy elite, mostly descended from the slave-era mixed race *affranchis* population (about 5 percent of the nation) and the 80 percent of Haitians who are severely impoverished, defined as those earning less than $2 per day (Farmer, 2006).

Many non-governmental organizations (NGOs) and other multinational organizations such as the United Nations and the Inter-American Development Bank have a strong presence in Haiti. Most have had minimal impact on the lives of everyday Haitians, as evidenced by their continued poverty. Moreover, NGOs have taken on most government functions, continually disempowering or sidestepping Haiti's government (Klarreich and Polman, 2012).

Climate change will exacerbate Haiti's environmental vulnerabilities. Haiti is located in a hurricane corridor, and as more carbon is put in the atmosphere through the burning of fossil fuels, oceans warm and hurricanes are more likely to increase in intensity and possibly frequency (Holland and Webster, 2007). In 2008 alone, four tropical storms and hurricanes hit Haiti in one month, causing floods, landslides, crop washouts, increased erosion, and 1,000 drowning deaths (Slagle and Rubinstein, 2012). What Haiti and her people need most are effective institutions that could lead to the qualities of effective nations – political and economic inclusiveness (Acemoglu and Robinson, 2012). I suggest that the *community school* – updated from its successful 20th Century incarnations developed by social worker Jane Addams and philosopher John Dewey, among others – might be a social policy initiative that could teach and model both democracy and sustainable social, economic, and environmental practices.

In times of dramatic change, new institutions are needed

One institution developed to meet new needs in a time of rapid social change in the US was the settlement house, exemplified by Chicago's Hull House, developed and run by Jane Addams and other early social workers. Hull House was envisioned as a site of community education and aid, a place where new immigrants to the US were able to access the skills and social relationships they would need to find success in the future. Imperfectly it helped build effective communities by engaging the gold mine of experience and skills that the immigrants brought with them from their countries of origin and incorporated these into the center's offerings (Longo, 2007).

Hull House's basic model of individual skill-building (human capital development) linked to efforts to make communities cohesive and effective (social capital development) was furthered by philosopher of education John Dewey, who developed a fully articulated model he termed 'the community school' (Benson *et al.*, 2009). The National Society for the Study of Education defines a community school as "a school that has two distinctive emphases – service to the entire community, not merely to the children of school age; and discovery, development, and use of the resources of the community as part of the educational facilities of the school" (Seay, 1953, p. 2). In laying out his pedagogic creed, Dewey stated, "through education, society can formulate its own purposes, can organize its own means and resources, and thus shape itself with definiteness and economy in the direction in which it wishes to move" (1897, p. 80). I extend his insight to suggest that a 21st Century version of the community school might serve a similar function in a context of climate change, poor governance, an extractive economy, and the racial and class remnants of colonialism.

The conditions of education in Haiti

Before the earthquake, only half of Haitian children attended school, because 85% of schools in Haiti were private and most parents couldn't afford the tuition. Only a quarter of those who attend school graduate from high school (CIA WF, 2013). Haiti needs more and better public schools at every level, but hampered by an ineffective and underfunded government and racialized class divisions, it is unlikely that Haiti's education sector will improve soon. Haiti's education system mimics its class structure: it provides quality high-tuition schools for children of the elite, while crowded schools with poorly trained teachers and few materials are the norm for the working class and poor. Though faith-based and secular NGOs and a few local groups have stepped in to provide free or low-cost education to needy students and to rebuild schools in the post-quake period, these efforts remain spotty, temporary, and inadequate and have not resulted in the development of a national school system that even begins to meet to needs of Haitian youth nor takes into account the precariousness of Haiti's environment.

What a Community School could do in poor communities

Community schools in Haiti could provide a central institution around which an effective community could be built. First, they could better serve one central function of education – the development of increased knowledge, skills, and capacities that sociologists term *human capital* – for both children and adults. Second, community schools can aim at developing social capital in impoverished neighborhoods. *Social capital* – a crucial resource in creating social change – is the result of social networks developed through joint participation in a project or organization, which then begin to shape norms and expectations. Networks developed through collective participation build trust among neighbors, and trust is the basis for community action that can lead to productive social change (Putnam, 2001).

As an institution that fosters learning, community schools can develop, model, and transmit methods for sustainable living and coping with climate change both in urban and rural contexts. Ecological educator David Gruenewald terms what is needed "a critical pedagogy of place" (2003, p.3) or "reinhabitation" – "learning to live-in place in an area that has been disrupted and injured through past exploitation" (Berg and Dasmann, 1977, p. 399). Often this includes intergenerational learning from elders. Reinhabitation combined with a critical pedagogy that stressed reflection on oppressive cultural and historical structures has the potential to begin to gradually address race, class, and gender inequities through ongoing conversation and reflection, or what revolutionary educator of the poor Paulo Freire (1989) termed praxis, defined as "a problem-posing education… [that] strives for the emergences of consciousness and critical intervention in reality." (p. 68). Combined with action to achieve justice and meaningful citizenship, critical place-based education has the potential to reclaim both the cultural and environmental 'commons' (Bowers, 2001).

Growing community schools from the ground up will not be quick or unproblematic, as elite interests will resist the growth of institutions that are inherently opposed to their continued domination. Haitian efforts to democratize or even mildly reform their country often have met with murderous opposition. Yet inaction in the face of continuing climate instability will ensure that Haiti's situation becomes increasingly dire. Even elites may not be able to weather the metaphorical and actual storms wrought by climate change, making social change in their interest as well. Moving to inclusive political and economic institutions has the best chance of saving a nation from continued strife and environmental degradation (Acemoglu and Robinson, 2012).

Environmental empowerment and education through Community Schools

One Haitian model that combines effective environmental protection, individual empowerment, and democratic development is the Lambi Fund, founded by Haitians, Haitian-Americans, and other North Americans in 1994. This NGO's mission is to promote economic justice, democracy, and alternative sustainable development in rural Haiti (Lambi Fund, 2012). A distinctive element of the NGO is its insistence that project suggestions come from villagers themselves, after a period of discussion and reflection. Once a group of people has met, formed an organization and outlined needs and goals in discussion together, Lambi Fund will consider providing economic support and expertise, depending on the merits of the plan and given that it fits within a model of environmental sustainability. While not specifically educational, Lambi Fund supports the creation of local institutions devoted to specific community improvement projects and insists that they be based on equal participation (including gender equality) and democratic deliberation and decision-making. This builds social capital, raises civic capacity, and empowers individuals – similar to the goals of community schools. Once one goal has been achieved, neighbors have the confidence to take on other projects. The Lambi Fund also stresses training and capacity building within the context of a particular project – thus the rural people who engage in these projects also gain human capital.

For example, one rural community group (Peasant Organization of Garat) decided they wanted to re-introduce pig-raising in their village; after some discussion, they asked Lambi Fund for 36 pigs, a veterinary technician, and organizational development training for the group. Some members would concentrate on breeding pigs to extend the benefit to other families. With the proviso that the pigs be raised in large, fenced-in areas so as not to cause further soil erosion through their rooting behavior, Lambi Fund agreed to finance this plan and train its participants. The project was a success and brought renewed income security to the farmers. They also learned about environmentally sound animal husbandry and saw how sharing young animals with neighbors increased the village's economy as a whole.

These farmers expanded their capacity to act as a civil society organization as well. At the same time this project was taking shape, the national government built a hydroelectric dam but did not extend electricity to this village. Now organized, the 500 members of this animal husbandry group persistently lobbied government officials until electric lines were brought to the village. This type of collective empowerment is the focus of the Lambi Fund and also the idea behind community schools. Their success in these endeavors is likely to empower them to ask one another what else they could accomplish by working together. As the work of Lambi Fund illustrates, the self-organization of communities that practice deliberative democracy and pursue environmentally sound economic uplift for their members provides a solid foundation for sustainable development and a demand for needs-based education.

Unlike many Western NGOs, Lambi Fund requires a lengthy period of self-organizing and discussion before a project grant is given. It does not try to intervene in communities with preconceived ideas of what the community needs. This was the lesson – to meet people where they are – that Jane Addams also eventually learned through her lifetime of involvement with the settlement movement. To meet people where they are, the community schools of the future would aim to truly serve children and their parents by providing knowledge about their culture, history, and current political and economic circumstances, especially their relationship to the environment and how they could improve that relationship. These 21st Century institutions must have as their central goal community and individual empowerment as democratic empowerment at the local level is the best tool humanity has found for creating societies that are inclusive, just, and future-oriented. Community schools are inherently concerned about the future, and because they are, they could well become the indispensable institutions in a future focused on stabilizing and renewing the environment on which we all depend.

References

Abbott, E., 2011. *Haiti: A Shattered Nation*. New York: Overlook Press.

Acemoglu, D. and Robinson, J., 2012. *Why Nations Fail: The Origins of Power, Prosperity and Poverty*. New York: Crown Business.

Benson, L., Harkavy, I., Johanek, H. and Puckett, J., 2009. The enduring appeal of community schools. *American Educator*, Summer, pp. 22–47.

Berg, P. and Dasmann, R., 1977. Reinhabiting California. *The Ecologist* 7(10), pp. 399–401.

Bowers, C., 2001. *Educating for Eco-justice and Community*. Athens, GA: University of Georgia Press.

Central Intelligence Agency, 2013. *World Factbook: Haiti*. [Online] Available at: www.cia.gov/library/publications/the-world-factbook/geos/ha.html, accessed 15 December 2013

Chavla, L., 2010. Has the US rice export policy condemned Haiti to poverty? *Hunger Notes*. [Online] Available at: www.worldhunger.org/articles/10/editorials/chavla.htm, accessed 11 December 2014.

Dewey, J., 1897. My pedagogic creed. *School Journal*, 54 (January), pp. 77–80.

Dupuy, A., 2004. Class, race, and nation: Unresolved contradictions of the Saint-Domingue Revolution. *The Journal of Haitian Studies*, 10(1), pp. 6–19.

Dupuy, A., 1985. French merchant capital and slavery in Saint-Domingue. *Latin American Perspective*, 12(3), pp. 77–102.

Emanuel, K., 2008. Hurricanes and global warming: results from downscaling IPCC AR4 simulations. *Bulletin of the American Meteorological Society*, 89(3), pp. 347–367.

Escobar, A., 2011 (orig. 1994). *Encountering Development: The Making and Unmaking of the Third World*. Camden, NJ: Princeton University Press.

Farmer, P., 2006. *The Uses of Haiti*, 3rd edn Monroe, ME: Common Courage Press.

Fox-Genovese, E. and Genovese, E.D., 1983. *Fruits of Merchant Capital: Slavery and Bourgeois Property in the Rise and Expansion of Capitalism*. New York: Oxford University Press.

Freire, P., 1989. *Pedagogy of the Oppressed*. New York: Continuum.

Gruenewald, D. A., 2003. The best of both worlds: a critical pedagogy of place. *Educational Researcher*, 32(4), pp. 3–12.

Holland G. J. and Webster, P., 2007. Heightened tropical cyclone activity in the North Atlantic: natural variation or climate trend? *Philosophical Transactions of the Royal Society* [pdf]. Available at: http://webster.eas.gatech.edu/Papers/Webster2007a.pdf, accessed 15 December 2014.

James, C.L.R., 2001 (orig. 1938). *The Black Jacobins: Toussaint L'ouverture and the San Domingo Revolution*. New York: Penguin.

Katz, J. M., 2013. *The Big Truck That Went By: How the World Came to Save Haiti and Left Behind a Disaster*. New York: Palgrave.

Katz, J. M., 2010. With cheap food imports, Haiti can't feed itself. *Huffington Post*. [Online] Available at: www.huffingtonpost.com/2010/03/20/with-cheap-food-imports-h_n_507228.html, accessed 15 December 2014.

Klarreich, K. and Polman, L., 2012. The NGO Republic of Haiti. *The Nation*. [Online] Available at: www.thenation.com/article/170929/ngo-republic-haiti#, accessed 13 December 2014.

LAMBI Fund, 2012. [Online] Available at: www.lambifund.org/, accessed 13 December 2014.

Longo, N. V., 2007. *Why Community Matters: Connecting Education with Civic Life*. Albany, NY: State University of Albany Press.

McDonald, B., 2011. Haitian divers hope to aid ailing reef. *New York Times* 1 September. [Online] Available at: www.nytimes.com/2011/09/02/world/americas/02reef.html?_r=0, accessed 12 December 2014.

Mintz, S. W., 1995. Can Haiti change? *Foreign Affairs*, pp. 73–86.

Oliver-Smith, A., 2012. Haiti's 500 year earthquake. In Schuller, M. and Morales, P. eds, *Tectonic shifts: Haiti since the earthquake*. Sterling, VA: Stylus Press, pp. 18f–22

Putnam, R. D., 2001. *Bowling Alone: The Collapse and Revival of American Community*. New York: Touchstone.

Seay, M. F., 1953. The community school: New meaning for an old term. *National Society for the Study of Education Yearbook*, 52(2). [Online] Available at: http://nsse-chicago.org/yearbooks.asp?cy=1953, accessed 14 December 2014.

Slagle, T. and Rubenstein, M., 2012. Climate change in Haiti. Columbia Climate Center, Earth Institute, Columbia University, 1 February 2012. [Online] Available at: http://blogs.ei.columbia.edu/2012/02/01/climate-change-in-haiti/, accessed 15 December 2014.

Than, K., 2010. Haiti Earthquake, deforestation heightens landslide risk. *National Geographic News* 14 January. [Online] Available at: http://news.nationalgeographic.com/news/2010/01/100114-haiti-earthquake-landslides/, accessed 15 December 2014.

United Nations Data, 2013. UNDATA Country profile: Haiti. [Online] Available at: http://data.un.org/CountryProfile.aspx?crName=Haiti#Economic, accessed 15 December 2014.

United Nations, 1987. Our Common Future, From One Earth to One World, Gro Harlem Brundtland Oslo, 20 March 1987 [Online] Available at: www.un-documents.net/our-common-future.pdf, accessed 15 December 2014.

World Bank, 2006. Haiti: Social Resilience and State Fragility in Haiti. Report 36069-HT [Online] Available at: www-wds.worldbank.org/external/default/WDSContentServer/WDSP/IB/2006/06/07/000160016_20060607092849/Rendered/PDF/360690HT.pdf, accessed 15 December 2014.

Zinn, H., 1995. A *People's History of the United States, 1492 – Present*. New York: Harper Perennial.

9 Of starving horses and growing grass

Resilience versus dependency in a Caribbean fishing community

April Karen Baptiste

After a hurricane, two farmers, husband and wife, find themselves left with little. They divide what was left between them equally. Each gets a horse and a field of grass able to feed that horse for a month if the horse is put to graze there immediately. Each horse can be hired out to earn money to buy enough food to feed one person for a day. The man asks the woman, "What are you going to do?" She says that she is going to hire out the horse each day for a month. She will take the money she earns and split it in two. She will use one half to buy half of the food she needs and the other to buy half the grass the horse needs. 'That way, by the end of the month, the grass in the field would have grown twice as much and there will be enough to feed the horse permanently.

"But," says the man, "After the first day, the horse will only get half of what he needs to eat and he will be hungry the next. On the second day, he will only be able to work for half the day, and earn you only half the money of the first day." And he continues with a laugh, "And if you then divide your earnings from him in half, there will be enough only to feed you a quarter of what you need and the horse a quarter of what he needs. By the fifth day, he will be hardly working and both of you will be starving. By the end of the month, you will both have starved to death before there is enough grass in the field to feed the horse in a sustainable manner."

The woman asks the man what he planned to do. He states, "I will let the horse eat its fill every day in the field, and hire him out to bring in a full day's pay, which I will use to buy a full share of food for myself, every day."

"But," replies the woman, "At the end of the month, there would be no more grass to feed the horse and you both will starve to death."

"No, my dear wife. By then, the government would have sent in emergency relief!"

Given the above story, one is inclined to say that the male farmer is perhaps being selfish or is not thinking long-term. We may even critique his saying, "by then, the government would have sent in emergency relief." What does it mean to rely on external support to enhance resilience as implied by the male farmer in the above vignette? Or is the female farmer's approach more appropriate for building resilience? Her approach is to use the resources she has to attempt to recover from the exposure and impact of the natural hazard. In so doing, however, her self-reliance may then lead to her demise. How do we address this tension? Should resilience be built internally? Should there be external support that is essential to the development and perpetuation of resilience? These are some of the questions are raised in this chapter.

Within both the natural science and the social science approaches, resilience is defined as the capacity of systems to absorb recurrent disturbances. Holling (1973) introduces the concept of resilience from the ecological perspective. The concept has recently been expanded within the social sciences from a geographical perspective. Here, the definition considers the degree to which a complex system is capable of self-organization and the extent to which the system can build capacity for learning and adaptation (Adger *et al.*, 2005). The capacity for learning implies that the system moves beyond the initial state prior to disturbance to a new, stable state, reflecting a balance between the new environmental demands and the system's adjusted responses. From these initial definitions there seems to be an underlying tone of an innate ability for the system, whether ecological or social, to achieve some type of steady state. However, this definition must be challenged: is resilience-building simply based on the innate characteristics of a system or can there be external sources that contribute to this attribute? Adger and colleagues (2005) further expand their definition by stating that resilient communities use diverse mechanisms to live with and learn from change and shocks.

For example, in looking at global climate change (GCC) for the Caribbean, there are a number of strategies that are used to deal with hurricanes. These include improving the monitoring of hurricane risk, establishing early warning systems, and encouraging mobilization among non-governmental and private organizations as part of first responder teams (Adger *et al.*, 2005). Hence, there is a need for some type of dependence on external forces to ensure that the system is building resilience to natural and man-made stressors. This dependence can take place within the frame of learning and adaptation, which, with respect to GCC, must be multi-dimensional.

In examining adaptation as a form of resilience-building, scholars have considered this as a social and political act (O'Brien, 2013), requiring consideration of the inherent structural and social disparities that exist within the society (Dulal, Shah, and Ahmad, 2009). Indeed, when thinking about adaptation as a form of resilience-building, equity is central: Questions such as "who gets access to adaptation measures and why?" must be addressed. Dulal and colleagues (2009) suggest that in order to address resilience adequately, the poor, indigenous populations, women, and children must be at the forefront of adaptation measures, as they are highly vulnerable to GCC effects.

Asset mobilization, both in terms of access and process, is another relevant dimension to understanding adaption as a form of resilience-building. Asset mobilization includes social (Orencio and Fujii, 2013; Shah *et al.*, 2013), human

(Orencio and Fujii, 2013), and financial capital (Fischer *et al.*, 2013), among other things. Access refers to the availability of resources, while process relates to the ways in which assets can be mobilized among entities. Governance structures at various scales are particularly important to understanding the ways in which different groups get access or fail to get access to assets. For example, in low-income communities provision of these social services are not strong, making it hard to effectively mobilize resources when needed.

A key dimension to understanding resilience-building is personal and political efficacies shaping human agency (Brown and Westaway, 2011). Individuals and social entities adjust to disturbances if they believe they are capable. Personal and political efficacies in resilience-building are shaped by many factors, two of which are explained here. The first is that of local 'ways of knowing' (WOK). Traditional communities rely heavily on WOK for day-to-day survival, which is a form of resilience itself, honed over generations. It represents repeated community response strategies to changes that are typically orally transmitted to succeeding generations of members of a community, building both individual and community capacity.

Moreover, personal and political efficacies among communities are also necessary for asset mobilization in the aftermath of the exposure to a natural hazard. This can range from neighbors and family members who rally to support each other, or it can include external sources such as first responders and other forms of assistances from non-governmental organization or community-based organizations. It is important to understand both from the community and governmental perspective the agencies responsible for making essential provisions available to individuals and communities enabling recovery from a stressor or event.

For the Caribbean, a significant amount of the research has looked at the institutional response to climate change adaptation and resilience-building, but has neglected to examine what happens at the level of the individual (Mercer *et al.*, 2012). This chapter fills this gap by focusing on the individual and highlighting the community or meso-level implications.

Climate change and the Caribbean – the Jamaican story

The growing literature on the effects of GCC in the Caribbean region indicates that it is highly vulnerable (Pulwarty, Nurse, and Trotz, 2010; Taylor *et al.*, 2012). The specific effects of GCC on Caribbean island states have not only been predicted but have also been observed recently: the air temperature has been trending upwards with an average increase of 0.6 degrees Celsius (Taylor *et al.*, 2012); the sea level rise has led to significant beach erosion, directly impacting tourism (Pulwarty, Nurse, and Trotz, 2010); increased coral bleaching has impacted biodiversity, and by extension livelihood activities (Alemu and Clement, 2014); and hurricanes and other storms have caused flooding and droughts, affecting local and national economies (Taylor *et al.*, 2012).

Natural resource exploitation, including farming, fisheries, and tourism are crucial to most of the economies of the Caribbean region (Pelling and Uitto, 2001), hence GCC is a direct and indirect threat to these economies. Specifically for fisheries, the sector of focus in this chapter, increasing sea surface temperatures, rising sea level, increased turbidity and damage from storms, and the effects of ocean acidification are relevant (Mimura *et al.*, 2007).

With respect to Jamaica, the agricultural sector accounts for 5.7 percent of the Gross Domestic Product (GDP) (Central Intelligence Agency (CIA), 2014). The fishing sector is important as a source of protein to both the domestic and export markets (Kong, 2006). For example, in Jamaica conch fisheries is one of the top export industries globally (Aiken, 1999). Studies are indicating that GCC is affecting Jamaica's marine environment, which then impacts fishing (Webber, n.d.). For example, Jamaica experienced six storm events, including three hurricanes, between 2002 and 2007 (Richards, 2008), which severely impacted coastal areas and decreased the amount of conch available, which negatively impacted the livelihoods of fishers. Understanding the vulnerability of coastal communities cannot be achieved without acknowledging intersectionality (Kaijser and Kronsell, 2014). Specifically, this intersectionality involves the multiple positions of fishers in society in terms of power differentials, proximity of environmental exposure to climate change events, legitimacy of different forms of knowledge, and high dependence on a scarce resource. These, acting together, determine the extent of vulnerability and resiliency among fishers and their potential for adaptation to the threat of climate change.

Approach to study

Within Jamaica, scientific studies have indicated that the south coast of the island will be particularly vulnerable to the effects of climate change (Food and Agricultural Organization (FAO), 2012; Richards, 2008). There are large, artisanal fishing villages located along the south coast, including one of the largest fish landing sites on the island, Old Harbour Bay (OHB) (FAO, 2004). OHB has a total population of 7,388 (Social Development Commission (SDC), 2009), of whom 632 are registered fishers (Government of Jamaica (GOJ), 2010).

A historical hazard impact and coping mechanism matrix for OHB indicates that the projected sea level rise will result in a considerable loss of land for the community (FAO, 2012). Given its size and significance to the domestic market, in addition to the current threats of overfishing and other environmental degradation, OHB demonstrates how compounded or intersecting vulnerabilities to climate change, enhanced by socio-demographic positionalities, are affecting marginalized communities.

A mixed-methods approach was used to get a sense of fishers' perceptions of climatic changes and the adaptation strategies that are used in this community. A closed-ended structured survey was conducted with a large sample of fishers (241), while semi-structured interviews were conducted with key informants from the community to add some depth to survey responses (14 interviewees; interview time 20–30 minutes).

The survey gathered perceptions of GCC effects; information on fishers' livelihoods; the type of adjustment strategies being used in response to the perceived effects; the reasons for the lack of adjustments in some cases; and general demographic information on fishers. As depicted in Table 9.1, the sample consisted primarily of males, with the highest education level being secondary. Most of the fishers in the sample were involved in the industry full-time and had monthly incomes ranging from $6000 to $24,999 Jamaican dollars [JMD] ($600 to $2,499 USD). Fishers were primarily engaged in net fishing near shore areas and had more than 30 years of experience in the industry. The majority were, however, not registered members of the local fishers' cooperative, though most were registered with the governmental fisheries division.

Table 9.1 Demographic of the survey sample of fishers from Old Harbour Bay

Variable	*Sub-component*	*Percent (n)*
Gender	Male	84.5 (202)
	Female	15.5 (37)
Education level	Pre-primary	6.2 (15)
	Primary/all-age	33.6 (81)
	Secondary	56.8 (137)
	Higher education	2.8 (7)
Employment status	Full-time	77.1 (185)
	Part-time	22.9 (55)
Monthly income (JMD)	80,000 and greater	3.0 (7)
	40,000–79,999	9.3 (22)
	25,000–39,999	7.2 (17)
	6,000–24,999	58.1 (137)
	< 5,999	22.5 (53)
Type of fishing	Net	65.3 (130)
	Pot	17.6 (35)
	Other	17.1 (34)
Area fished	Near shore	58.8 (114)
	Deep water	32.0 (62)
	Near banks	9.3 (18)
Length of time involved in fishing	30 years or more	28.2 (68)
	21–30 years	24.5 (59)
	11–20 years	24.1 (58)
	10 years or less	23.2 (56)
Member of fisherman's cooperative	Member	10.3 (22)
	Not a member	89.7 (192)
Registered with Fisheries Division	Registered	69.2 (157)
	Not registered	30.8 (70)

Results

Perceptions of the impacts of climate change

From the survey, a majority (80.9 percent, or 195) claimed to have observed some type of GCC impact in the previous five to ten years, exampled below through interviews.

> a lot of changes mi (I) see. Erosion from the sea, erosion from the land. Mi (I) see sea take back nuff ah de (a lot of the) land. Mi (I) see all kinda environment changes over these 40 years ... Storms nowadays more than when I was a little boy ...
>
> (Interviewee 3)

> ... storms and storm surge, it has never been so prevalent. You know you getting more storms and more hurricanes now that the whole environment situation has evolved ... You even having less rainfall ... months when you nah [don't] expect no rain, you expect good fishing you having rain, so the situation is so contrary now...
>
> (Interviewee 4)

Fishers also indicated that the above changes would have negative impacts on their livelihoods. As indicated in Figure 9.1, they believe that there would be more sickness in the fish, changes in migratory patterns and species diversity, and increases in fishing expenses.

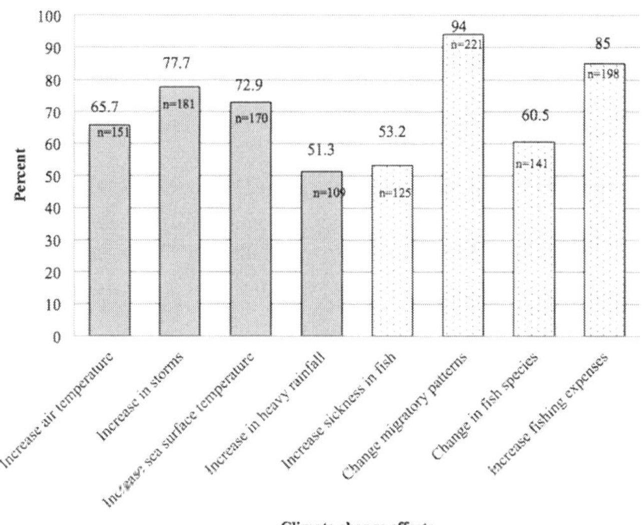

Figure 9.1 Distribution of perceptions of the effects of climate change on the physical environment and livelihoods among fishers in Old Harbour Bay[1]

Interviewees 1 and 4 lamented how their fishing trade had been impacted by climate change in the time since they got into the industry:

> yuh [you] see when mi [I was] smaller, still storm come unno but, here now, yuh [you] see [the] last storm whey gone [that pass], no fisherman, a hold here, no fisherman a no pot out there [none of the fishermen did not catch, nor did they have any catch in their pots]. Yuh [you] see [the] last storm, [the] last storm mash we up …
>
> (Interviewee 1)

> … there are certain times during the year where you will look forward to good catch, special species of fish traveling and so forth, you find that that doesn't happen. … You find that it is still limited because what has happened is that global warming has been taking place all over the world. So it has affected the fishing industry not only in Jamaica but worldwide in that …, you have a lot of erosion in the sea bottom, of the habitat where the fish lives.
>
> (Interviewee 4)

Fishers generally see climate change not just as a threat to the physical environment, but also to their day-to-day survival in the industry.

Adjustment strategies

Of the interviewees that have noticed changes in climate, 27.3 percent say they have engaged in some type of adjustment activity in response to the livelihood threat presented by GCC. For those fishers who indicated that they did adjust, four main strategies were recorded: going further out to sea (42.2 percent), decreasing the number of days at sea (35.9 percent), changing fishing grounds (28.1 percent), and changing the type of fishing technique used (17.2 percent). Additionally, interviewee 2 explained:

> Well, adjustment goes by keeping close … to weather reports and the signs that we know in the ocean … We also depend on those that are using the technology to warn and to give a more precise estimate of what is expected. Sometimes it work, sometimes it does not …
>
> (Interviewee 2)

Interviewee 2 further explained how pot fishers adjust. He said

> The only thing that we can do to prepare those traps and protect them is by removing from rocks, and carry them to mud … We [also] trim the buoys, smaller buoys, because the bigger the buoy the easier for the waves to help lift the traps. So things like those we can do but we cannot take our traps on the south side back on land.
>
> (Interviewee 2)

Interviewee 4 also indicated that some fishers are engaged in adjustment measures. She said:

> So what happen is that people are more pro-active … with regards to that [GCC effects] so they are protecting their livelihoods better now. So you find that people are easier to bounce back now because we have learnt resilience and how to cope as a community.
>
> (Interviewee 4)

However, survey results indicated that the majority of fishers are not engaged in adjustment strategies (72.7 percent), with several reasons provided for this lack of adjustment (Figure 9.2).

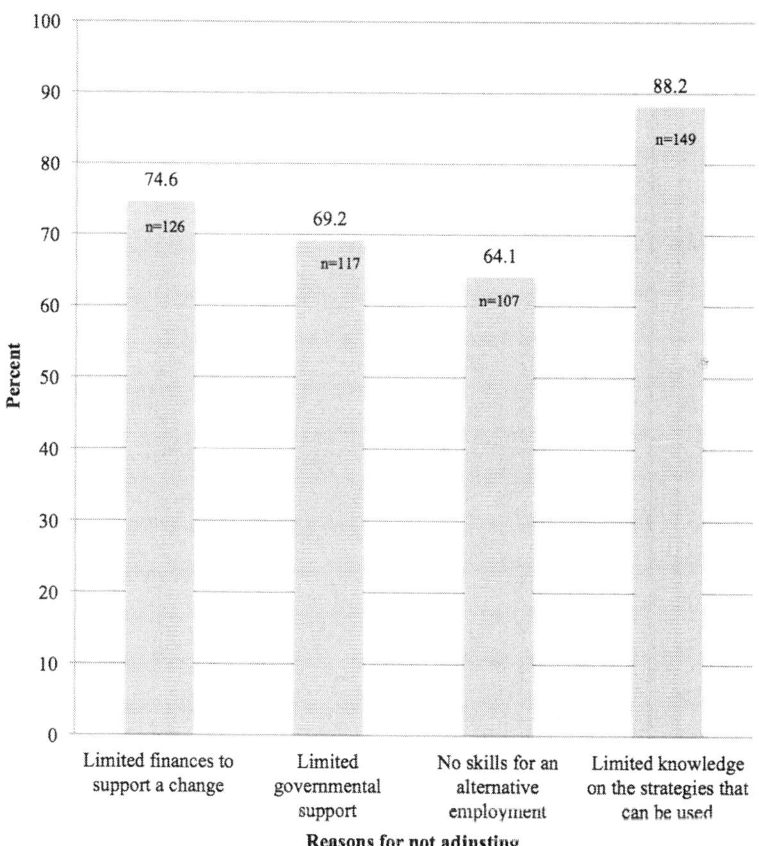

Figure 9.2 Distribution of reasons for not adjusting to climate change impacts among fishers in Old Harbour Bay, Jamaica[2]

Interviewee 2 indicated the difficulties in adjusting to the threats to their livelihoods with reference to the frequency of storm events:

> ... let me say if you try to rebuild, and another one [storm] come and hit you again, especially when a third one [storm] come and hit you, then you are so frustrated that even if there is fish out there, you cannot afford to keep up because the fishing equipment are very expensive. So that is another factor that we have to take into consideration.
>
> (Interviewee 2)

Knowledge regarding GCC causes and effects also plays a role. Fishers gained knowledge from two main sources: formal institutional training and traditional knowledge. In terms of formal training interviewee 4 stated:

> You have agencies such as C-CAM that comes in and ... do training with persons in the community on how to manage whatever resources ...you have, because the time you sit down and wait for help coming from outside you lose everything already... you have agencies coming into the community and teaching you disaster risk management, giving courses and telling you how to cushion the effect of some of these things when they do happen.
>
> (Interviewee 4)

The experience of interviewee 4 with formal training was not typical among the fishers. Many more reported having access to local WOK. Two fishers summarized:

> I is a fisherman, I is a navigator. Yuh [you] see when yuh [you] born as a fisherman, de [the] fisherman nah make [is not made], de [the] fisherman [is] born. And see when yuh [you] born as a fisherman, yuh [you] have fi know nuff tings bout di sea [you must know a lot of things about the sea]. Mi [I] look pon de [on the] stars and know time and something, ... we just have inna we knowledge, inna we brains [We just have the knowledge in our brains].
>
> (Interviewee 1)

> There are a lot of natural signs that we can see in the ocean and we can also tell what's coming. For instance when there would be a lot of rain, there is a certain grass that drifts upon the ocean. And by that we can prepare and know that there will be a lot of rain ... The stars, the moon, the sun, we can use them... But we as fish catchers has got an in-born thing in us, so that we can perceive.
>
> (Interviewee 2)

These quotes suggest that at least some fishers are incorporating local WOK in the adjustment measures they are taking in response to climate change.

Implications

Resilience has two requirements. First, that which is resilient needs to survive in the immediate term. Simultaneously, it must be engaged in actions that guarantee its long-term survival. That was the debate preoccupying the farmer couple whose story began this chapter. Thus, in the context of the fishers in OHB, resilience requires more than immediate adjustments to the changes that come with climatic events (Adger, 2003; Pelling, 2011). Being reactive, while necessary, must be part of a broader and long-term response to societal problems. It is this type of response that guarantees a system's overall ability to be able to adjust to impacts (Pelling, 2011).

What then is the resilience factor in the adjustment measures being taken in OHB? One way to determine this is through examining the extent to which the goals are achieved (Moser and Boykoff, 2013). While goals may not have been explicit in the findings, these were implied given the adjustment measures. Specifically, most fishers indicated that the strategies used were primarily to retain livelihoods, to guarantee immediate survival. By going further out to sea, reducing the number of days out at sea, or changing the fishing grounds, fishers sought to increase their catch. By these means, they hoped to maintain or increase earnings.

To what extent did these short-term, reactive adjustments succeed in achieving their goals? Specific to increasing fish catch, fishers indicated that the fish catch had been decreasing. Often they lamented incurring expenses for fuel and equipment maintenance and not being able to recover revenue for these expenses with the limited catch that they bring back. As such, it could be asserted that these adjustment strategies only partially achieved their stated goals of maintaining livelihoods. The conundrum facing OHB fishers was how to feed a horse while the grass is still growing.

Less reactive measures were also mentioned as being adopted. These include moving boats and other fishing equipment to protect them in times of unpredictable weather events, trimming buoys so that high waves would not lift the fish traps, and paying closer attention to the weather bulletins. In addition, local traditional knowledge had been brought to bear in predicting weather events and understanding their effects. These measures all represented permanent changes in behavior aimed at protecting livelihoods and in the last case, life itself. They clearly had a strong resilience element in them since they were reproducible in response to every climate event and capable of reducing the effect of these events.

Notably, only 27.3 percent of those surveyed reported adopting any adjustment measures and even smaller numbers report adopting measures which strengthen resilience. Whether what has been found represents evidence of a strong, resilience-related adjustment by the community or not is unclear. One scenario can be proposed, which is that resilience-related adjustment measures may have been in the process of spreading through diffusion of intervention (Katz, Levin and Hamilton, 1963). Thus, those fishers who were engaging in long-term adjustment measures that were more anticipatory were acting as change champions for these and promoting acceptance of these measures within the fishing community. This may eventually become part of the culture and value system of the community, enhancing the resilience-building process.

Alternatively, and more realistically, given the lack of success of even the short-term, reactive adjustment measures the 27.3 percent adjusters claimed to have engaged in, there would be few wanting to imitate them. One can link this firstly to the high percentage of those who claim to see the effects of GCC and make no adjustment. They are, it is being suggested here, acting rationally since any short-term adjustments only result in a marginal improvement in their situation. This can be surmised from the many statements from interviewees that they rely heavily on external financial resources for purchasing equipment or recovery after adverse climate effects, and suggests strong community dependency on external resources. This is the opposite of resilience, which involves the community as a self-sustaining entity being able to respond successfully on its own to threats to its livelihoods.

Like the female horse owner in the story, trying to be resilient and self-reliant is an exercise that leads to early and inevitable extinction. The irony that trying to be resilient leads to extinction raises the question of whether self-reliance itself is resiliency. The more rational choice given the dependency development model is to eat, drink, and hope for outside help. One way in which resilience can be built that may also be akin to outside help can be through the combination of efforts among members of the community. For example, the female farmer in the vignette might combine her efforts with that of her husband's. For the fishers of OHB, they may want to combine their resources and WOK to build community resilience. This must be considered carefully though, as in the case of low-income communities often the resources are limited and even with combined efforts they may not properly meet the needs of the aggregate.

As expected by the male farmer, another way to look at outside help for the long-term benefit of OHB has come in the form of protection of the fisheries related eco-system through the Portland Bight Protected Area. This protected area is the body of water off the south coast of Jamaica between Portland Ridge and Hellshire Hills, and includes the near shore region of OHB (Caribbean Coastal Area Management Foundation, n.d.). The area forms a habitat for a number of endemic species and includes a fish sanctuary, which contributes to improving the livelihoods of fishers. Having a fish sanctuary not only allows the near shore to be protected, but contributes to the fish stock, which is a central concern for fishers. This initiative was started by an NGO from outside the community and is endorsed and sanctioned by central government. Though the NGO is primarily responsible for managing the protected area and the fish sanctuary with the assistance of marine police, there is also some community and self-policing where fishers report illegal activity. Local knowledge is being employed here to enhance the resilience created by a long-term adjustment measure originating from outside the community.

It is important to note that there are some contradictions in relying on outside help. One might argue that this is not resilient, as it increases vulnerability in having to rely on someone/something else. However, we must look at the historical context of entities like Caribbean islands that have been exploited over time. There must be some expectation for reparation in order to ensure that entities are able to realistically attend to the threats they face.

Conclusion

SIDS are increasingly vulnerable to the effects of climate change given their geographical location, size, and availability of resources for adaptation. The need for adaptation is a direct result of the excessive carbon emissions of advanced industrial countries causing global warming. Often there is subtext surrounding SIDs dependency that does not take into consideration the historical context of exploitation of these states, which has inhibited the response capacity. The resources to deal with the problem reside outside these communities and in the wealthy countries whose wealth was generated by the creation of the climate problem.

In the vignette at the beginning, the female horse owner is hungry from the first day, and before the end of the week is starving. In a worst-case scenario, the male horse owner begins to starve at the end of a month, during which he hopes "… that something will turn up." OHB has opted for the second option and, true to form, the NGOs and help from the central government has come in.

The chapter argues that this dependence on external support is a form of developing resilience attributes, not only for small communities like OHB that have been used and neglected, but also for Caribbean states who have been exploited by the developed world and are now suffering the consequences of large scale production and industrialization. There is a tension between resilience being defined as something that is internal and an attribution that must be developed by the entity or ecosystem internally, and the dependence on external factors to enable the resilience. Whether this dependence will be enough to save the livelihoods of OHB and other vulnerable Caribbean fisher communities is a question well worth pondering.

Notes

1 These responses are for those who indicated that they believe that the named effect will take place, hence the totals will not add to 100%, nor will they add to the total sample size.

2 These percentages are for those who responded "yes" to the questions related to not engaging in adjustment strategies, hence the totals will not add to 100%, nor will they add to the total sample size. The percentages are valid percentages, meaning that missing values are not included in the overall percentage.

References

Adger, W.N., 2003. Social capital, collective action, and adaptation to climate change. *Economic Geography*, 79, pp. 387–404.

Adger, W.N., Hughes, T.P., Folke, C., Carpenter, S.R. and Rockstrom, J., 2005. Social-ecological resilience to coastal disasters. *Science*, 309(5737), pp. 1036–39.

Aiken, K., 1999. The queen conch fishery on Pedro Bank, Jamaica: Discovery, development, management. *Ocean & Coastal Management*, 42(12), pp. 1069–81.

Alemu, J.B. and Clement, Y., 2014. Mass coral bleaching in 2010 in the Southern Caribbean. *Plos One*, 9(1), pp. 1–8.

Brown, K. and Westaway, E., 2011. Agency, capacity, and resilience to environmental change: Lessons from human development, well-being, and disasters. *Annual Review of Environment and Resources*, 36, pp. 321–42.

Caribbean Coastal Area Management Foundation (C-CAM), n.d. *The Portland Bight Protected Area.* [Online] Available at: www.ccam.org.jm/pbpa/the-portland-bight-protected-area, accessed 15 September 2014.

Central Intelligence Agency (CIA), 2014. *The World Factbook: Jamaica.* [Online] Available at: www.cia.gov/library/publications/the-world-factbook/geos/jm.html, accessed 20 May 2014.

Dulal, H.B., Shah, K.U. and Ahmad, N., 2009. Social equity considerations in the implementation of Caribbean climate change adaptation policies. *Sustainability*, 1, pp. 363–383.

Fischer, A.P., Paveglio, T., Carroll, M., Murphy, D., and Brenkert-Smith, H., 2013. Assessing social vulnerability to climate change in human communities near public forests and grasslands: a framework for resource managers and planners. *Journal of Forestry*, 111, pp. 357–365.

Food and Agricultural Organization (FAO), 2004. *Fishery and Aquaculture Country profiles. Jamaica.* [Online] Available at: www.fao.org/fishery/countrysector/FI-CP_JM/en, accessed 15 May 2013.

Food and Agricultural Organization (FAO), 2012. *Agriculture disaster risk management plan: Old Harbour Bay, St. Catherine*. Rome: FAO.

Government of Jamaica (GOJ), 2010. *Statistical Trend (2001-2010)*. Kingston, Jamaica: Government of Jamaica.

Kaijser, A. and Kronsell, A., 2014. Climate change through the lens of intersectionality. *Environmental Politics*, 23(3), pp. 417–33. doi: 10.1080/09644016.2013.835203

Katz, E., Levin, M.L. and Hamilton, H., 1963. Traditions of research on the diffusion of innovation. *American Sociological Review*, 38(2), pp. 237–52.

Kong, G.A., 2006. Consideration of socio-economic and demographic concerns in fisheries and coastal area management and planning in Jamaica. In: U. Tietze, M. Haughton, and S.V. Siar, eds, 2006. *Socio-economic Indicators in Integrated Coastal Zone and Community-based Fisheries Management: Case Studies from the Caribbean*, Rome. Food and Agricultural Organization, pp. 45–64.

Mercer, J., Kelman, I., Alfthan, B. and Kurvits, T., 2012. Ecosystem-based adaptation to climate change in Caribbean Small Island Developing States: Integrating local and external knowledge. *Sustainability*, 4, pp. 1908–32.

Mimura, N., Nurse, L., McLean, R.F., Agard, J., Briguglio, L., Lefale, P., Payet, R., and Sem, G., 2007. Small islands. In: M.L. Parry, O.F. Canziani, J.P. Palutikof, P.J. van der Linden and C.E. Hanson, eds, 2007. *Climate Change 2007: Impacts, Adaptation and Vulnerability. Contribution of Working Group II to the Fourth Assessment Report of the Intergovernmental Panel on Climate Change*. Cambridge, UK: Cambridge University Press, pp. 687–716.

Moser, S. and Boykoff, M., 2013. Climate change and adaptation success: the scope of the challenge. In: S. Moser and M. Boykoff, eds, 2013. *Successful Adaptation to Climate*

Change: Linking Science and Policy in a Rapidly Changing World. New York: Routledge, pp. 1–33.

O'Brien, K., 2013. Review of Mark Pelling, 'Adaptation to Climate Change: From Resilience to Transformation' (Abingdon: Routledge). *Progress in Human Geography*, 37(5), pp. 729–31.

Orencio, P.M. and Fujii, M., 2013. An index to determine vulnerability of communities in a coastal zone: a case study of Baler, Aurora, Philippines. *Ambio*, 42, pp. 61–71.

Pelling, M. and Uitto, J., 2001. Small island developing states: Natural disaster vulnerability and global change. *Environmental Hazards*, 3, pp. 49–62.

Pelling, M., 2011. *Adaptation to Climate Change: From resilience to transformation*. New York: Routledge.

Pulwarty, R.S., Nurse, L.A. and Trotz, U.O., 2010. Caribbean islands in a changing climate. *Environment*, 52(6), pp. 16–27.

Richards, A., 2008. *Development trends in Jamaica's coastal areas and the implications for climate change*. [Online] Planning Institute of Jamaica. Available at: www.pioj.gov.jm/portals/0/sustainable_development/jamaica_climate_change_paper.pdf, accessed 11 August 2014.

Shah, K.U., Dulal, H.B., Johnson, C., and Baptiste, A., 2013. Understanding livelihood vulnerability to climate change: applying the livelihood vulnerability index in Trinidad and Tobago. *Geoforum*, 47, pp. 125–37.

Social Development Commission (SDC)., 2009. *Community Profile: Old Harbour Bay, St. Catherine*. Kingston, Jamaica: Government of Jamaica.

Taylor, M.A., Stephenson, T.S., Chen, A.A., and Stephenson, K.A., 2012. Climate change and the Caribbean: review and response. *Caribbean Studies*, 20(4), pp. 169–201.

Webber, D., n.d. *Climate Change impacts on Jamaica's biodiversity*. Presentation to National Environmental Protection Agency. [Online] Available at: www.nepa.gov.jm/neec/front_page/CCF/presentations/Dr.%20Dale%20Webber%20-%t20Climate%20Change%t20Impacts%20on%20Jamaica's%20Biodiversity.pdf, accessed 1 October 2014.

Part II
Earth

Place Development Farming Food Pollination

Figure II.1 Earth

10 The Memory of Land … The Law of All Belonging to Earth

Paul K. Haeder

Great amnesia
civilization's womb ripped
earth, totalitarian agriculture
singeing earth, flowing rivers
dammed, cesspools of filth burned
earth drawn & quartered

hunter-gatherer sing
name *orca* "black fish," raven
grandfather, heart of monster
Nez Perce borne of coyote

we court magic
two hundred thousand years
then fertile crescent terror
claims earth … upright
believer in one god, creation
flogs all creatures

plague of farming
starvation, toiling, flagrant
disregard of sister, brother
beetle, salmon
burns earth music
buries bird poetry
three million years foraging
recalling gatherings
monarchs, elk, humpbacks

plow, death rows
stagnation, diseases shared
dark villages, sky emptied of stars
passenger pigeons

civilization, seeds of war
festooned to metal
forgetting sacred vows
from river, lake, mountain, valley

Seattle said, "Every part of this earth is sacred
… every shining pine needle, every sandy shore,
every mist in the dark woods, every meadow,
every humming insect …
holy in the memory and experience of my people. "

we are crashing
burning, evidence
shattering before this culling
"entire plagues upon
vermin," planters, civilizers profess
pesticide plagues upon
the land, fellow hominids, bellies
aching, great revolution
of agriculture, feeding war
erasing
brother-sister-aunt-uncle
mountain-river-grassland-forest
father-eagle, mother-deer

11 How Dare You

Irene Hardwicke Olivieri

Figure 11.1 How Dare You. Oil on wood, 2005, 34" × 19"

This is my painted plea against hunting. I mixed up a fiery red for the background, symbolizing my passion for animals and the outrage I feel about hunting. The woman's braids surround the animals, protecting them. The text on her body describes the lives of these animals, in hopes of inspiring love and respect for them from the viewer. There are tiny hunters in coffins at the lower part of the painting. Vultures hover nearby.

12 Bringing goddesses down to Earth

Bandana Purkayastha

This chapter presents a reflection on the intersections between spiritual imageries, environment, sustainability, and practices within changing socio-political contexts. I focus on a Bengali Hindu religio-cultural celebration—Durga Puja—that symbolizes some of the deepest values of sustainability, that is, the harmony of humans (and their actions) with a larger universe. Sustainability is not merely about human actions to improve the current environment and initiate long-term change, though sustainability certainly includes such action. Sustainability is attained through consciousness and action that reflects the indivisibility of humans-animals-plants-inanimate objects within a holistic universe; at the same time the message highlights the importance of diversity and heterogeneity. This idea of sustainability is closely linked with deep environmental messages that invite us to understand the environment to be a part of us, not something that exists apart from us, which we can thoughtlessly control or destroy without destroying ourselves.

The chapter is divided into two parts: the first focuses on the symbolism of Durga Puja, especially its focus on a mother goddess and the deep environmental messages that are presented through this celebration; the second part focuses on actual practices and the changes in practice that simultaneously undermine and uphold some of Durga Puja's environmental messages in the contemporary world. I end the chapter by linking this celebration and its message to sustainability in today's world.

It is important to point out a few caveats at the beginning of this chapter. Most spiritual terms in Hinduism cannot be wholly translated to English because the English language simply does not have terms to convey the social imaginaries (Chandran, 2009). Equally important, there are very few, if any, Hindu celebrations that are celebrated in exactly the same way and for the same reasons across the Indian subcontinent. Hence, in this chapter I use some beliefs and practices with which I am most familiar: a Kolkata-Bengali form of Durga Puja.

Durga Puja: the tableaux of imageries

The mother goddess, Durga, who is worshipped in Bengal, India and across the world by Hindu Bengalis, is most often presented as the central figure in a tableau of five goddesses and gods along with their attendant animals and plants, and a prone male figure near her feet. This tableau provides an interesting array of imageries that emphasize some core principles of sustainability: an appreciation of the indivisible connection of humans to a larger universe; an understanding that the principle of indivisibility, which is maintained through actions, preserves life forms while resisting forces of wanton destruction; and, ultimately, an emphasis on everyday action to transcend entrenched separations and promote inclusiveness.

The mythological story that I am most familiar with is of Durga as a slayer of Mahisasur. According to this story, Mahisasur or Asur—sometimes described as a demonic male—had acquired significant powers through a boon granted by the god Brahma. Armed with these untrammelled powers, Asur then proceeded to terrorize and wreak havoc on heaven and earth. The gods recognized that Asur had to be stopped from disturbing the harmony of the universe. Yet, individually, they failed to stop Asur. So the gods—Agni, Siva, Vishnu, Indra, and others—bestowed their choicest weapons on Goddess Durga. Armed with formidable weapons in her ten hands, life-preserver Durga confronted Asur and defeated him after a prolonged battle (cf. Calcuttaweb, 2014).

The imagery of *mother* Durga, certainly emphasizes the power of females ("Shakti or strength") because she alone can defeat Asur. Yet, there is also a deeper message about the need to go beyond simple binaries. Durga includes the power of the male gods; she is at once life preserver and Asur-destroyer. Similarly, Asur cannot be reduced to a simple evil demon (opposite to god). He has received his powers through his disciplined worship of the god Brahma. The popular tableau depicts the moment when Durga's trident has pierced Asur's body; he lies near Durga's feet to symbolize his defeat. Numerous images of Durga are now available on the web; Graner's (2013) blog provides an interesting set of images that are easy for non-initiated readers (also search online for images of "durga puja kolkata"). While each image differs in the details and artistry, most depict the moment of Durga's triumph. Asur's destructive violence harms many entities, but Durga's use of violence is qualitatively different because it begins and ends with defeating Asur. This aspect indicates a core principle of sustainability—*the need to confront and defeat the powers of destruction in order to restore the harmony of the universe.*

How is this harmony conceptualized? In the tableau, the images of Durga's children are juxtaposed against the central images of Durga and Asur, who are engaged in battle. The children are not engaged in the battle but simply appear on either side of their mother as a larger statement about a universe that is at once filled with conflict and calmness (as reflected in the postures of the children). On Durga's left is her daughter Saraswati, the goddess of learning; next to her appears Durga's son Kartik, the god of war. On Durga's right appears Lakshmi, the goddess of wealth and on Lakshmi's right, Durga's son, the elephant-headed Ganesh, the bestower of wishes. The tableau of the mother and her children (each of whom is invested with distinct powers) and the array of animals and plants emphasize another aspect of sustainability: a *unity* composed of a *diversity* of roles, powers, and abilities of diverse life-forms within *one* family. This diversity is the basis of harmony and interrelationships in the universe.

In this tableau, plants,, animals, and birds attend Durga and her children emphasizing the message of the indivisibility of humans and these life-forms. These animals and birds are not pets, i.e. maintained according to human terms within homes. These are 'bahan,' i.e. they carry, attend, and support the goddesses and gods freely. The bahans emphasize a message about the interdependence of the gods and animals (and humans). If the gods and goddesses are not complete without the bahans and plants, neither are humans. The additional message is a reminder not to embark, like Asur, on the path of wanton destruction of these interrelated life-forms.

Durga is shown astride a lion, which represents her regal valour and power, and its oneness with the goddess. Saraswati, is depicted with a manuscript ('puthi') and a veena (a stringed musical instrument) in her hands. These symbolize different types of knowledge and learning. Saraswati is shown with a lotus and a swan. In Bengal (and elsewhere), the beautiful lotus blossoms appear above the mud and waters of the ponds in which the plant is rooted; the petals appear to be impervious to all that would tarnish the blossoms. Similarly, the swan is popularly attributed with the ability to filter muddy water within its mouth, and separate the valuable from the dross. The message is that learning and knowledge are not simply about books and music or the arts, it is about the ability to link learning-music-beauty to discernment and to develop the ability to create beauty and harmony amidst earthly limitations.

Lakshmi, the goddess of wealth and prosperity, is also shown with a lotus (to accentuate the same message about creating beauty and harmony, but now with wealth). Her attendant is an owl, a bird that can see clearly through darkness. In other words, prosperity is depicted as a good thing as long as it does not get shrouded in the darkness of selfish acquisitiveness. Durga's son Kartik, the god of war, is shown with a peacock—a bird that dazzles with its beautiful colors, but has a harsh voice. We can draw upon a message about the superficial glamour versus the harsh realities of war. The other son, Ganesh, is shown with an elephant head. Ganesh is the remover of all obstacles, the benefactor, and a good-natured, wise figure. His embodiment challenges conventional ideas of 'beauty,' and calls attention to the blurred boundary between humans and animals. Ganesh is worshipped first—before his mother and siblings—to emphasize the importance of the message he embodies. Ganesh's attendant, the lowly mouse, depicts the critical importance of the lowliest creatures within a larger environmental chain. The veiled figure next to Ganesh—an actual banana plant draped in a sari—is a depiction of greenery and the ultimate source of all life forms. These imageries encapsulate, in a single tableau, the key values of sustainability—acknowledgement and deep understanding of human-animal-plant life forms along with a conscious appreciation of the beauty and harmony of the universe. It also emphasizes justice, discernment, ever-expanding knowledge and learning, goodwill, and responsibility as part of a plan for ethical and sustainable practice.

Durga Puja is widely celebrated as a *sarbojanin* puja, or one that is held in public spaces within temporary edifices called pandals that are dismantled after the puja. As the images on the web show, these pandals are works of unparalleled craftsmanship. Since the puja typically occurs in public spaces, it is open to everyone irrespective of class, caste, religion, gender, age or other social hierarchies. Durga Puja is celebrated in autumn, a time when the summer rice has been harvested in rural Bengal, a time of plentiful food. With the harvest completed, in rural areas, this has been a season of family visits: migrant workers return home from cities, married daughters return to their natal families with their children for their annual visits. After the fourth day of this puja, people visit each other renewing family and community bonds as a reigniting of human relationships to a period of celebration. At present, all over Bengal, in cities and rural areas, schools and offices are closed for several days during the Durga Puja.

For those who celebrate the puja, it reminds them of the oneness of the universe and their position—as humans within an interrelated universe—as part of this oneness. The worshipper is expected to understand this principle: the gods and goddesses are part of us; the images have no significance apart from the values they depict. The images of Durga are *created* each year; traditionally these images were created out of clay or silt from the nearest rivers (see The Source Project, *A Love of Mud* 2015). The images are rarely worshipped in temples, for reasons I describe later. As part of the celebrations and rituals of worshipping Durga, the images are consigned to rivers at the end of the puja, and the clay images are expected to disintegrate and meld with the river waters. The annual cycle of creation of the images, of investing these images with values, and then letting go of the images-as-props, further underscores the permanence of the *messages of oneness and ethical practice*, with or without the images at other times of the year. Along with the recognition of Durga as an invincible goddess, she is worshipped *and* loved—as an intimate, a daughter who has returned home each year to her own family. The worshipper and the community is 'the home;' they have to welcome her by making her one of their own. So this second layer of messages—emphasizing *intimate* everyday loving relationships—show these values are not distant esoteric principles that are practiced by ascetics, but are meant to be part of everyday living. Thus intimacy and love, which are key to sustainable practices, are reflected in this puja as well.

Social practice and sustainability

To what extent have these lessons translated to practice? As I describe in this section, the meanings of several practices have changed over time, nonetheless, the puja has been a pivot of many collective practices that emphasize the interconnections of humans and other life forms. The evolution of Durga Puja into a *sarbojanin* puja encapsulates a social history of resistance (of people in Bengal) as well as one of inclusiveness (especially an emphasis on keeping the puja open for everyone regardless of caste, class, or creed).

Radha Kumar (1990) reminds us that the practice of this particular form of community based Durga Puja, gained widespread popularity during the late nineteenth and early twentieth century amidst the rapid expansion of British colonial power in India, beginning first with the area of Bengal where the British had their capital city: Calcutta. The British used the imagery of the effeminate, traditional Bengali (and, later, Indian) men to justify the need for the upright, rational British to rule and civilize Bengal (and later, India). This discourse about effeminacy, traditionalism, and general civilizational backwardness of Bengalis, was used both by the political machinery and Christian missionaries who wished to convert Hindus so they could worship the "true (male) God" that they espoused (Rolfson, 2005).

In Bengal, the cultural and political resistance that developed against the British addressed both of these discourses. Confronted by the masculinist discourse of the state and state-supported missionary efforts, Bengalis resisted British colonialism culturally, socially, and politically by revitalizing the worship of Durga and goddess Kali (Kumar, 1990; Narayan and Purkayastha, 2009). The strong mother imageries were used to draw boundaries between the cultures of the British and their valorisation of male gods and the Bengalis, for whom gods were neither male nor female and could be worshipped in the female forms as embodiments of power. It also acted as a form of resistance to the British charge that powerless Bengali and Indian women had to be rescued by the British from their men (Sinha, 2000). Seen from this vantage point, Asur could be interpreted as a/the religio-cultural representation of British power, a destructive *colonial* power that had to be challenged, confronted and defeated. In practice, it became a fulcrum of a religio-cultural-political movement that encouraged people to resist political and social subjugation within their everyday spheres of life.

As a cultural renaissance intensified during this period, the worship of Durga adopted other interesting features that are important for this discussion of sustainability. A key tenet of sustainability is inclusiveness. By the early twentieth century, Durga puja began to be organized widely as a sarbojanin puja, in order to break down the barriers of religion, caste, gender, age, class and all other boundaries that are codified into buildings and spaces that herald a particular religious lineage. The location in public space makes it easier for people to join in, as worshippers or simply spectators, and mingle with a variety of people.

The idea of links between places, people, and life forms near and far are further emphasized through the rituals. Just as the images of Durga were traditionally made out of materials—clay and straw or of the pith of banana plants—that can be restored to the earth (or, in modern terms, recycled), the rituals of Durga Puja rituals require earth materials: water, earth, and fire. The water used in this ritual practice is a collection of waters from five great rivers from different parts of India, highlighting the links between the local puja and the distant places from which the waters are collected. The earth that is used for the rituals has to similarly combine earth from different sources. One source stands out: the puja requires the earth from the doorstep of a prostitute. Here again, the people who are shunned in daily practice are ritually included in the performance of the puja. Whether most people have followed these lessons of inclusion in everyday life is, of course, an open question.

Durga Puja is a festival, so other festive activities spring up surrounding the actual site of the puja. Artisans and craftspeople develop avenues for earning income locally by selling their wares to people who visit the pujas. So a variety of people enact different roles, buying, selling, worshipping the images or simply enjoying the entire scene. The materials required for the puja were drawn from the local context and acted as a stimulus to the local economy while emphasizing the work of different strata of people in the community. Similarly, the immersion of the clay images each year, and the creation of images from the silt of rivers, contributed to the idea that we would now describe as reusing and recycling.

Reflections

Durga puja remains very popular in Bengal. However, the scale of the pujas—the size of the images, the pandals, the decorations, and the size of the crowds—have gotten bigger leading to some challenges of balancing sustainability in the era of neoliberalism.

A key change is in the creation of the images. As the pace of life has increased manifold from the early twentieth century, craftspeople who make the images are simply not able to keep up with the demand for the highly artistic images. Earlier the images were made from clay (typically gathered from rivers) and the clay was used to create the face, the arms, legs and parts of the body that would be visible, while the colors were made from natural ingredients. When the images were immersed in the water, they became a part of the river's water and silt, a ritual return to the origin as well as a lesson in sustainability that rests on our willingness to let go of tangible possessions.

If we stop and think of what exactly this image making entails, it will be clear that image-making combines spirituality and creativity. The artist-craftsmen's 'dharma' (which can be translated as 'essential purpose,' and/or religion) was to create images that expressed the quality of the goddess or god. The degree of skill required for this task can be best explained with an analogy. Many museums in North America and Europe now house stone images of Buddha from the earliest times. For anyone who looks at the images in historical sequence, it is apparent that it has taken a few centuries for artist-craftsmen to create the image we are most familiar with: the meditative yet benignant Buddha. It is not easy to capture that expression. So too with the Durga images. Even though these images are made each year, it is not easy to create images that encapsulate the qualities of each goddess or god. It requires very highly skilled artists/craftsmen. The skills for this type of artistry have required lifetime practice and were traditionally imparted through the generations within families. Producing these images required time.

In contemporary times, the skilled image-makers cannot keep up with the demand for these images and have turned to the use of moulds and plaster of Paris to mass produce the body parts of the images. The paints they use are commercially available synthetic paints. While some images continue to be made the traditional way, these are expensive. As a result of this shift, the immersion of images that emphasize deep environmental messages have, ironically, become a significant source of environmental pollution (see e.g. Mishra, 2009) though a few pujas now are consciously trying to go back to the environmentally conscious roots (Dutta 2003).

Another significant change, at least in large metropolises like Kolkata, is that corporations mostly sponsor the larger sarbojanin Durga Pujas, so the pujas reflect commercial extravaganzas. Intricate pandals are created; many are so elaborate and beautiful that they rival more permanent, world famous buildings of brick and stone. Huge amounts of electricity are consumed to light up numerous pandals and adorn the streets. The images are even more intricately adorned than before. Where music is allowed, the noise blasted through speakers is significant. Not surprisingly, the pujas have become formal tourist attractions. It is not clear whether all the messages of the puja are evident when one looks in from the outside as a tourist and only sees the images. Nonetheless, the main tableau is there for everyone to see.

So, what are the implications of sarbojanin Durga Pujas that are no longer reliant either on the local community or local produce? Some of those local circumstances cannot be recreated, but our understanding of 'local' has changed; it now includes larger expanses of the world. Pujas are now 'glocal,' i.e. global and local simultaneously, so there is the possibility of our learning about the glocal community to which each of us is connected. At the same time, the central idea that these pujas have to be celebrated in public spaces has allowed the puja to flourish in ways that combine the spiritual and the non-spiritual, the religious and the secular.

Durga Puja: Some reflections on intersectionality

Scholars of intersectionality focus on the exercise of power at different levels of society: at the international level, at the national level, at the local level. They describe how these intersecting levels of power create and maintain gender/class/race/other social hierarchies (e.g. Collins, 1990; Purkayastha, 2012). Corporations, large non-governmental organizations, governments, and groups at the upper levels of these intersecting social hierarchies exercise power in ways that create and sustain inequalities locally and internationally (e.g. Armaline *et al.*, 2015; Shrivastava and Kothari, 2013). The discussions about Durga Puja offer many examples for understanding intersectionality. I will highlight three lessons.

We can begin with the history of Durga Puja in Bengal and the messages that linked Asur's violence with British colonialism that I described earlier. Colonialism reaches across state boundaries to subjugate nation states. It survives through institutions and practices of extreme coercion, control, and large scale routinized violence against the people it seeks to control. It uses ideologies that justify the subjugation of people in order to maintain the privileges of those who are at the top of the hierarchies. In Bengal, the revival of Durga Puja as a form of cultural nationalism and resistance against colonialism was preceded by a period when the colonizers systematically destroyed local industries to ensure their own profits and ruthlessly subdued all those who protested. Much like the state of the gods before they turned to Durga to control Asur, the colonial powers unleashed large-scale violence, yet the earliest resistance efforts were not successful in stopping such destruction. Later, beginning in the early twentieth century, the use of 'feminine tactics' of non-violent resistance led to more successful forms of resistance against the colonial powers. While Durga is depicted in her violent encounter with Asur, the totality of Durga's symbolism is about feminine power that is guided by the good of *the interconnected universe*, not merely of humans or gods. At the same time the puja's rituals offer alternatives to practices that entrench race/class/gender and related social hierarchies. The rituals act as a form of resistance to ideologies that make untrammelled, violent power *and* social hierarchies seem normal. So the lesson of intersectionality inherent here is about the realities of power and violence, but also the need to resist such powers actively. This perspective of realities and resistance is based on a perspective that is based on the indivisibility of the universe.

We can also think about intersectionality by reflecting on the transformation of Durga Puja in parts of Bengal today. As I described, capitalist ideologies of consumption and tourism are shaping the form of sarbojanin pujas today. The corporatization of the Durga Pujas has led to significant use of resources, a delinking from the locally rooted community enterprises, and has promoted the mass production of images that no longer use sustainable material and contribute to significant levels of pollution. All of these changes are inimical to the messages of sustainability that constitute the principles inspiring the pujas. At the same time the changes reflect the control of powerful entities. Durga Pujas that thrive as tourist-attractions emphasize décor and artistry; the messages and practices of inclusiveness are drowned within these new emphases on the people as consumers of art rather than people as actors in the effort to address social hierarchies. Additionally, the co-optation of the pujas means the ideologies of resistance to contemporary capitalism, i.e. the type of ideologies that had linked Asur to British colonialism in earlier historical periods, are muted, at least for now.

A core message of Durga Puja's is about indivisibility of life forms of the universe and the need to recognize and sustain these links through on going practice. Ranged against these expectations of sustainable practice are the powerful effects of contemporary capitalism:

> [t]he most serious of these problems show signs of rapidly escalating severity, especially climate disruption… an accelerating extinction of animal and plant populations and species … depletion of increasingly scarce resources … These are not separate problems; rather they interact in two gigantic complex adaptive systems: the biosphere system and the human socio-economic system.
>
> (Armaline, Glasburg and Purkayastha, 2015)

Many scholars and activists, such as Vandana Shiva (e.g. 2013), have pointed out that capitalism operates by forcing people and states to obey dictates. The type of wanton violence and destruction that I described as 'Asuric' in the sections above, are similar to the destruction of the earth's soil, water, seeds, and food by powerful global entities, like the World Bank, corporations, and governments (see Shiva 2013). These Asuric practices have introduced massive climate change and induced human suffering, but this suffering is borne by people at the lower strata of social hierarchies. Thus, if we juxtapose Durga Puja as a set of practices that emphasize interconnections, a set of practices that seek to dismantle hierarchies, we can contrast it to contemporary capitalism and witness both the power and privilege of those who profit from wanton violence and destruction and the plight of those who suffer.

Durga Puja and the lessons about sustainability as everyday practice

Despite the changes in form and scale, I would conclude by emphasizing that the central message of the pujas endures. Since the tableau does not change, the most evident message about Durga the life preserver and the beloved daughter continues to be a relevant social message. For people who actually participate in these pujas, especially away from the tumult of the big pujas in the larger cities, this is still the time to stop the routine of everyday life and focus on the messages and celebrations. It is a time to renew oneself and reflect on the universe. People celebrate: they eat different types of food each day of the puja. They use different kinds of flowers and adornments. They visit family and friends and pay homage to their elders. They go to public spaces to celebrate, and then they celebrate within homes with friends and family again.

Even though I described a Hindu religious practice, the message of this puja does not have to be *narrowly religious*. Indeed, the development of sarbojanin puja was meant to address those narrow socio-religious boundaries. As I have alluded to earlier, the boundaries between 'sacred spaces' and secular ones (e.g. temples vs. schools or church halls), between who is Hindu or Muslim or Christian or Jew, who is Brahmin and who is not, is diluted in this form of spiritual practice. At this level, it is one more way of promoting inclusiveness. Equally important, the images that are likely to evoke questions in people's minds about the presence of non-human life forms as part of a puja tableau open up opportunities to discuss the indivisibility of the universe. In the end, I don't think the message of the puja needs to be tied to one specific faith: the principles are open for everyone to practice whichever way they please, including strengthening or expanding their spiritual practice by renewing their attention to sustainability principles. Indeed, sustainability itself can be imagined as a 'universal' spirituality that transcends any 'human religion.'

Spiritual practices rarely exist apart from their socio-political contexts. Each year some Durga Puja creates Asur in the image of some contemporary character who is seen as a force of evil. Thus, the use of a spiritual image to promote social messages is not extinct. Equally important, with a swing away from globalization and homogenized cultural practices in India, at least among some sections of society, there is a renewed questioning of the 'modern' form of the Durga image—especially the potential to pollute the rivers. So there is a renewed attention to the environmental principles of the puja: the need to constantly pay attention to our impact on a wider universe and be responsible about preserving life instead of destroying it.

The imageries of Durga Puja emphasize the values that have been key to a secular understanding of sustainability, values that reflect the consciousness of an interconnected universe. Such an interconnected universe has to be understood through discernment (through knowledge and appreciation of its beauty), enjoyed through prosperity (but a prosperity that goes beyond the prosperity of some at the expense of others) where the objective of action is to remove obstacles to benignant practice and work towards harmonious and just existence. The combined lessons of the puja emphasize our everyday action to transcend entrenched separations and promote inclusive-ness. In this manner we are symbolically able to *bring goddesses down to Earth*.

References

Armaline, W., Glasberg, D. and Purkayastha, B., 2015. *The Human Rights Enterprise*. London: Polity Press.

Burke, M.L., 1983. *Swami Vivekananda in the West: New Discoveries*, 6 vols. Kolkata: Advaita Ashram.

Calcuttaweb, 2014. Durga Puja. [Online] Available at: http://calcuttaweb.com/durgapuja, accessed 15 August 2014.

Chandran, M., 2009. Whose English is it? Translating in a neocolonial world. In *Apperception: Journal of Visva-Bharati, Santinikentan*, Volume IV, pp. 38–52.

Dutta, K., 2003. *Calcutta: A Cultural and Literary History*. Oxford: Signal Books.

Graner, S., 2013. Durga Puja – West Bengal's biggest festival. [Online] Available at: http://fellowsblog.kiva.org/fellowsblog/2013/10/23/durga-puja-west-bengals-biggest-festival, accessed 1 September 2014.

Kumar, R., 1990. *A History of Doing*. New Delhi: Zubaan Books.

Mishra, B.K., 2009. Idol immersion pollutes Ganga. [Online] Available at: www.timesofindia.indiatimes.com/home/environment/pollution/Idol-immersion-pollutes-Ganga/articleshow/24220624.cms, accessed August 30 2014.

Alvares, C., Ziauddin S., and Nandy. A., 1994. *The Blinded Eye: Five Hundred Years of Christopher Columbus*. New York: Apex Press.

Narayan, A. and Purkayastha, B., 2009. *Living Our Religions: South Asian Hindu and Muslim Women Narrate Their Experiences*. Stirling, VA: Kumarian Press.

Purkayastha, B. and Narayan, A., 2009. Bridges and chasms: orientalism and the making of Indian Americans in New England. In *Asian Americans in New England*, edited by Monica Chiu. Durham, NH: New England University Press, pp. 124–149

Purkayastha, B., 2012. Intersectionality in a transnational world. Symposium on Patricia Hill Collins. *Gender & Society*, 26, pp. 55–66.

Rolfson, C., 2005. *Resistance, Complicity and Transcendence: A Postcolonial Study of Vivekananda's Mission in the West*. Unpublished MA thesis, Queens University, Ontario, Canada.

Sinha, M., 2000. Refashioning Mother India. *Feminist Studies* 26, r\o. 3 (Fall 2000), pp. 623–644.

Shiva, V., 2013. *Making Peace with the Earth*. London: Pluto Press.

Shrivastava, A. and Kothari, A., 2012. *Churning the Earth: The Making of Global India*. New Delhi: Penguin Books.

The Source Project, 2012. *A Love of Mud: Kolkata and the Durga Puja*. [Video] Available at: www.vimeo.com/50520389, accessed August 29, 2015.

13 The farmer and the witch

Replanting the seeds of indigeneity

Nala Walla

The seeds of indigeneity

Just as any store-bought apple will always sprout a unique wild variety when planted (Pollan, 2002), so every person on this globe—even the most domesticated among us—contains the feral seeds of our own indigenous origin. Though they may be deeply buried, so deeply that we may be unaware they exist, these seeds are of incalculable value to anyone interested in the germination of sane and sustainable cultures.

Mayan author and teacher Martin Prechtel (2012, p. 53) shares his extraordinary insights about these seeds of indigeneity:

> The world is populated with … [p]eople who've lost their seeds. They are not bad or useless people, but … [t]he real people they used to be, like the seeds, have vanished … to hide in an inner world inside modern, citified people. In some small, never-looked-at place in the forgotten wilderness of their souls, invisible to the forces that would invade and take over, their indigenous seeds of culture and lifeways live exiled from their everyday consciousness.

The quiescent kernels of indigeneity are resting patiently within all of us, waiting for our variously industrialized and wounded bodies to step outside our climate-controlled routines, into the nourishing rain and soil, so these seeds can flourish once again.

Yet, the "simple" act of spending time outdoors, working again with soil and seeds, animals and trees, has been enormously complicated by oppressive systems designed precisely to break human connection with earth, with each other, and with the wilderness embedded in our own psyches. A profound sense of meaninglessness and depression often results from this disconnection, as described by depth psychologist and wilderness guide Bill Plotkin (2013, p. 160):

> Affective depression is, at root…the blockage of the wild, indigenous, emotive, erotic, and fully embodied dimension of our human wholeness. The best therapy for depression begins with the resuscitation, animation, and liberation of [our] Wild Indigenous One.

But accessing our indigenous wisdom is much more than just an excellent strategy for healing our personal psychological wounds. Such liberation involves the deep shift in consciousness needed in order to perceive solutions to seemingly intractable societal and ecological problems. These solutions may have been right in front of us all along, but it has been difficult for us to see them, embedded as we are within a paradigm of exploitation, separation and division. Writer and herbalist Stephen Harrod Buhner (2014, pp. 24–25) emphasizes the importance of our capacity to see beyond entrenched assumptions, to a deeper reality of connectedness with the whole of nature:

> [T]here is every reason to view this capacity as a crucial evolutionary adaptation, a capacity hardwired into all living organisms, and which serves a specific purpose … Despite our culture's willful ignorance, deeper perceptual experiences and paradigmatic shifts in cognition are spontaneously emerging with more frequency, and much more strongly, into the human species. For using this different kind of perception and thinking is the way out of our predicament, the way to solve the problems that those older kinds of thinking have caused. It is an evolutionary necessity.

The convoluted histories which taint our relationship with landscapes, both inner and outer, render earth-based work an extremely powerful catalyst for healing between individuals and families, between nations and races, as well as for the living planetary ecosystems of which we are all part. Our ability (and responsibility) to rebuild our connections with natural communities—human, animal, vegetal, bacterial—is underlined and potentiated by the severity and depth of our wounding. Though this type of paradigm-shifting work may not be easy, it can be extremely rewarding, and can be regarded, as both Prechtel and Buhner do, as an "evolutionary necessity."

Whether we become advocates for youth naturalist programs, dig a garden in an urban pea-patch, create permaculture programs in prisons (Thomas, 2015; Vosper, 2015), or organize large-scale holistic land management (Savory, 2015), opportunities abound to reclaim our birthright as wild creatures on an awe-inspiring planet. Indeed, our ability to respond creatively and decisively to rising sea levels, to civil wars, to nuclear pollution, is directly dependent upon our ability to reconnect with our inner wildness, regarding it as a wellspring of wisdom, rather than an unruly riot which must quickly be tamed.

The farmer and the witch

As I write, the colorful Halloween holiday, with straw-stuffed scarecrows and spooky lil' ghosts parading across homes and storefronts is approaching all over the Northern Hemisphere. It's my favorite time of year.

Crooked-toothed icons of witches on their brooms are plastered everywhere, and I can't help but marvel at how, even after centuries of efforts to hunt and exterminate her, 'The Witch' nevertheless continues to capture our imaginations.

Even through the thick synthetic cloak of modern culture, our subconscious selves dimly recognize the witch—that earthy woman stirring her pot of herbs and flying through a magical nighttime sky—as our ancestor. Despite pervasive miseducation, and rampant dilution of her cultural history, the witch endures.

The means by which the long and rich history of witch culture has been eroded include all the typical mechanisms of exploitation we are familiar with today: terrorism, colonialism, genocide, propaganda. The medieval witch hunts themselves served as the proving grounds which developed and refined the above mechanisms, when combined forces of Church, State and media experimented with global violent crusades whose purpose was to sever the connection of the peasantry to the land (Federici, 2011). Only slightly different in style and scope today, these techniques remain favorites of belligerent governments and corporations around the world that wish to remove all resistance to exploitation.

Current cartoonish portrayals of witches—virtually devoid of any real meaning—are a testament to the 'success' of these terror and slander campaigns, which have destroyed most of the detail about how ancestral pagan cultures actually functioned, and the extensive knowledge they contained. In just a few hundred years, common representations of the witch shifted from a revered, medicine woman embodying a culinary, shamanic, and healing tradition, to a warty, cackling buffoon in a pointy hat who exists only in picture books.

A similar fate has befallen another figure who, in the public view, once possessed extensive knowledge about the land: 'The Farmer.' The infantilized image of the witch mentioned above is reminiscent of popular depictions of farmers, ranchers, and herders as clumsy hicks who are, at best, unsophisticated and out of touch with the slick urban "reality" of modern life, and, at worst, stupid and irrelevant to the river of progress.

As with witch culture, the details of once-hearty and self-reliant agrarian communities have been glossed over in the creation of the current degrading stereotypes. I was ashamed to find on Wikipedia a whole list of pejorative slang used to refer to rural people—the very people who negotiate our relationship to the land and are responsible for our sustenance: boor, bumpkin, churl, hayseed, hick, hillbilly, lob, redneck, rustic, and yokel.

These slurs wound on several levels, translating not only to a philosophical disrespect, but an actual biting of the hand that feeds us, as well. Even worse, they demonstrate the thoroughness with which we modern people have internalized our own oppression, colluding with the severing of our original connections with the land, slashing at the lineages of our own indigeneity.

Though references to farmers in the West today usually assume 'white,' 'Christian,' and 'male,' both the farmer and the witch—with their millennia-long lineages, and bountiful knowledge of food, animals, herbs, and handicrafts—are characters which grace the family trees of diverse ancestries. Men and women worldwide have pagan and agrarian roots of which we can be proud, yet despite rich historical links, the potential solidarity between the average modern, industrial citizen and figures such as witches and farmers has been cauterized, allowing for ignorant and dangerous stereotyping to spread.

Grandmother witch

To prepare for our harvest feast, my son and I are headed to the local market in our little town. He always loves coming here, helping to fill our basket with an assortment of the succulent fruits and veggies available this time of year. But as I put my hand on the door, I feel a small jolt of fright as I notice the illuminated witch-in-silhouette, flying across the face of the waning moon—and it's not because I am 'scared of witches.' Rather, I shudder to think about what falsehoods, what shallow slanders, this image will be conveying to him about his own ancestors?

For all its tiring over-generalizations, it can at least be said that this green-faced portrait *is* an accurate representation of how desperately little knowledge remains about my son's own mixed heritage. How the outlines of his original Indigenous Body have been buffed and muted into a puffy caricature. I wonder how bewildering the Witch concept will likely be to his budding Jewish identity, since her image was influenced by and conflated with the anti-Semitic images developing in Europe during the same period as the witch hunts? *Wow, Mama, look at how long that witch's nose is!* Will I really have to explain to him that since the entire populace of Europe was once wiccan, some had big noses, and some little? And how will I counteract the confusing fact that witches are pictured almost exclusively as women? *You mean there's such a thing as a boy witch, Mama?* I'm merely trying to get some groceries, yet I've inadvertently exposed my son to a triple whammy: sexism, classism and anti-Semitism all rolled into one.

One of the eeriest things about this minstrelized Witch is how well-suited she is to the bland palate of modern industrial society in general, which is in such poor digestive health it can hardly stomach anything more than fluff, even as it starves for meaning and connection.

It may come as a surprise to many readers that people of European ancestry were (and arguably still are) subjected to the same processes of pauperization, industrialization and commodification that are currently occurring in so-called 'developing' countries. In fact, we are so accustomed to seeing 'white' people in a privileged, oppressor role, we assume it must have always been this way. We forget to inquire how Europeans got so disconnected from the their lands? Is it possible that people of European descent—is it possible that *white* people—also have indigenous roots?

Like existing indigenous peoples all over the globe, pre-conquest Europeans were earth-centered, pagan peoples—a term derived from Latin *paganus*, meaning 'not cultivated' or 'wild'—and intimately connected to a living, breathing land that they revered as the source of all life. Similar to tribal people worldwide, ancient European tribes had no formal money systems, and had no need for them, as they inhabited a gift culture based on careful stewardship of the commons—that great interlocking web of physical, cultural, and spiritual relationships. Lo and behold! *Europeans* once displayed the same connective qualities and behaviors we currently attribute to *indigenous people*.

Please allow me to propose a journey of kinship and solidarity with a larger family of pagan cultures: if the old European clans' practice of 'wicca' or 'witchcraft' (a more modern term) was similar to that of indigenous tribes worldwide, then can we reclaim and revalorize the term 'witch' as a loose description of any intact, nature-centered culture?

In the Dark Ages, the witch-hunting authorities themselves certainly did not limit the label 'witch' to European pagans, and they still do not. Snared in that same net—-a net cast broadly enough to encompass almost any subversive activity, as "conveniently and strategically vague" (Federici, 2005) as the word *terrorist*—were colonial subjects from Africa to the Americas, at whom were hurled the same accusations of flesh-eating, fornication with the devil, and infant-stealing, and who suffered the very same torture rooms, pyres, and gallows that so efficiently broke the communities of their European counterparts overseas. And the witch-hunting violence continues to this day, for example in the contemporary murder in London of accused Congolese 'witch' Kristy Bamu (La Fontaine, 2012).

Previously just a name for European pagan culture, the brand 'witch' was appropriated and became a slur used to describe anyone viewed a threat to authoritarian control—black, brown and white alike. Just a handful of generations ago, then, before mechanization, before colonization, before Christianization, *we were all witches*.

Amazingly, even after centuries of terrorism heaped upon the witch on at least four continents—despite her constant demonization, degradation, minstrelization, and Disneyfication—-her image continues to haunt the collective soul, even penetrating the bubbliest halls of pop culture. Bovenschen *et al*., (1987, p. 87) describe the irony and importance of the witch's 'staying power':

> In the image of the witch, elements of the past and of myth oscillate, but along with them, elements of a real and present dilemma, as well. In the surviving myth, nature and fleeting history are preserved … In turning to an historical image, [we] do not address the historical phenomenon, but rather its symbolic potential…To elevate the historical witch … to an archetypal image of female freedom and vigor would be unimaginably cynical, given the magnitude of her suffering. On the other hand, the revival of the witch image today makes possible a resistance which was denied to historical witches.

I reconsider the witch cartoon on the front door of our local market: at least this image can serve as a segue for conversation with my son. Maybe, as Bovenschen *et al*. (1987, p. 85) suggest, the omnipresence of this image evidences a collective 'return of the repressed.' Perhaps she is being reclaimed for purposes of liberation, as seen with the label 'queer' in the LGBTQ rights movement, for example? To be certain, the sheer persistence of the witch to this day is indicative of an archetype not easily forgotten. It may well be that the witch everywhere, because she is our grandmother.

The cop in the head

As my son and I wait in the checkout line, I overhear a woman describing an argument with her friend, exclaiming "Geez, what a *witch*!" I cringe at the harshness of this internalized oppression, as she not only denigrates a fellow woman in this small community, but also slanders her honorary grandmother. One of the main symptoms by which people in advanced stages of colonization can be recognized is that they have been recruited to participate in their own degradation and destruction, mostly unwittingly.

Using a marginalized person or group (such as 'witches,' 'terrorists,' or 'Jews,'etc.) as a scapegoat upon which to blame virtually anything is an all-too-common human response to stress. And it is one that elite classes have long encouraged, since it successfully diverts attention away from the real source of the stress: the concentration of wealth and power into the hands of the very few. And because scapegoating is but a mere temporary release-valve for tensions, the original problem eventually boomerangs back upon the thrower, destroying families, communities and ecologies in the process. Today 'isms' are being hurled on a massive scale in the form of rampant racism, classism, sexism, homophobia, xenophobia and anti-Semitism, all overlying an anthropocentrism nearly as ever-present as the air that surrounds us.

Indeed, the breaking of the power to resist subjugation and appropriation of resources is the original and primary goal of all the isms. The campaign of terror against witches was designed with this exact intention in mind—to attack the women who were the foundations of pagan, peasant communities, as well as the backbone of the resistance to the 'enclosures', the medieval beginnings of the unrelenting privatization that continues to this day.

In *Caliban and the Witch* (2002), scholar Silvia Federici reveals how persecutions of witches in Europe in the years leading up to the industrial revolution were overwhelmingly aimed at poor and working-class women, stereotypically represented in ragged clothing (not unlike today's popular culture depictions of the lazy and tattered cowpoke). Old women who retained their abilities to subsist on the land base were especially singled out for targeting, since they were the most likely to embody the cultural knowledge and heritage of ancient ways like raising crops and animals, herbal remedies, midwifery, community ritual, and so forth—skills which preserved the health and independence, and thus the power of the peasantry to resist exploitation.

This disturbing strategy by which community strength is efficiently broken by sexist targeting of women leaders was 'perfected' in this era. As described by feminist theorist Maria Mies (1986, p. 81):

> Recent feminist literature on the witches and their persecution has brought to light that women were not passively giving up their economic and sexual independence, but that they resisted in many forms the onslaught of church, state and capital. One form of resistance were the many heterodox sects in which women either played a prominent role or which in their ideology

> propagated freedom and equality for women and a condemnation of sexual repression, property, and monogamy. Thus the 'Brethren of the Free Spirit', a sect which existed over several hundred years, established communal living, abolished marriage, and rejected the authority of the Church. Many women, some of them extraordinary scholars, belonged to this sect. Several were burnt as heretics ... Some argue that the witches had been an organized sect ... where all poor people gathered and already practised the new free society without masters and serfs.

Unfortunately, modern attempts to manifest a 'new free society' are still being hampered by sexist infighting. The consequent scourges of mistrust and abuse remain primary factors in the weakening of community resistance movements which oppose the separation of people from their lands and means of subsistence.

Today, the mechanization of industrial agriculture ensures that a minimum of people know how to grow food or medicine, the remainder being completely dependent upon service and high-tech for their work and their sustenance. During any current election year, the thoroughness with which modern people have been disciplined to accept roles as 'workers' can be heard in the constant clamor for "more jobs!" By contrast, the early sixteenth century European peasant would rather risk the gallows than submit to wage labor. Hence the irony that the wristwatch—once a symbol of slavery and an artificially imposed time, disconnected from the natural rhythms of the land—has become a modern status symbol (Federici, 2004).

The horror of separation from the land created the stressful conditions ripe for scapegoating. As mistrust was sown within pagan communities, peasants began accusing each other and cooperating with their own marginalization. This is the terrorized and disturbed ground in which the isms took root, and continue to 'flourish' today. In modern, industrialized peoples for whom a subsistent, nature-connected life is already long-gone, these isms have become the preferred method of social control: an internalized, instead of overt, oppressor with whom we cooperate in the effective policing of ourselves. Much tidier, and a lot cheaper than inquisitions and bombing, we become, as Brazilian activist Augusto Boal describes in *Theater of the Oppressed* (1993) our own "cop in the head."

During the harvest season where I live in the northwest United States, I see examples of this self-inflicted oppression everywhere, as people routinely consume and propagate over-simplified, 'pin-up' versions of witches and buck-toothed, grinning farmers with their pants falling down. In an astonishing ignorance of our own pagan and agrarian past (and future!), we conspire in the turning of both witches and farmers into cackling, guffawing minstrels.

Yet, we are beginning to understand that large-scale human estrangement from the land is threatening the extinction of our and many other species. Instead of taking crude potshots at farmers or witches, perhaps it is wiser for those of us who have lost our connection to the land to seek out the people who have been safeguarding it for centuries against all odds? Perhaps we might recuperate this wisdom—preserved within each of our indigenous lineages—and do our best to enact it, learning more about our food systems, our ancient healing customs and remedies, about working with animals, plants, and the cycles of the moon?

In the shallow images of the farmer and the witch lie the remnants of our very own ancestral cultures, and therefore, they deserve to be paid some much deeper attention. As we embrace them with an attitude of openness and curiosity, can our historical traditions and lifeways reveal potential solutions to serious cultural and ecological problems? Could our heritages, for example, contain a key to reversing climate change?

Bring back the buffalo: how animals can reverse climate change

Perhaps this is the first time that you've encountered the hopeful idea that animals can help heal large-scale weather and ecosystems, but I hope it won't be the last. What follows is one example of how stunningly straightforward reversing climate change can be.

The research of Allan Savory has not yet made it into breakfast-table conversation in mainstream, industrial society, as the television stuffs us instead full of pop culture and trivia, yet he and his colleagues in the field of holistic resource management (HMI, 2015) have discovered something of extreme importance for anyone interested in climate change: a method for swiftly and drastically reducing atmospheric carbon levels that uses no technologies other than livestock.

Livestock? You mean ranchers and cowpokes—those backwards, lazy, know-nothings—can help reverse climate change?

All grasslands—prairies, savannahs, steppes, and so forth—originally co-evolved with dense herds of grazing animals whose natural ranging behaviors provided the mowing, mulching, fertilizing, soil aeration, and seed dispersal functions essential to the health of these ecosystems. For decades, in a misguided attempt to stop 'overgrazing,' standard land-management policies worldwide have removed herds—and the herding peoples whose lives were intertwined with them—from these lands. The result has been a drastic *acceleration* of desertification and therefore, of climate change, as well as displacement and pauperization of countless indigenous people (Schwartz, 2013).

What does desertification have to do with climate change?

As enormous amounts of carbon previously contained in grassland soil is released into the atmosphere (think of the American Dustbowl), Savory emphasizes that desertification is as big or bigger of a contributor to global warming as burning fossil fuels (Savory, 2013). Savory's efforts have been assisting people on 40 million acres in Africa, Australia, Europe and the United States to bring back the herds, recreating, out of barren desert, both healthy grassland ecologies *and* right livelihood for pastoral peoples. Simply by returning the animals to desertified places, and helping to mimic their natural movement patterns in the landscape, soil and range management scientists estimate that we could again achieve *preindustrial levels of atmospheric carbon in less than 40 years* (Sacks *et al.*, 2013, p. 15).

Amazingly, pastoral skills are now being revealed as an integral part of reversing climate change, as carbon moves out of the atmosphere back into grassland soils (White, 2014). It seems that a restoration of respect for these skills—some of the very same skills witches worldwide gave their lives to protect—is critical to restoration of the land itself. If we are serious about reversing climate change, animal husbandry will necessarily become again, a respectable occupation. Imagine shepherding as the preferred profession for the hip and fashionable, the next 'cool' thing to do!

Indeed, many people are being inspired by the example of Joel Salatin, dubbed 'World's Most Innovative Farmer' by TIME magazine in 2011. Salatin is rapidly becoming a well-known example of how using the simple, low-tech strategies of holistic management is not only good for soils, animals, and humans, but can be economically viable, as well. Salatin's (relatively) small 550-acre *Polyface Farm* in Virginia, USA had over $2 million dollars in yearly sales (Gabor, 2011), an impressive accomplishment for a farm of this size. Polyface's success, completely independent of the enormous subsidies given to many US agribusinesses, casts doubt upon the assumed 'necessity' of ever-escalating investments in hi-tech and government subsidization, and points in a more hopeful and healthier direction.

As more and more people embrace the instinctual impulse towards reverence of the land that is the source of all sustenance, reestablishing a holistic and sustainable relationship to it, all kinds of unanticipated resolutions to ecological impasses like the example above will arise. A huge accomplishment will be to perceive the stereotypes we hold for what they are: examples of internalized oppression, and a disrespect of our own ancestors, the witches and the farmers. It is time for every citizen of this precious planet to identify as a creature indigenous to earth, and to reclaim a history full of herbalists, shepherds, and agrarians. Can we imagine a world where our educational systems encourage our children to cultivate 'green-collar' careers in fields such as holistic ranching, dairying, and farming? Where "bring back the buffalo!" becomes the rallying cry for the sustainability movement?

Reconnecting with the land: the key to ecological restoration

Since moving to a rural island over fifteen years ago, my own experience with farmers—especially small farmsteaders seeking to steward their lands organically and sustainably—-has consistently contradicted the stereotypes I grew up with in suburbia. Far from naïve simpletons, most small-scale farmers and ranchers I know are astoundingly savvy and resourceful. In order for their farms and gardens to survive as businesses, today's agrarians are required not only to become proficient with a hundred skills having to do with raising plants and animals (including entomologist, plant pathologist, vehicle mechanic, veterinarian, and so on) but, as fellow citizens of the Information Age, they often are expected to maintain websites, intern programs, and community outreach calendars, as well as possess enough shrewdness to navigate a veritable gauntlet of health and food regulations, cutthroat subsidies and strategic marketing climates. One local farmer in our valley earned an MBA before starting his farm, and one local butcher is also a working surgeon. It would be very difficult to consider them 'simpletons.'

The farmers and ranchers in our county are part of a larger national trend of young people and white-collar professionals who cherish having their 'hands in the dirt,' and are voluntarily trading in their high-tech futures for trowels and tractors, returning to our neglected farms, fields and forests (Markham, 2011). In search of deeper nourishment, they are spitting out the thin gruel that our larger exploitative society tries to pass-off as sustenance, and rejecting the dominant cultural memes of our time that denigrate working with ones' hands. For many, this means leaving urban environments and moving back to the land, in a small, but encouraging reversal of the demographic shift towards urbanization that has been in place since the beginning of land privatization.

And many others are digging right into the urban environments where they live, in the process, healing trampled land as well as tired stereotypes about where our food comes from and who grows it. The urban agriculture movement in the USA is headed-up by many people of color, and helps heal the ironic and inaccurate idea that all farmers are white. An example: Through its creative and inspiring New Roots program, the International Rescue Committee (2015) is helping refugees to share their farming expertise with their families and neighbors. These innovative programs are often located in "urban food deserts" where residents otherwise have little access to fresh food.

The opportunity to witness and work alongside other people of color who are expert farmers, right in their own neighborhoods, is healing for those whose land was taken from them, including African Americans, for whom the very idea of farming has been tainted with the traumatic legacies of slavery, sharecropping, and racist government policies (Thomas, 2015). Urban gardens give people of color a way to reestablish agrarian skills without having to move away from the safety of their own communities into rural areas, which they, often correctly, perceive as racist and hostile.

Other organizations encourage people of color to work in rural areas, such as the Fresno-based African American Farmers of California, which trains African Americans in essential skills such as irrigation and operating farm equipment on their Central Valley farm, and then helps them to sell their produce at farmer's markets all over California (Scott, 2013). John Boyd—founder of the National Black Farmer's Association—worked for decades to expose the widespread discrimination and abuse against blacks by the US Department of Agriculture, and eventually won back the farm that was taken from him. Though Boyd agrees that growing food "in your own backyard" is a huge step towards reconnecting with the land, he urges fellow black Americans to take a "second look" at farming because, "when we lose our land, we are also losing a part of our history" (Thomas, 2015).

In the UK, Wilfred Emmanuel-Jones founded the Young City Farmers program based on a similar sentiment. After leaving his inner-city upbringing to realize his dream of owning a farm on the rural Devon/Cornwall border, he wanted to help others do the same. "Exposing 'hardcore urbanites' to the rural environment…can trigger a deep seated affinity with the land…it opens up a huge amount of options to someone who may have thought they were headed for life's dustbin heap" (The Black Farmer, 2009).

Sacred farming

In both urban and rural environments, people everywhere are breaking with conventional notions that have dictated how and where they interact with the land, and getting involved however they can. Perhaps the time has finally come for us to recognize that it is not required to become landlords of large swaths of land in order to access a meaningful relationship to earth; is not necessary to first become a paragon of virtue before we can begin healing familial patterns; we need not have all the answers before attempting to reshape our culture to be friendlier, more humane, more connected. We can begin wherever we are, just as we are. Prechtel (2013, p. 313) describes this start-small attitude as a type of "sacred farming":

> [A]ny worthy culture has to sprout right out of the slag heap of the world's present condition … These cultures…start in many ugly places in ways hardly noticed at first … For we, as "sacred farmers "… know we must learn to metabolize our grief into a nutrient … compost the failures of civilization's present course, and cultivate … a future worth living in, all smack-dab in the middle of modernity's meaningless waste.

After observing their contemporaries growing ever more hunched and pale in front of computer screens, people of all stripes are choosing to buck the technological tide by embracing traditional skills —starting small dairies, organic gardens, natural building co-ops, wildcrafting herbal medicines, and relearning classic occupations such as tanning, smithing, orcharding, shepherding, masonry, and boat building. Simultaneously, people are also recreating a world where the time-honored arts that grease the wheels of social and inner harmony—dance, storytelling, music and theater—are celebrated and integrated into everything we do. Innovations which incorporate nature into the healing arts are becoming more and more common, as well. All of the above, and more, qualifies as sacred farming.

In revaluing these timeless and enduring skills, we are growing real roots into our communities, and into the ground, gaining a visceral understanding how the fate of the trees, the animals, the plants, the waters are bound up with our own. Working amidst a tearful rain of human gratitude, we are making it possible to sprout forth the seeds of indigeneity that have been dormant within our bodies since our cultures were uprooted, perhaps hundreds, or even thousands of years ago.

These trends towards re-skilling instinctively recognize that when we are connected in a tangible way with the Earth are much more likely to act in reverence and stewardship of it. As Wendell Berry (2003, p. 85) elucidates:

> In a state of total consumerism—which is to say a state of helpless dependence on things and services and ideas and motives that we have forgotten how to provide ourselves—all meaningful contact between ourselves and the earth is broken. We do not understand the earth in terms of either what it offers us or of what it requires of us, and ... people inevitably destroy what they do not understand.

Only when we risk rekindling a messy love affair with our estranged beloved Earth will we gain the inspiration and the courage required to act resolutely when confronted with challenges such as melting sea ice, species extinction, massive pollution and 'permanent' war. Thus, a human race moving robustly into a healthy, ongoing future, is destined to be a life which involves a reclaiming of our indigenous heritage—the basic right, and the basic pleasure of working with wood, with soil, with Earth. For more and more people, a healthy life will be determined by how much dirt we have under our fingernails.

The wheelbarrow activist

Sometimes, on a windy October night like this one, I can actually catch a glimpse of the ghostly forebodings of my immigrant forefathers wafting around: *Gotta get into a good school. You don't wanna end up a dirt-farmer, like your poor grandfather! You're smart enough to be a doctor or a lawyer!* In these voices, which are threaded deep into the fabric of my personality, I can hear the echoes of a long history of exile from the land. Even after over a decade of living elbow-deep in a food-forest, I can still perceive the cop in *my* head trying to convince me that working with the land is despicable, suitable only for 'peasants,' or, more pointedly in a hyper-phobic and racist America, for 'Mexicans.'

In response, I heft my wheelbarrow full of leaves and manure into our garden, and blanket the beds for their winter slumber. I laugh with my toddler as he affectionately labels the pile "big poop!" and encourage him help dig with his tiny shovel. I thank the cleansing winds as those voices catch an updraft and blow out to sea, and replace them with gratitude for the chance to work with earth—a freedom for which our ancestors sacrificed their lives, and for which people everywhere are still fighting—from Indian farmers resisting seed exploitation by Monsanto, to villagers in Bolivia and Greece blocking privatization of the water supply, to Amish farmers battling for the right to drink raw milk from their own cows, to modern herbalists preserving their grandmother's healing recipes despite increasing regulatory pressure from Big Pharma, to urban farmers markets which sell food grown exclusively by African-Americans.

Like the green leaves that can always be found pushing their weedy heads through cracks in the sidewalk, no matter how many times they are torched, weed whacked and herbicided, the unceasing sprouting of wild human ingenuity consistently thwarts every authoritarian attempt to pave it over. For modern people to recognize and repair the damage done to our connection to Earth is perhaps the pivotal task of our generation. It is for this reason that I stand in solidarity with farmers and witches all over the globe, and reclaim them as titles of distinction and pride. I am a Farmer, and I am a Witch.

A future worth living

In putting these words on 'paper,' I hope to contribute to the enormous task of piecing back together what Prechtel calls the *tribal shards* of original human culture, shards from which we can reconstruct the blueprint for an ample and sturdy cooking vessel. Only in a pot as miraculous as this, made up of pieces recovered from deep within each of us, can we simmer up the deliciously innovative responses needed to sate the rowdy ecological and social crises currently seated at our dining table and "begin remembering our Indigenous belonging on the Earth back to life" (2012, p. 10). As we reach out to the banished farmers and witches exiled within us, we will welcome also the wild solutions we need to transform travails into triumphs, and give birth to 'a future worth living in.'

References

Berry, W. 2003. *The Art of the Commonplace: The Agrarian Essays*. Berkeley, CA: Counterpoint Press.

Boal, A. 1993. *Theatre of the Oppressed*. New York: Theater Communications Group.

Bovenschen, S., Blackwell, J., Moore, J., and Weckmueller, B. 1978. The contemporary witch, the historical witch and the witch myth: the witch, subject of the appropriation of nature and object of the domination of nature. *New German Critique*, (15), 83–119.

Buhner, S.H. 2014. *Plant Intelligence and the Imaginal Realm: Beyond the Doors of Perception Into the Dreaming of Earth*. Rochester, Vermont. Bear & Company.

Federici, S. 2004a. *Caliban and the Witch: Women, the Body, and Primitive Accumulation*, New York: Autonomedia Press.

Federici, S. 2004b. Audio Lecture with Silvia Federici, recorded live at Fusion Arts, 30 November, Little Red Notebook. [Online] Available at: www.silviafederici.littlerednotebook.com/

Gabor, A. 2011. Inside Polyface Farm, Mecca of sustainable agriculture. *The Atlantic*, 25 July. [Online] Available at: www.theatlantic.com/health/archive/2011/07/inside-polyface-farm-mecca-of-sustainable-agriculture/242493/

HMI-Holistic Management International 2015. HMI. Heathy Land, Sustainable Future. [Online] Available at: http://holisticmanagement.org/

International Rescue Committee. New Roots in America [Online] Available at: www.rescue.org/new-roots-america

La Fontaine, J. 2012. Witchcraft belief is a curse on Africa. *The Guardian*, 1 March. [Online] Available at: www.theguardian.com/commentisfree/belief/2012/mar/01/witchcraft-curse-africa-kristy-bamu

Markham, L. 2011. The New Farmers. *Orion*, 27 October. [Online] Available at: https://orionmagazine.org/article/the-new-farmers/

Mies, M. 1986. *Patriarchy and Accumulation on a World Scale: Women in the International Division of Labour*. London, UK: Zed Books.

Plotkin, B. 2013. *Wild Mind: A Field Guide to the Human Psyche*. Novato, CA: New World Library.

Pollan, M. 2002. *The Botany of Desire*. New York: Random House.

Prechtel, M. 2012 *The Unlikely Peace at Cuchumaquic: The Parallel Lives of People as Plants: Keeping the Seeds Alive*. Berkeley, CA: North Atlantic Books.

Sacks, A.D., Teague, R., Provenza, F., Laurie, J., Itzkan, S. and Thidemann, K. 2013. *Restoring the Climate Through Capture and Storage of Soil Carbon Through Holistic Planned Grazing*. The Savory Institute. [Online] Available at: http://savory.global/assets/docs/evidence-papers/restoring-the-climate.pdf

Scott, W. 2013. Scott Family Farms: AAFC Black Farmer Training [Online] Available at: http://scottfamilyfarms.net/services

Schwartz, J. 2013. *Cows Save the Planet and Other Improbable Ways of Restoring Soil to Heal the Earth*. White River Junction, VT: Chelsea Green.

Savory, A. 2013. How To Green the World's Deserts and Reverse Climate Change. TEDx talk, 4 March. [Online] Available at: www.ted.com/talks/allan_savory_how_to_green_the_world_s_deserts_and_reverse_climate_change.html

Thomas, M. 2015. What happened to America's black farmers? *Grist*, 24 April. [Online] Available at: http://grist.org/people/what-happened-to-americas-black-farmers/

Thomas, P. 2015. *Pathways to Resilience*. [Online] Available at: http://pathways2resilience.org/#

Vosper, N. 2015. *Empty Cages Design*. [Online] Available at: www.emptycagesdesign.org/permaculture-prisons-links-resources/

White, C. 2014. *Grass, Soil, Hope: A Journey Through Carbon Country*. White River Junction, VT: Chelsea Green.

14 Dia de la Tierra

Jose Gonzalez

Figure 14.1 Dia de la Tierra

15 A personal journey to a universal approach

Permaculture

William Hooper

My childhood home abutted three narrow acres of woodlot. One edge, adjacent to a four-lane road, had been stripped down to the red clay in a thirty foot swathe during the road construction. By the time I was in grade school, pine tree saplings were tenaciously rooted in the clay, transmuting it back to soil.

When I was no more than ten, some other young boys and I decided to take heavy sticks and attack these small trees. I think it was the culmination of a 'let's pretend' game in which we fought valiantly against mountain trolls, enemy agents, or some similar 'bad guys.' So we hefted our sticks and went at the small trees. Now, thirty years later, I can talk about 'performative masculinity' and 'socialization to equate boyhood with destruction,' but at the time it was only the joy of exertion and the thrill of seeing things change by your own hands.

We never again played such a game, not out of guilt, but because it never came up again. I don't remember any trouble or even realization from grownups that we'd done this thing. We weren't destructive kids, as kids go; we loved the woods and spent countless hours in them, and this was an aberration without any overarching purpose.

I had many years of watching the strange unevenness at the forest's boundary, echoes of one casually destructive afternoon. There was guilt, but there was also affirmation that even small bursts of concentrated activity could evoke great differences in the environment, over time. I just needed to focus my efforts in the right way.

What was the right way?

My mother practiced 'organic gardening' when it was an exotic notion. The strangeness of the concept, magnified by our conservative upper-middle-class white bread suburb, meant my mother's compost piles and controlled lawn-burnings fell, in terms of respectability, somewhere between nudism and devil worship.

The shaded side of our garage was a raised bed covered by composting piles. I was asked why my family maintained a garbage dump. Yet after several years, the hardpan red clay of that side yard evolved into black loam so thick with earthworms that a handful would literally wriggle out of your grip. This, a stone's throw from neighbors' yards where tanker trucks labeled "LAWN-GRO" would trundle up, so that men in hazmat suits could spray the grass. Afterwards, you could watch earthworms in those lawns crawl to the surface and die, and within a few days, the lawns resumed a color like underripe limes and grew thick as shag carpeting, so that every Saturday shrieking ride-on mowers could mow it pool-table flat again, because ... because that's what you did. Our yard was the oddity, able to be sustained by our own waste and labor.

Why were we the odd ones?

I remember our first watermelon patch. We planted the big black seeds, tended the green-yellow vines twisting out of the soil, and carefully protected the growing melons. Following Mark Twain, we listened as the unripe *pinkpinkpink* of the thumped melon became the ripening *pankpankpank* and finally the deep *punkpunkpunk* of a fruit ready to eat.

One afternoon, I picked the first ripe watermelon, placing it on the patio table and slicing it open. The hard, dark green rind of the small melon split to reveal crimson flesh. Standing on the concrete of the patio, in the fading light of the late summer twilight, staring over the grass lawn and adjacent gardens and forest fading off at the rear, I took my first bite. Watermelon flesh crunched, crisp, peak of ripeness, juicy and cool and dripping-down-your-chin perfect.

Decades later, I don't know that I've ever tasted anything sweeter.

As I'd effected negative change with one afternoon's burst of destructive energy, I now made good from months of gentle, occasional intervention.

Typical responses or justifications for non-intervention

Global climate change cannot be avoided at this point; no matter what we do, we're still dealing with some level of effect, and we're unlikely to take serious action until the system is so far from equilibrium that even 'some level of effect' means intermittent catastrophe. At the same time, despite unimaginable accumulated wealth on a macroscopic level, the distribution of wealth grows increasingly, worryingly, lopsided in much of the world, and little of that wealth funnels towards relieving basic human suffering. Hierarchical structures rooted in racism, sexism, and classism still dominate humanity, control most of the obvious mechanisms of power, and subtly poison much of our interaction. The whole edifice currently rests on a foundation of cheap, plentiful fossil fuels, a foundation that unravels as we speak.

This age requires active effort to avoid a constant barrage of hopeless imagery and soul-withering facts. In contrast to all the materials that might run out, reasons to despair are infinitely renewable. Several defense mechanisms spring up readily to shield one from these realities (Passerini 1992).

Denial can be found in those who wrap their assertions in a veneer of cherry-picked data mistaken for healthy skepticism, or those who confuse contrarianism with insight, or those whose circumstances shield them from uncomfortable realities. For instance, middle-class white folks like myself often enjoy, and encourage, high walls (physical and metaphysical) between their immediate lives and deeper problems. Powerful and profitable forces work to encourage denial, because an easy path to denial is to consume instead of worrying.

Pollyannaish optimism doesn't deny problems per se, but gives them no serious attention, due to limitless, generally baseless confidence that smart/self-interested/altruistic/rich people will solve them before they seriously impinge on our lives. Underpinning this outlook is a shallow, backwards glance at history – obviously, all previous problems have been sufficiently solved for us to be here, now, discussing it – which falls under the same logical breakdown as "I don't need to worry about ever dying, because I've never died before," usually combined with an unstated but unshakeable faith in gee-whiz technological solutions that are always just about to emerge.

Hopelessness grows in the soil of vast problems and blooms in several ways: for some, inescapable despair, an emotional oubliette from which no escape seems possible. The suffering this brings can literally lead to death, or an even more devious alchemy, where the psychic imprisonment itself is a kind of defense. After all, if the situation is hopeless, you know in your heart that you cannot make a difference. Since you can't make a difference, there is no point in trying. If there's no point in trying, you're freed from feeling like you ought to try. Hopelessness leads paradoxically to the 'freedom' of indifference.

Embittered cynicism: those too active to be indifferent, and too resilient to merely despair, attack others – particularly those trying to fix the root issues. "Why bother," says the cynic; after all, nothing ever changes. Once you're as cleverly wise and world-weary as I am, you'll see that the dead end of cynicism is the place to be. In a culture hostile to vulnerability, cynicism presents a swaggering, tough-seeming, 'manly' alternative to the apparent weakness of admitting fear and hopelessness. The media exploit the cynical, allowing them to vote with their dollars for prettified nihilism. A manifestation of this comes in parts of the 'doomsday prepper' community, for whom a world-overturning calamity offers the possibility of being freed from the injustices and discomforts in their current situation. Inevitably, the fantasized apocalypse wipes out their credit card debt and dead-end job, but never leads to death from dysentery or starvation, at least not for them or anyone they like.

Rage is the response to the world's injustices and hurts for some. Anger pushes back against the world, threatens it right back, and breaks the paralysis of hopelessness with a thundering counterattack of righteous indignation. Anger certainly taps a deep well of energy for many people, and it would be historically foolish to deny the central role of anger in fomenting many of the revolutions, positive and negative, that shape our current world. In my experience, though, anger by itself burns like a bonfire; it produces great heat and light, but does little other than to consume whatever is burning. So you burn with rage, only to find that nothing outside you permanently changes, and you're exhausted. Markets exist to exploit this anger, too – a whole range of shock journalism and clickbait articles (to stoke the anger), feel-good superficial greenwashed consumer goods and no-effort 'slacktivist' behaviors (extremely low-effort actions taken because they flatter the ego, not produce change). They exist to generate a profit from the movement and generation of this anger. But ultimately, anger must be wedded to and transmuted into positive action in order to make change; by itself, anger just makes you angry.

What, then, can be done?

When I was twenty-five, I found myself staying overnight at the home of a friend of a friend, in rural Massachusetts, far from my home in Alabama. I appreciated the hospitality of the older man and his wife, and thanked them for putting me up. As I was leaving, the man mentioned getting rid of some old books, and invited me to scrounge before he gave the lot to Goodwill. I flipped through the box, and one text in particular caught my eye. It was black with a cover illustration that looked like an architectural diagram of a house and forest – it reminded me of my younger days of drawing maps for Dungeons and Dragons games. The curious book was *Introduction to Permaculture* credited to Bill Mollison. I opened the book. And it changed my life.

> Permaculture is a philosophy of working with, rather than against nature; of protracted and thoughtful observation rather than protracted and thoughtless labor; and of looking at plants and animals in all their functions, rather than treating any area as a single product system.
>
> (Mollison 1997, p. 1)

I found, reading over the next few days, the repeated feeling that many of the half-formed, niggling questions, the ones that eat at the edges of consciousness and make it hard to sleep late at night – that someone else had grown those questions to full form, brought them into the light, and built a framework for developing answers. Mollison presented a vision not merely of how things could be, a dream of some future world, but an extensive practical blueprint for moving towards that dream. A way forward, even if you were working just in your own little corner of the planet, paths towards creating sustained and meaningful transformation of whatever small slice of planet you lived on.

Permaculture provided a uniting narrative and toolkit for addressing the concerns I had about the environment. It was grounded in actual work, not merely big dreams, and drawn from lived experience, not just predictions of abstracted models.

> What is Permaculture? … Consciously designed landscapes which mimic the patterns and relationships found in nature, while yielding an abundance of food, fiber, and energy for provision of local needs…[and] the use of systems thinking and design principles that provide the organizing principles for implementing [that] vision.
>
> (Holmgren 2002, p. xix)

It would take another five years for permaculture to become a cornerstone of my actual lived experience, but from the moment I opened that text, the ship of my life began to tack into a new and different wind. Permaculture, if not the singular answer, gives the right framework to evolve answers. To build recognition of problems sufficient to avoid denial, give hope sufficient to climb out of despair and grow out of cynicism, and give voice and power to righteous anger by providing engines by which anger can be made into change, and growth, for the better.

Permaculture, as a word, originally came from combining "permanent" and "agriculture". Over time, reflecting a shift in the movement away from simply seeing permaculture as a novel set of strategies for sustainable agriculture, and towards applying it as an intentional transformation applied across human activity as a whole, the portmanteau now combines "permanent" and "culture" (Mollison 1997).

With such a broad goal, it can be difficult to define what constitutes permaculture. Such a vast topic can be approached accessibly by contrasting it with its better-known opposites. The other end of the continuum would be the standard model of industrialized, consumption-driven, hierarchical culture that exerts hegemony in much of the world today. For example, food production as an industrial activity, often performed thousands of miles from its ultimate destination. To produce and process and ship and store food, our diets have become homogenized and independent of local climate, season, or needs. All of this leads to the cluster of problems previously discussed, as well as the various defense mechanisms deployed to not have to address them.

Permaculture is built on the belief that we can take some of the best of the past, and wed it to the best of the present. Assessing these issues, seminal permaculturist David Holmgren argues for four potential futures for humanity (Holmgren 2002).

First, the 'Techno-Fantasy' outcome, in which – any day now! – some technological salvation appears, allowing us to continue up the slope of increasing per-person consumption forever (whether this be through some magical unlimited energy source, or via some transmuting of humans into computer programs, depends on the person doing the dreaming). Without even digging into the implicit assumptions of this possibility, such as the assumption that this would make people happy or solve any of the deeper structural problems currently shackling us, it falters because it kicks the can down the road. Maybe we will find techno-Jesus to save us from ourselves. But until that (maybe) happens, we've got work to do. I might also win the lottery, but I still think it's wise to save for retirement and not rack up debt right now. My personal interactions with nuclear physicists, geneticists, neuroscientists, and so forth generally reinforce the suspicion that the people proclaiming most loudly that techno-utopia is at hand have minimal connection to the people actually wrestling with the complexities of fusion power, artificial intelligence, and so forth. This someone-else-solves-everything outcome is one of the places denialists and Pollyannaish optimists pin their hopes (Passerini 1992).

Secondly, Holmgren (2002) describes the "green-tech stability" outcome. In this version of the future, the industrialized world sets its thermostats a little lower, installs LED lightbulbs, puts solar panels on everything, and manages to limp along more or less as-is, with slightly less energy consumption per capita but without major disruption to our 'way of life.' This future sits more in the bedrock of reality than the techno-magic dream; industrialized nations are increasingly investing in renewable energy, their total populations are leveling or dropping, and energy usage in much of the first world on a per-capita basis is lower than it has been historically. Mark Jacobson of Stanford University and Mark Delucchi of UC Davis developed a well-regarded and well-researched plan in which humanity's energy needs could be met by completely renewable resources by 2030 (Jacobson and Delucchi 2009). However, the 'green tech' response still minimizes climatic chaos and its impact on food production (particularly in the non-industrialized world), the incredibly unified political and economic will required to make such a shift wholesale, and the reality that it would still do nothing to address the underlying dynamics of power, hierarchical control, and corporate control that plague us right now. It assumes, ultimately, no real change in what we do or how we do it; we just swap out some of the 'under the hood' pieces and keep rolling along. This 'good-enough' outcome is the other place denialists tend to congregate. It's a movement in the right direction, but we must go further, and embrace the next and most viable future.

The third possibility is called the 'Energy Descent,' or permaculture future, also known as 'descent with dignity' (Holmgren 2002). In this scenario, total worldwide energy usage drops to sustainable levels, from a combination of shifting energy production itself to renewable sources and greatly scaling back our individual consumption. This doesn't mean a future of subsistence agriculture bare-bones survival for everyone, or a Luddite destruction of any device more complicated than a bicycle. Instead, it means a world in which intentional choices about energy usage lead to optimizing for long-term utility. For instance, turning oil into reusable plastic parts, into communications satellites that will stay in orbit for millennia, into applications that will continue to produce benefits indefinitely, instead of burning it. Emphasizing local decision making, flattened hierarchies, and cooperative structures for governance and commerce. Most centrally, a redesign of the food system, towards a future in which 80% of all food consumed is produced within your local area (Hopkins 2014), using methods that sustain and build soil health rather than stripping it, and do not depend at base on constant petrochemical-driven intervention.

The fourth is the 'Atlantis' scenario, where energy usage collapses due to system-wide failure, depletion of available sources, and accelerating climate change (Holmgren 2002). To evoke it, just imagine that neither you nor your neighbors can get food at the grocery, medicines at the pharmacy, or electricity at the power line, at first for weeks at a stretch, then eventually, forever. The cynics and the hopeless and many of the angry end up feeling this is the inevitable outcome. Let's agree to disagree with them, and focus on solutions.

As noted by Holmgren (2002), permaculture can be misread as a hodgepodge of techniques for gardening, and this often arises when 'permaculture' gets identified with particular techniques in its repertoire (e.g. hugelkultur mounds, keyline swales, herb spirals, etc.). These are an essential aspect of the whole, but they grow out of deeper principles. Likewise, it can be misconstrued as a generically countercultural stance, and indeed what I've written so far could readily be read as such.

At its heart, permaculture's focus isn't thumbing its nose at the status quo, or just producing more vegetables; it is creating viable systems for meeting human needs, particular needs for food and energy, in a way grounded in intentional ethical stances, rather than first focusing on productivity, convenience, or contrarianism (although those frequently result from the ethics). From the ethical core, all the other principles and techniques flow. Although Holmgren (2002) uses different words, he lays out the ethics of permaculture as *Earth care*, *people care*, and *fair share*.

Earth care gets listed first, but the three tenets do not form a hierarchy of importance. Instead, they are three legs of a stool, all supporting the whole concept, working in concert, and equally necessary. Earth care means care for the planet itself, but depth as well as pitfalls come with such an uncontroversial idea. As we all depend on the planet for our lives, caring for the planet can be justified as self-interest, but permaculture reaches higher than utilitarian pragmatism. Biodiversity, and the species creating it, cannot be measured merely in immediate utility or convenience to humans, but instead have inherent worth (Holmgren 2002).

A central earth care theme in permaculture is devotion to soil health. Everything ultimately grows from, depends on, and returns to the soil. The fundamental importance of healthy soil is seen in an old gardeners' joke – "What do I grow? Soil. As a side effect, fruits and vegetables." This gives rise to most of permaculture's techniques, including no-till approaches, berming and swaling (creating mounds and troughs in the land to capture and direct water), composting techniques, and so forth – it all goes back to soil care.

People care embodies an equally important aspect of permaculture: the valuing of, caring for, and providing for human needs, ranging from the material to the spiritual-emotional (Holmgren 2002). The inclination for many ecologically-focused folks is to make care of people subservient to care of the earth, but this falters on several fronts. It implicitly reinforces the folly that humanity exists separate from the earth; this outmoded dualism produced the environmental issues we now contend with. It forces a false dichotomy in which people must prioritize one or the other, a self-damaging and needlessly hierarchical approach. Finally, it devalues the unique capacities humans have for making positive changes. We appear to be the only species capable of consciously observing and predicting changes years in advance. We can intervene, with deliberation and care, to achieve better outcomes. The planet itself produces biodiversity and resilience, but it does so without intention or planning, and without contingency plans; we can model what we see in it, but leverage knowledge and intention to accomplish those ends more universally, and despite the damage we've already done when acting without careful forethought.

By providing pragmatic tools for local-scale production of food, by emphasizing reduced consumption and a community focus over isolating hyper-individualism, and by providing a framework emphasizing accepting feedback from the system and acting based on it, permaculture provides tools not merely for propagating the status quo for a few more years, but provides transformative direction. When you grow food with your own hands, and share it directly with neighbors doing the same, artificial distinctions based on color or creed or wealth become more difficult to maintain. You eat, they eat; you need food, they need food; you all share a table, and 'they' become 'we.' It is no accident that nearly all religious celebrations, across the planet, involve a communal meal.

Fair share, the final notion, helps focus the other two. We must produce a surplus, in order not only to provide for ourselves, but also to provide for others (Falk 2011). Part of producing a surplus is planning to always produce a little more than our own needs. A common metric is to plant 10% more than your own family will eat, to create a 'tithable' portion for neighbors, soup kitchens, folks whose own crops fail, and so forth. At the same time, if we limit our own consumption, we reduce how much we need to produce to begin with, and living below your means is a cheap way to be rich.

From this foundation, permaculture builds principles, techniques, strategies, and a way of planning and doing. It does not proffer a single 'right answer' to all problems, but equips us with the language and tools to positively transform and evolve our own world and our own, not only for the improvement of ourselves, but for the betterment of the whole system in which we are, always and forever, inextricably a part.

My own work within permaculture allowed me, in the spirit of making use of the resources at hand, to use the raw materials of my own earlier life. From the creeping dread and hopelessness about the world, to the early experiences with forest and garden, and destruction and creation, through the college studies in chemistry and biology and history, to the adult experiences grappling with vocation, masculinity, fatherhood, and husbandhood, permaculture provided a language and skillset and goal-orientation to combine all of those parts into fertile soil, in which grows hope, direction, and positive effort. It holds the possibility of similar transformation for other individuals, and the world we share.

References

Falk, B. 2011, Permaculture ethics, lecture notes from Permaculture Design Course at Whole Systems Design, Moretown, on 2 August 2011.

Holmgren, D. 2002, *Permaculture: Principles and Pathways Beyond Sustainability*, 4th edn, Hepburn, Victoria: Holmgren Design Services.

Hopkins, R. 2014, *The Transition Handbook*, Cambridge: UIT Cambridge Ltd.

Jacobson, M., and Delucchi, M. 2009, 'A Path to Sustainable Energy by 2030', *Scientific American*, November 2009, p. 58.

Mollison, B. 1997, *Introduction to Permaculture*, revised edn, Tasmania: Tagari.

Passerini, E. 1992, *The Curve of the Future*, Dubuque: Kendall Hunt Publishing.

16 Climate change and sustainable agriculture

Why inclusive farmers' markets matter[1]

Ryanne Pilgeram

The promise of the Green Revolution—high food yields—has meant a fundamental shift in how food is produced. Huge monocropped farms that rely on chemical fertilizers and pesticides are its hallmark, and these hallmarks of conventional, industrial agriculture have been linked to everything from increased rates of cancer because of pesticide drift and runoff, to massive 'dead zones' in the ocean, to a substantial drop in the number of honeybee colonies, to developmental abnormalities in amphibians, to desertification, and, of course to climate change (Horrigan *et al.*, 2002). Of particular importance to those concerned with climate change, the Johns Hopkins School for a Livable Future notes that, "Agriculture is directly responsible for about 20% of human-generated emissions of green-house gases … [and] Changes in land use contribute about 14% of the total human-generated emissions of greenhouse gases, and much of this land development is for agricultural purposes" (Horrigan *et al.*, 2002, p. 446). This should not be surprising given the high use of fossil fuels in conventional agriculture: "From the tractors that break the ground for planting, then return to do the planting and harvesting, to the transport and processing, to the further transport to the supermarket, and all the way to your drive to make the purchase (unless you bicycle and cut a calorie or two off the process), energy is used and carbon produced" (Ross, 2009, p. 78). Thus, the conventional agriculture system is invested in a treadmill of production completely dependent upon fossil fuels.

Beyond these environmental issues, the Green Revolution has also fundamentally changed the US and global economy by shifting agricultural production away from small, family run farms to large industrial operations that require tremendous investments in everything from fertilizers and seeds to farm equipment. These shifts have put an increasing demand on cheap immigrant labor and created a glut of certain food items, such as corn, which have not only affected how and what we eat (try finding a processed food without corn in it) but have also affected how and what others eat globally. One particularly salient example is how traditional methods and types of corn are being farmed in Mexico. Because cheap, subsided US corn is imported into Mexico, it is often cheaper to buy corn than farm it, pushing and pulling people from the land into the cities. Not only has this changed the Mexican economy, there is also evidence that strains of Monsanto's genetically modified corn are contaminating the heirloom varieties of corn that have been traditionally grown in Mexico, meaning that traditional foods and the means of growing them may forever be altered.

Of course, these examples extend well beyond Mexico or the US; industrial agriculture by almost all measures is a disaster for the environment and for human communities, something that has become increasingly clear as more and more people globally and domestically are seeking alternatives. Perhaps the most obvious of these alternatives on a domestic scale are US farmers' markets that connect farmers and consumers to locally produced food. In fact, scholars and activists alike have described farmers' markets as important sites for combating the effects of global warming, since food grown for farmers' markets often comes from smaller, less fossil-fuel dependent farms, requires significantly fewer off-farm inputs, and uses less fossil-fuels to be transported and packaged (Horrigan *et al.*, 2002, p. 446). Thus farmers' markets may serve as important sites for repairing the environmental damage done by conventional agriculture by contributing to an alternative and more sustainable food system. What is more, in addition to the potential positive effects for climate change, farmers' markets are often vibrant parts of their communities. At farmers' markets, people come together in public spaces not only to buy food, but also to share ideas about that food and the farms that grow it. Because of this, farmers' markets have been described as "keystone institutions in rebuilding local food systems" (Gillespie *et al.*, 2008, p. 70).

Yet, despite the importance of farmers' markets in creating sustainable alternatives to industrial agriculture, the fact remains that most of the research on sustainable agriculture has used an environmental and natural science perspective, where the challenges and promises of sustainable practices in promoting a more environmentally sustainable and just planet are the central focus (Goodland, 1995; Kates, 2001; Wackernagel *et al.*, 2002). The social aspects of sustainable agriculture, or 'social sustainability,' and farmers' markets are far less researched. In fact, the discourses about sustainable agriculture have been framed in such a way that certain topics or critiques are simply not addressed and do not seem to be a part of the dialogue. As Allen (1993, p.145) notes in her book on sustainability, "social issues, when raised, are often safely vague and framed in terms of 'socially acceptable' as it refers to environmentally and economically sustainable institutions and practices." For Allen, "this begs the question, socially acceptable for *whom?*"

Sustainable agriculture is touted as "an opportunity to improve the economic, environmental and social sustainability of agriculture" (Barbercheck *et al.*, 2006); however, while there are strict environmental regulations before "organic" can be stamped on a box, this stamp says nothing about the social relations of producing that product. Using the Northwest Farmers' Market as a case study, this paper explores how consumers and farmers involved in sustainable agriculture put the goals of social sustainability into practice. Thus, rather than assuming that farmers' markets and sustainable agriculture are "good for everyone," this project examines how privilege operates in the space and practice of sustainable agriculture.

As this suggests, this interrogation of privilege within sustainable agriculture is based on an intersectional understanding of social justice. Recognizing that race, class, and gender continually overlap with and reinforce each other, this paper examines the issue of social justice from this broad and intersectional perspective, analyzing the complex ways that various privileges and assumptions across these categories manifest themselves at a Northwest farmers' market. If sustainable agriculture is as important in addressing climate change and other environmental degradations as we have been led to believe—and based on the research, there is little doubt it is—then we must also question how to produce and distribute this food in ways that emphasize social justice. Social sustainability asks that we examine the laborer in the field of organically grown tomatoes as well as the consumer sitting at the table enjoying the fruits of those labors. Thus, in this paper I explore how the space of the Northwest Farmers' Market is constructed to privilege certain peoples and identities and connects those privileges to larger systems of structural inequality.

Methods

This project specifically examines the spaces and social relations at the Northwest Farmers' Market, a farmers' market that sells sustainably grown foods and other products in a medium sized town in the Pacific Northwest that wins national awards for being one of the "greenest" in the US. I explore the space of sustainable agriculture using three different qualitative methods—participant observation at the market, structured interviews with consumers, and key informant interviews and farm visits with farmers.

In 2008 I observed the market for a full season (approximately six months, from mid-April through mid-October). At the height of the market season, I observed at the market three times per week with each observation lasting approximately two hours. I observed on different days and during different time periods to get a better sense of the ebb and flow of consumers and farmers. Thus, one day I observed during the set up period, one day during the middle of the day, and another time when the market was being packed up. I also participated in the market as much as I could—buying produce and trying to get into the mindset of a consumer. This was not terribly difficult: as a white, able-bodied heterosexual woman who often visited the market with my male partner and young baby, I looked very much the typical shopper at the market. The more time that I spent at the market, the more word seemed to have spread that the woman wandering around with a notebook and audio recorder was just one of the many unusual things at the market that became 'normal.'

The farmers' markets and whiteness

What stands out at the market is the food, both because of the space it occupies and because it is the reason for the market. In the presence of this cornucopia of fresh produce, it becomes easy to pay attention to the processes and soil that helped to produce it rather than the human elements. The market is constructed in particular ways, ways that emphasize these environmental processes and inputs over the human element. For example, the farmers use signs to announce that their farms are "all organic" or "bio-dynamic" and have compost buckets at the market that are labeled "organic only." Except for signs that suggest the farms are "family" farms, no one announces with pride the working conditions or social relations on their farms; there are no signs that announce: "our employees have health insurance!" Furthermore, though the Northwest Market Handbook is clear to point out the list of requirements for labeling food as "NATURALLY GROWN" versus "NO SYNTHETICS APPLIED" versus "CERTIFIED ORGANIC" as well as the consequences for not doing so adequately (which includes "disqualification to participate in Market"), the requirements are all tied to environmental regulations about how the food is grown (emphasis theirs). There are no rules about the working conditions on the farms or any other element that might fall under the guise of "social sustainability." On a basic level this means that each farm can independently decide how they want to construct the social relations of their farms, but in a society with deeply entrenched social inequalities, even with a deep commitment to social justice it can be difficult to ameliorate these inequalities. Without an awareness of these issues, it becomes nearly impossible.

A number of scholars have pointed out that farmers' markets are spaces that tend to privilege middle-class whiteness. Demographic studies report similar findings, noting that the average farmers' market consumer is a woman in her late thirties to early forties with an above average income and education (Walton *et al.*, 2002; Onianwa *et al.*, 2006). Class, in particular, is an area that is often discussed by scholars who note a classed dynamic to sustainable agriculture, arguing that the consumption of sustainable agriculture seems to be related to a high education and income level (Goodman and DuPuis, 2002; Hinrichs and Kremer, 2002; Slocum, 2007).

For example, Slocum (2007, p. 522) notes that, "those involved in alternative food tend to be economically and/or socially middle class. They have the wealth to buy organic, the inherited or schooled knowledge about nutrition or the environment and they are politically liberal to left." Goodman and DuPuis (2002, p. 29) reiterate this point, arguing that "organic food consumption is presently a middle-class privilege—a 'class diet,' if you will." Similarly, other scholars have argued that farmers' markets are often spaces of white, middle-classness (Onianwa *et al.*, 2006; Slocum, 2007; Guthman, 2008). Slocum (2007, p. 521) argues that, while "it should be said that there is something white about alternative food practice, that 'something white' is not equivalent to 'something negative.'" She sees whiteness in sustainable agriculture as something to be understood but not condemned outright because she argues, "whiteness coheres in alternative food practice in the act of 'doing good'" (p. 526). Furthermore, she notes that, "the desire for good and sufficient food and jobs and thriving economies is not white" (p. 521). Nor, I would add, is it only a middle-class prerogative. Consistent with research findings, I found whiteness central to the construction of the Northwest Farmers' Market. In particular, my observations at the market focused on understanding how race/ethnicity[2] affected interactions in this space—rather than counting bodies—in part because there were so few obvious non-white people at the market. One of the most salient examples of this was the images used to market the farmers—images that were often somewhat nostalgic images of white children or families on the land. These images reflect both the racial/ethnic make-up of the farmers and consumers (but not necessarily the farm labor), but also reflect the ways whiteness may be written into ideas about sustainable agriculture. As Slocum (2007, p. 521) notes, "Whiteness is hegemonic in the US; it is dominant regardless of the number of bodies in a certain place." Rather, this is about an image, an idealized representation that relies on more than numbers—it's about space, geography, what people do, the roles they have, in the market.

For example, during the course of my research, I made two observations that were contrary to a vision of racial/ethnic inclusiveness. If it were only about counting bodies, these observations would suggest that the market was more inclusive of non-white people; however, based on the power dynamics involved, it became clear that it is not enough to discuss race/ethnicity without contextualizing it. It is difficult to weigh the importance of these observations, because, on the one hand, they were relatively rare, but on the other hand, given the small number of non-white people at the market (particularly Hispanic people), these observations stuck out to me.

The first observation happened during the early morning hours while I watched the vendors setting up. I watched a Hispanic man haul vegetables from a truck to the stand where two Hispanic women were arranging the vegetables in the display. While the Hispanic workers were busying themselves with the task of setting up the stand—a stand I always found particularly beautiful because of the ways the vegetables were stacked in such large, yet neat piles—two young, white women busied themselves with writing out the price list for the stand and giving directions to the Hispanic man and women. As I watched the Hispanic man hauling heavy boxes from the truck and the Hispanic women laboring to unload box after box under the direction of the young white women, I found myself thinking: why have I never seen the Hispanic workers during any of my other observations? Then, before the market was opened, the man and woman got in the farm truck and drove away, leaving the white women in their summery cotton dresses to work the stand. Over the course of the summer I watched this scene take place whenever I observed the market before it opened.

This wasn't the only stand at which non-white people labored behind the scenes. I observed something similar at the only flower stand that occupied an area in the core area of the market. A white, middle-aged woman usually ran the stand. Fragrant peony blossoms and trailing houseplants surrounded her as she quickly assembled stems of flowers into bouquets for waiting consumers. Yet during the set-up process she is not alone; two Hispanic men, who do the heavy labor of hauling bucket after bucket of flowers from a large van to the stand, assist her. The morning I observed, they were behind schedule. The woman running the stand was agitated and the language barrier between her and the men, who appeared to only speak Spanish, only exacerbated her frustration. There was an awkward, aggravated dance between the men and the woman as she shouted "al frente, al frente," trying to convey that she wanted a specific flower from the front of the van. I felt uncomfortable observing these interactions that normally take place behind closed doors in workspaces.

However, what I watched next affirmed the ways that people who are racially/ethnically marginalized—particularly Latinos in this context—can literally become invisible in the space. The stand was being set up quite late; it was graduation morning at the local university and they likely had orders to fill before setting off to the market. But before they could even finish setting up, customers were eager to buy bouquets. The first consumer, a white, affluent-looking woman in large sunglasses, immediately began demanding that certain flowers be removed from the bouquet she had selected, pointing to the flowers she liked and indicating that she wanted them in her bouquet. At the same time, two Asian American women came to the stand and began selecting armfuls of flowers. The flower stand is one of the smallest at the market. Buckets overflowing with flowers line a narrow aisle less than five feet wide. Yet, all four women were crowded into this space, and all of them seemed hurried. The proprietor seemed caught off guard by the rush and flustered that the stand wasn't completely set-up. In spite of all this, somehow in this tiny space, the two Hispanic men continued to set up. With buckets in hand, they ducked between the women and everyone carried on as if the men were not even there. No one backed up to let them through; no one made eye contact with them. Only the woman running the stand seemed to acknowledge their presence, as she occasionally yelled something like, "where are the scissors?!" Soon, the customers were taken care of, the stand was set up, and the Hispanic men left in the van as the woman put the finishing touches on the stand.

Both of these examples suggest that, though the labor of Hispanic people may be necessary for some of the farmers, this labor is hidden in a variety of ways. This is not to say that every farm engages in these practices; the farmers themselves, farmers who are mostly white, run most of the stands. Yet the observations at these two stands suggest that racial/ethnic labor is hidden from the buying public. These stands whitewash their wares by using Hispanic laborers for set-up, but not as the public faces of the farm itself. Furthermore, customers (at least the ones at the flower shop) are implicated in this as well by not seeing this labor even when it is right in front of them. I was shocked at how the men at the flower stand seemed to work completely unseen in a space no bigger than a small bathroom.

This same morning I watched another Hispanic man help set up a produce stand, but his relationship with his coworkers at the stand was markedly different than at the other stands I described. It is worth noting that this man also looked different from the men and woman who were setting up the other two stands. He was taller and lighter skinned than other laborers setting up the stands, and the fact that he was dressed similarly to his white, middle-class coworkers suggested a class differential between him and the other Hispanic workers I observed. However, at this stand, rather than doing the heavy, manual labor alone, everyone working at the stand came together to load the van and to set up the produce. Then, when the stand was set up, they sat down together on a bench behind the display and rested together as they enjoyed a brief break before the opening. Even after the market opened, they all worked together selling produce and replenishing the food. I was struck by how different this was from my other observations. This was clearly a team of people who worked together quite well, and from my observations seemed to be treated equitably. He was not treated like a 'Hispanic farm laborer' in the ways the people setting up the two other stands were. Importantly, if a consumer were to visit the market, more than likely they would never see the first two cases. If they arrived even 10 minutes after opening they would only see the beautifully set up stands run by white, male/female dyads.

Of course to understand these dynamics, one must examine the larger structural conditions that they are a part of. As Bonilla-Silva (2014, p. 233) contends, "because of a number of important demographic, sociopolitical, and international changes, the United states is developing a more complex system of racial stratification…suggest[ing] three racial strata will develop, namely, whites, honorary whites, and the collective black and that 'phenotype' will be central factor determining where groups and members of racial and ethnic groups will fit—lighter people at the top, medium in the middle, and dark at the bottom." Thus, Bonilla-Silva's argument contextualizes the racial/ethnic dynamics that allowed for the lighter skinned, tall Hispanic man to become a public face of a farm and explains the contexts where the darker-skinner Hispanics were made invisible in this space. Furthermore, it explains the relationship between the Asian American women buying flowers, the white woman running the stand, and the Hispanic men setting up the flower stand.

These racial/ethnic stratifications are further exemplified when looking at land ownership. The racial/ethnic make-up of landownership means that there are simply far fewer non-white farm owners than there are white farm owners. For example, African-Americans own less than one percent of the land in the US and Hispanics own less than two percent, in part due to partitioning sales, non-participation in farm programs, and systemic discrimination by the USDA, as was noted in a 1997 class action lawsuit against the agency (Gilbert *et al.*, 2002). Importantly, control over farmland does more than just allow people of color opportunities to farm. Research suggests that when people of color own farms they are more likely to sell their food in underserved communities of color, and that people of color are more likely to shop at farmer's markets where they feel they have a connection to the farmers (Suarez-Balcazar *et al.*, 2006). Moreover, access to healthful food is a social justice issue that is tied to racial inequality. For example, the poor and racial/ethnic minorities are more likely to live in 'food deserts' where access to healthful food is limited; African American neighborhoods are eight times more likely to have liquor stores that feature high-priced, prepackaged food as compared to white and integrated neighborhoods (LaVeist and Wallace 2000).

Understanding the dire situation for obtaining healthy food in one's community (for example, only one grocery store—and 36 liquor stores—to feed 20,000 residents), The West Oakland Food Collaborative started in 2001 in Oakland, California. According to the group, the cooperative emphasizes community self-sufficiency and links black farms with the black community (Alkon and Norgaard, 2009). In their 2006 study, Suarez-Balcazar and colleagues found that African-Americans in a low-income Chicago neighborhood preferred produce from their farmers' market because the market was located in their neighborhood and, equally importantly, the farmers selling the food were African-American themselves. Yet, they worry that "working class neighborhoods are more likely to attract low-income farmers who might lack the government support necessary to maintain a more sophisticated farming enterprise" (p. 9).

Both the West Oakland Food Collaborative and the Chicago study suggest that land ownership matters, and that whiteness in farmers' markets may be tied to inequality in land ownership. Of course, understanding the distribution of sustainable food is complex. For example, Webber and Dollahite (2008, p. 187) note that, "farmers practicing sustainable agriculture are at a distinct disadvantage in reaching these [low-income] households given the subsidies and 'economies of scale' that conventional, petro-chemical-based and subsidized agribusiness enjoy." These two studies, however, suggest that when a person of color owns land and grows food on that land, it is more likely to end up on the plate of a person of color. Yet, despite the necessity of labor from people of color and women to the success of both conventional and sustainable agriculture, "women and ethnic minorities have not had equal access to land, capital, or decision making in the food and agriculture system" (Allen, 1993, p. 148).

My observations, which focused on understanding how race/ethnicity was constructed in this space, support similar findings of earlier research and supplement these assertions with ethnographic evidence. Ultimately, this research on race/ethnicity and farmers' markets suggests that sustainable agriculture is a space of whiteness. It also suggests that efforts towards a more inclusive sustainable agriculture will require a broader coalition of people to have a voice. Initiatives like the ones in Chicago and West Oakland, as well as in other cities such as Detroit's "Food Justice Task Force," indicate that this is not an impossible project, but it is one that will require mindfulness on the part of people who are invested in making sustainable agriculture a meaningful agent to combat climate change.

Conclusion

Having grown up on a cattle ranch in Montana that was lost as a direct result of the 1980s farm crisis, I've seen the ways that industrial agricultural has broken people who believed in it most. It also made it clear to me that we cannot separate the environmental from the social. Real people apply pesticides and herbicides; real people eat those foods; real people live downstream. Of course, we all live 'downstream' when it comes to climate change. Shifts in US sustainable agriculture that are more inclusive will also have direct consequences within the US. The burdens of industrial agriculture are not equally shared—the concept of environmental racism is best understood when we look at who is applying pesticides and herbicides in our fields. Perhaps the clearest example of this in Northern Idaho is the fact that just about the only thing translated into Spanish are the OSHA videos and flyers that farmers are required to show farm laborers about the dangers of farm work.

Sustainable agriculture and farmers' markets offer real promise for changing our food system, but an important part of this promise is working to make the goals of social justice and social sustainability more visible, and working to make them realities. The beauty of the sustainable food movement is that its history is being written now. But we are at a crossroads; without an emphasis on social justice, this part of sustainable agriculture will simply be written out the equation.

My findings at the Northwest Farmers' Market suggest that it is too early to proclaim farmers' markets as 'good for everyone.' Often, despite good intentions, there are class, racial/ethnic, and gender dynamics in these spaces that create inequity. The more people who buy sustainably produced food, who grow this food, who have a place at the proverbial table, the greater the opportunities for positive changes on a host of environmental issues, not the least of which includes climate change. Part of working to change these systems is an understanding that access to sustainably grown food is linked to larger structural systems of domination in the US and globally. For example, patterns of residential racial segregation in the US mean that people of color are significantly more likely to live in food deserts where they have little access to grocery stores, let alone farmers' markets (Block *et al.*, 2004; Allen, 2006; Allen and Sachs, 2007; Allen, 2008). These same structural barriers also explain why so few people of color own farms, sustainable or otherwise. These barriers to ownership are salient given the amount of farm labor that is done by Hispanic men and women in the Northwest.

Throughout my research, I was struck by the difficulty of putting 'social sustainability' into practice. The farmers and consumers existed within a system where inequality is deeply structured. The farmers need to sell their food at a premium if they are going to pay themselves anything for their labor. These practices led to a particular class of consumer being privileged in this space. These same elements lead to farmers who typically come from a particular social location, often one of privilege.

While finishing this project, I was often asked to share it with 'the community' when completed, with the hope that perhaps this project would help address some of the places where sustainable agriculture seems inequitable. I continually wonder who the appropriate audience for this plea is. Ultimately, change in the food system needs to come from a structural level. Of course, I recognize the scope of—and the political battle involved in—this kind of suggestion. If we are truly invested in changing the food system, it must change on a structural level; of course, the paradox is that structural change begins with actions of collective individuals.

Notes

1 Portions of this chapter previously appeared in: Pilgeram, Ryanne. 2012. "Social Sustainability and the White, Nuclear Family: Constructions of Gender, Race, and Class at a Northwest Farmers' Market." *Race, Class and Gender* 19: 37–60.

2 In line with Grosfoguels (2004, p. 315) assertion that, 'the traditional distinction between race and ethnicity is considered highly problematic,' I use the term 'race/ethnicity' rather making a distinction between race and ethnicity as separate categories.

References

Alkon, A.H., and Norgaard, K.M., 2009. Breaking the food chains: an investigation of food justice activism. *Sociological Inquiry*, 79(3), pp. 289–305.

Allen, P., ed., 1993. *Food for the Future: Conditions and Contradictions of Sustainability*. New York: Wiley.

Allen, P., ed., 2006. *Together at the Table: Sustainability and Sustenance in the American Agrifood System*. University Park, PA: Pennsylvania State University Press.

Allen, P., ed., 2008. Mining for justice in the food system: perceptions, practices, and possibilities. *Agriculture and Human Values*, 2, pp. 157–161.

Allen, P. and Sachs, C., 2007. Women and food chains: the gendered politics of food. *International Journal of Sociology of Food and Agriculture*, 15, pp. 1–23.

Barbercheck, M., Hinrichs, C., Karsten H., Mortensen D., Ostiguy N., Richard T., and Sachs, C. 2006. Organic farming offers societal benefits, but it needs support (editorial). *The Centre Daily Times*, State College, PA.

Block, J.P., Scribner, R.A., and DeSalvo, K.B., 2004. Fast food, race/ethnicity, and income: a geographic analysis. *American Journal of Preventive Medicine*, 27, pp. 211–217.

Bonilla-Silva, E., 2013. *Racism Without Racists: Color-blind Racism and The Persistence of Racial Inequality in America*. Lanham, MD: Rowman & Littlefield.

Brown, A., 2001. Counting farmers' markets. *American Geographical Society* 91, 655–674.

Gilbert, J., Sharp, G., and FeZin, M.S., 2002. The loss and persistence of black-owned farms and farmland: a review of the research literature and its implications. *Southern Rural Sociology*, 18, pp. 1–30.

Gillespie, G., Hilchey, D.L., Hinrichs, C.C., and Feenstra, G., 2008. Farmers' markets as keystones in rebuilding local and regional food systems. In: C.C. Hinrichs and T.A. Lyson, eds, *Remaking the North American Food SFood System: Strategies for Sustainability*. Lincoln, NE: University of Nebraska Press. pp. 65–83.

Goodland, R., 1995. The concept of environmental sustainability. *Annual Review of Ecology and Systematics*, 26, pp. 1–24.

Goodman, D., and DuPuis, E.M., 2002. Knowing food and growing food: beyond the production-consumption debate in the sociology of agriculture. *Sociologia Ruralis*, 42, pp. 5–22.

Grosfoguel, R., 2004. Race and ethnicity or racialized ethnicities? Identities within global coloniality. *Ethnicities*, 4(3), pp. 315–336.

Guthman, J., 2008. If they only knew: color blindness and universalism in California alternative food institutions. *The Professional Geographer*, 60, pp. 337–396.

Hinrichs, C. and Kremer, K.S., 2002. Social inclusion in a Midwest local food system project. *Journal of Poverty*, 6, pp. 65–90.

Horrigan, L., Lawrence, R.S., and Walker, P., 2002. How sustainable agriculture can address the environmental and human health harms of industrial agriculture. *Environmental health perspectives*, 110, pp. 445–456.

Kates R.W., Clark, W.C., Corell, R., Hall, J.M., Jaeger, C.C., Lowe, I., McCarthy, J.J., Schellnhuber, H.J., Bolin, B., Dickson, N.M., Faucheux, S., Gallopin, G.C., Grubler, A., Huntley, B., Jager, J., Jodha, N.S., Kasperson, R.E., Mabogunje, A., Matson, P., Mooney, H., Moore, B., O'Riordan, T., and Svedin, U., 2001. Environment and development – sustainability science. *Science*, 292, pp. 641–642.

LaVeist, T.A. and Wallace, J.M., 2000. Health risk and inequitable distribution of liquor stores in African American neighborhood. *Social Science & Medicine*, 51, pp. 613–617.

Onianwa, O., Mojica, M. and Wheelock, G., 2006. Consumer characteristics and views regarding farmers markets: an examination of on-site survey data of Alabama consumers. *Journal of Food Distribution Research*, 37, p. 119.

Ross, C., 2009. The Second Green Revolution. *Race, Poverty & the Environment*, pp. 78–81.

Slocum, R., 2007. Whiteness, space and alternative food practice. *Geoforum*, 38, pp. 520–533.

Suarez-Balcazar, Y., Martinez, L.I., Cox, G. and Jayraj, A., 2006. African Americans' views on access to healthy foods: what a farmer's market provides. *Journal of Extension*, 44, pp. 1–11.

US Department of Agriculture, 2009. *Farmers Market Growth: 1994–2008.*

Wackernagel, M., Schulz, N.B., Deumling, D., Linares, A.C., Jenkins, M., Kapos, V., Monfreda, C., Loh, J., Myers, N., Norgaard, R., and Randers, J., 2002. Tracking the ecological overshoot of the human economy. *Proceedings of the National Academy of Sciences of the United States of America*, 99, pp. 9266–9271.

Walton, D., Kirby, C., Henneberry S. and Agustini, H., 2002. *Creating a Successful Farmers Market.* Oklahoma, OK: Kerr Center for Sustainable Agriculture.

Webber, C.B. and Dollahite, J.S., 2008. Attitudes and behaviors of low-income food heads of households toward sustainable food systems concepts. *Journal of Hunger & Environmental Nutrition*, 3, pp. 186–205.

17 Coming home to our bodies/healing the Earth we share

Madronna Holden

Abandoning the body: the roots of our crises

I once heard a member of World War II Allied Forces relate his experience liberating a Nazi concentration camp. For him, the evil of the camp was embodied in the smell of dead and decaying human flesh that penetrated his senses, linking him with the camp's inmates. It took over a month after leaving Germany to banish the odor of human agony from his skin and clothes. Yet when he asked villagers living near the camps how they stood the stench, they replied, "We smelled nothing." In abdicating this aspect of their embodiment, these villagers abdicated their connection to others with a body. In denying their senses, that is, these villagers denied their conscience.[1]

If we do not belong to that most intimate of homes—our own flesh—we belong to nothing else and in such failed belonging, we lack both the personal presence and the connection to others that grounds moral choices. From the fundamental communality between one's own body and other bodies—from "sharing one skin," to use the indigenous Okanagan (of the US Pacific Northwest) phrase for community—flows compassion and care. As elder Janet McCloud (1994) from Puget Sound put it, our body *is* nature and thus "our first teacher"—modeling our proper relationship to other lives, human and more than human.

Ecofeminists affirm this idea in asserting that all humans are necessarily embodied and embedded in ecological systems–exposing the destructive results of disembodiment and its accompanying displacements (Plumwood, 1993). McCloud also saw the pioneer search for "something else, somewhere else" as a symptom of "deep spiritual poverty," paralleling the observations of ecofeminists that patriarchal societies rest on the manipulation of deficiencies—which see bodies as less than minds, women's bodies as less than men's, and nature-centered lives of all species as less valued than "civilized" ones (Plumwood,1993).

To counter destructive social hierarchies and their psychological impoverishment, ecofeminists envision an embodiment that places humans in a web of natural belonging constituted by a "democracy of all life" (Shiva, 2005). This natural belonging is the legacy of enduring place-based cultures throughout the globe (Holden, 2010), such as the Lower Chehalis in Washington State. When

I first met Lower Chehalis elder, Henry Cultee, he was 84 years old and living in his "fishing shack" on Grays Harbor, where his ancestors had fished for generations. They knew this place as *Samamanauwish*—which was also Cultee's inherited name, following the tradition by which his people named themselves for their places on their land. Cultee's parents "put a power on him" that was known as "rooted to this ground"—to which he owed his long life and personal wellbeing. His people told him, "Don't lie to your life: the eyes of the world are looking at you." This was the standard by which the Chehalis gauged their ethical choices, urging them to generosity, reciprocity, and care (Holden, 2009).

Pioneer displacement held them to no such standards. Those so displaced were "without a heart," according to Okanagan teacher Jeanette Armstrong (2005)—and thus blind to the destructive effects of their actions on the world from which they detached themselves. Such displacement anxiety, as Armstrong termed it, went hand-in-hand with a hunger that threatened to "eat up" the whole world, sending entire species and native peoples to the land of the dead—a story Henry Cultee told me—expressing an apt metaphor for the results of compulsive consumerism.

"We are a restless race," one descendant of a pioneer family in Washington State told me, relating her ancestor's habit of abandoning his children in an unfinished house open to the winds blowing across Grays Harbor to take off on one 'scheme' after another. In his 1860 letter to Cornwall, England from nearby Grand Mound, Washington, pioneer Samuel James observed: "The Americans … generally calculate to … do a little work on a piece of land, and then watch the first opportunity for selling … and thus the great multitude of them are always on the move" (1980, p. 39). Those who "were always on the move" treated the land and its inhabitants as a "one night stand," in the words of Wendell Berry (1999, pp. 37–40). Henry Cultee averred that non-Indians engineered some useful things, but they also had a tendency to "chew right through a mountain rather than go around".

These pioneers sidestepped the considerable moral concern of their actions with the narrative of Manifest Destiny, according to which native peoples faded away before the march of pioneer progress. Native peoples certainly faded away before pioneer *vision*. Their "invisiblizing," "backgrounding," or "systematic non-seeing," as ecofeminist Val Plumwood terms it, transformed indigenous lives from subjects with lives of their own into objects that served pioneer goals (1993, pp. 54–70).

Pioneer backgrounding of native peoples paralleled "fantastic" observational errors of women's bodies documented by James Hillman, who delineates how Western philosophers, psychologists, and scientists cast themselves as "mind" (privileged observers) to the bodies of women (as their "data"). As Hillman (1972, pp. 216–225) observes, "Seeing is believing," and in turn, "believing is seeing," and thus these observers saw the "structural inferiority of the woman's body" in which they believed. Hillman concludes that Western consciousness will never be whole nor recover its connection to the *animus mundi* (soul of the living world) until it heals this history of misogynously mistaken perceptions of female bodies.

As did the scientists who saw in women's bodies what was not there, and pioneers who failed to see in native villages what was there, Nazi doctors muted their senses and their consciences together in the process Robert Jay Lifton terms "psychic numbing." Lifton's interviews with these men revealed that, in the part of their lives in which they were situated in their bodies, these doctors upheld their moral responsibilities as fathers and husbands, as well as doctors. But in their work in concentration camps, they set mind over body, living in a psychic "elsewhere." They well understood that, had they been present to the touch, smell, sight, and sound of the pain they caused, they would not have been able to continue their actions (Lifton, 1986, pp. 442–5). Lifton further observes that the Nazi doctors were hardly alone in expressing the egregious potentials of displacement from life in a body. He gives examples of US scientists, psychologists, and doctors who also practice psychic numbing to suspend their ethics in the treatment of patients and the carrying out of questionable research experiments (pp. 442–5; 501–4).

The numbing of men's bodies is pervasive in contemporary team sports, in which those bodies become instruments with which to injure other bodies. Michael Messner exposes the tragic irony with which athletes who formerly wished to honor their bodies are trained to objectify them, to "play through the pain." Many take up drug and alcohol use to numb themselves to their bodies' signals and push them beyond their limits.

Lifton's proposal for immunizing society against the moral dangers of psychic numbing entails prioritizing the embodiment of knowledge and experience, such that empathy for others becomes a guide in professional practice. In the Eurocentric worldview, however, an obstacle to this remedy is the emphasis on transcendence of the body as well as earthly life. The masculinist strains of Judeo-Christian tradition objectified forests, streams, animals, slaves—and women—in order to license their domination.

Pit River (Northern California) elder "Wild Bill" shared his own perspective on Eurocentric objectifications in response to a linguist's attempt to elicit a word for "object" in his native language. He noted that by "object" one could only mean "dead person," and those living in a world of objects "don't believe anything is alive…they are dead themselves" (Callahan, 1979, pp. 240–1). Wild Bill thus echoes Lifton in observing that objectifying others parallels deadening oneself. In her feminist reading of the history of Western logic, Andrea Nye (1990, pp. 1–5) explores the dual objectification of self and other deriving from a thought process detached from the senses so that "natural life is only dead skin that falls away before the hard bone of logic … unrelated to conflictual relations between men and women, between men, or between men and the natural world." Thus, a logician might be "a dogmatist in religion, a fascist in politics, a sadist in love"—all supposedly irrelevant to his thinking process. Beginning with Parmenides and his abhorrence of bodies in general and women's bodies in particular, Nye traces the history of Western logic to the twentieth century Nazi sympathizer Gottlob Frege, who put himself "out of touch with all reality … the reality in the streets, the disappearances, the deaths, the concentration camps."

His abstracted logic's "final perfection" is based on the "thinghood" of both self and other (pp. 1–5; pp. 163–171).

This is what Val Plumwood terms the "logic of colonialism," and indigenous Okanagan term "talking talking inside the head": in a thinking process severed from body, earth, and community. To affect such logic, one must not see what one's eyes see, just as one must not feel what others feel. This logic can only be practiced in the rarified air of the disembodied mind, a mind that not only separates itself from the body but separates humans from one another: from "sharing a skin" in community. This separation is fully in line with the capitalist notion that one must "earn a living," that the human body is undeserving of food and shelter without paying a cost. In contrast, embodiment elicits reciprocal rather than dominating relationships with others, prompting us to ensure that our neighbors are fed if we are (Holden, 2008). In Henry Cultee's tradition, one didn't ask, "Are you hungry?" One just brought out the food, avoiding the implication that the host was in charge of food meant to be shared with all.

The worldview that severs mind from body undermines the occasion for honoring human rights to bodily wellbeing, instituting physical punishment to coerce labor and obedience. The widespread violence toward women in the contemporary world, the massacres perpetrated on indigenous peoples in global colonial history, and the physical punishment of children derive from a common root: a worldview that authorizes mortification of the body as a means of social control. Notably, Augustine originated the concept of "original sin" during the historical period in which the Christian hierarchy became intertwined with empire. For Augustine, the fact that humans are born with the stain of sin on their souls makes women's bodies especially culpable, since it is through them that original sin is passed on at birth. This doctrine emphasized the separation of soul from body and heaven from earth even as it licensed physical punishment of the "sinful" impulses of human bodies. Original sin's implication of the flawed character of earthly life rationalized human suffering and authoritarian social arrangements and urged the acceptance of both (Fox, 1983). As Plumwood observed, this coincided with the belief in a "dualized nature" which consisted of "the suffering body deprived of agency, and the mastering external rational agent" (Plumwood, 1993, p. 38).

Enacting such cultural beliefs, US missionaries and boarding school administrators used beatings, hunger, and forced labor on their native charges with the rationale that punishing their bodies was the cost of saving their souls. Mortifying the body as a means of social control in turn parallels modern science's punishing technologies for controlling the natural world. There are pointed similarities between the experimental method of modern science ushered in during the Renaissance, and the torture of the witches, whose murders by the millions went hand-in-hand with the rise of capitalism (Federici, 2004). Francis Bacon, originator of the scientific method geared to create an "Empire of the Human," used the language of the witch trials to describe the forced interrogation of nature—not surprisingly, since he was an influential lawyer and politician during the witch burning era. Bacon's argument that any procedure leading to human domination of nature is good per se has an insidious side, as well as insidious persistence. Hillman (1972) observed that nineteenth century French psychiatrists used a diagnostic method on female patients formerly used by the Inquisition on witches. In his *Gendered Atom*, Theodore Roszak (1999) explores the worldview that turns its objectifying gaze similarly on women and nature, so that the depiction of nature as atom/thing parallels the patriarchal objectification of women's bodies, and licenses the rape of them both.

The scientific method that dissects a formerly-living body to force it to reveal itself operates under the illusion that a dissected body is equivalent to a body that is whole, embedded in its natural ecosystem—and alive. This method views natural life emptied of its story in the same way that plastic-wrapped meat in a supermarket case is emptied of the story of its animal origin. That packaged meat yields no knowledge of the singular being whose living flesh it once was; it exhibits only the absence of the relationship between its consumers' animal body and the animal body that sustains it.

The notion that forced dissection and exhibition leads to knowledge is as false as Freud's assertion that he understood the consciousness of the girls whose bodies he faultily observed (Hillman, 1972). Whatever the consciousness of medieval women accused of witchcraft, their torturers would never know it. Lisel Mueller's (1996, p. 11) powerful poem, 'The Unanswered Question,' attacks exhibition's imperial impulse, asking the reader to give up affiliation with the curiosity-seekers who viewed the last survivor of her Tasmanian culture put on display in a cage in London. Instead Mueller challenges witnesses to place themselves in bodily sympathy with this exposed woman, pondering any word of the lost language that might pass through the bars between them.

The Tasmanian woman's exhibition is doubly egregious as an example of what ecofeminist Maria Mies terms the "White Man's Dilemma," in which members of colonializing cultures seek to recover their lost bodily feeling through the experience of exotic places and peoples, even as they continue to destroy them (Mies and Shiva 1993, pp. 132–63). The oppression inherent in the exhibition of other bodies is echoed in contemporary pornography. John Stoltenberg's workshops place men in poses assumed by women in pornographic magazines, and these men reveal the sense of personal violation in being put on display in these ways: even fully clothed, they cannot endure being thusly exhibited for more than a few minutes.

The exhibition of women's bodies follows a centuries-old tradition in Western art, depicting women as arranged before the appropriating gaze of the viewer. This dynamic is taken up by modern media, which empties women's bodies of their agency as surely as the gaze of colonists empties indigenous lives and landscapes of theirs (Berger, 1972). Indeed, contemporary media atomizes women's bodies as its exhibits them, portraying them as consumable in the same way that capitalism portrays nature as consumable. In *Deadly Persuasion*, Jean Kilbourne (1999) analyzes media depictions of women's bodies as segmented things—legs, buttocks, headless torsos—as in the notorious ads showing the lower half of a woman's body protruding from a garbage can, wearing an advertised pair of shoes.

Further, media objectification portrays silenced or voiceless women, girls as young as four or five years of age, as sex objects, and women's bodies as passive objects of violence. In the words of one such woman: "I saw myself piece by piece. I didn't see the connections … I didn't see me … I hated my body because I hated myself. I doubted my body because I doubted myself" (p. 259).

Kilbourne labels the media depiction of women's bodies as "killing," echoing Pit River elder Wild Bill's link between objectification and "dead people" and Carolyn Merchant's (1980) historical analysis of the "death of nature" with industrialization's objectification of the natural world. Kilbourne notes that media images of women's bodies are growing more phantom-like in their passive immobility and their airbrushed absence of pores. Stuart Ewen's historical chart of media images of the female body shows how women's bodies have been literally vanishing, which we might also surmise from the new, size 'zero' clothing. "When your body says more", one commercial cited by Ewen states, "Say less" pp. 180–183). Images of female voicelessness, ads in which women's mouths are covered up or even sewn shut, go hand-in-hand with socialization, resulting in girls' failure to "speak up for themselves or to use their voices to protect themselves" (Kilbourne 1999, p. 139).

Here women of depth and flesh are missing, as the proffered image parallels Parmenides' escape to the realm of pure concept and Frege's insanity. As real women lose agency in this process, the engine of consumption gains momentum. The fact that many women are starving themselves to death while the US overconsumes to the extent that its five per cent of the world's population consumes a quarter of its resources, exposes the connection between the oppression of nature and of women. As explored in Katherine Gilday's (1990) documentary, *The Famine Within*, and Peggy Orenstein's *Schoolgirls* (1994), women's bodies have become the symbolic battleground for the control of nature. In that arena, women's bodies and nature—both of which nourish others—have no rights to nourishment themselves.

Authentic embodiment: restorative visions

Given the cultural devaluation of nurturers, of women and nature together, feminists may be understandably wary of nurturing maternal metaphors. But the remedy is not, as some have suggested, to join men in the realm of pure rationality or to objectify the earth as "it" rather than "she" (Mellor 1997, pp. 7–13). Such tactics only strengthen the alienation of both women and men from their bodies and natural life, with the destructive consequences we have seen. Instead, we need to restore the social power of nurturers, following the example of certain indigenous societies in which gender *ground* replaces the restrictive gender *roles* that make being born in a woman's body a liability under patriarchy. Gender grounds are social locations of mutual support for each gender—for female and male and those who walk between (as 'bridge' or flexibly gendered persons in many indigenous societies). Such grounds are determined by their members and characterized by inclusion rather than prohibition and exclusion. They often have a mythic dimension, imparting meaning, authority, and responsibility to life in the body of a woman, a man, or a flexibly-gendered person. For instance, an Apache girl connecting with Changing Woman in her puberty ceremony expresses the power of creation as she enters the ground of her womanhood. Typically, societies with strong gender grounds for women honor the value of nurturance, whether expressed by human individuals or by the natural systems that sustain us.

Gender ground is also the ground onto which elders welcome the young to guide, care for, protect and encourage them. In an example from Africa, Malidoma Somé (1997) relates how a Dagara "male mother" (mother's brother), worked to "take the anger out of a young man"—in this instance, himself—caught between the Eurocentric and traditional Dagara world. Orenstein (1994) documented the contrasting situation of girls in the contemporary US in the absence of a circle of women elders to welcome them onto a ground where they might find protection, nurturance, and purpose—and honor for growing into the body of a woman. Instead, Orenstein found that adolescent boys initiated these girls into womanhood, grooming them for compliance and shaming them for their achievements. The girls' mothers, isolated from one another and constrained to their roles in the nuclear family, were ineffective in offering their daughters meaningful support. Their daughters thus came to womanhood alienated from their bodies and lacking personal purpose, in decided contrast with the Apache girl whose initiatory dance on the ground of her womanhood is considered a potent prayer for the whole of her people.

Indigenous women's gender ground characteristically links nurturance with social power, prioritizing community care for other lives and future generations. We have something to learn from the man who refused a mine owner's bribe to subvert his mother's work protecting native forests in the Chipko movement in India with these words: "Money I can get anywhere, but my mother's dignity and respect comes from the village community and we can never sacrifice that" (Mies and Shiva, 1993, p. 249). Gender ground carries the potential of the 'Idle No More' movement, founded by four mothers to fight Canadian tar sands and oil pipeline development, supporting indigenous rights and environmental justice for all Canadians in concert with advice from indigenous elders (Van Gelder, 2013).

Strong gender ground for women not only creates the motive and affiliation in working for such remedies, but releases the energy that contemporary media would instead have women spend on remaking their bodies. Debold, Wilson, and Malavé (1993) observe that young girls' sensual connections to the natural world express their vitality, passion, and joy—immunizing them from the cultural story in which adolescent women trade their full-bodied erotic relationship to the natural world for reductive sexuality controlled by a man. Embodied connections to the natural world put young women in touch with authentic personal desire, engaging their life purpose and passion.

By contrast, the disembodied view of the world not only "kills things by viewing them as dead, it imprisons us in that tight little cell of an ego" (Hillman, 1989, p. 100), which separates each of us from one another and from the consequences of our actions. In plucking a chocolate bar from a grocery store shelf, we have no experience of the child that might be enslaved on an African plantation (as many today are) to grow and harvest that chocolate. Between biting into that chocolate and the growing of its ingredients lies such long and complex lines of labor, transportation, and packaging that there is neither the recognizable earth of Africa nor the hands of the harvester for the consumer to perceive. In parallel disconnection, vehicles with self-enclosed interiors separate their passengers from awareness of the natural costs of both their manufacture and their use. Full bodied connection to the natural world counters the "anesthetizing" technologies of "plastic, styrofoam, cold metal," whose loss of feeling allows our hands to become "brutal" (Hillman and Ventura, 1992, p. 212).

Imagine, by contrast, a technology modeled on the values of the indigenous Chinook at the mouth of the Columbia River, who carved a heart and lungs on their ocean-going canoes, sailing them as living bodies upon living bodies of water. If such sensual attentiveness guided our contemporary choices, it might usher a profound shift away from the technology of control and segmentation and toward the technology of partnership and care. Here is a challenge that authentic embodiment presents: replacing Bacon's technology of control (which bears too many affiliations with the technology of torture) with technology that expands its users' sensual reach to other living bodies. Such technology would alert us to the results of our actions—for instance, increasing our awareness of the living soil and natural lives that produce our sustenance. As Courtney White's recent work (2014) indicates, such awareness may lead to food-growing techniques that restore water tables and sequester carbon as does all healthy soil, thus addressing climate change by nurturing the land even as we feed ourselves.

Authentic embodiment consists not only in listening *with* the whole body, but *to* it. Awareness of the body's messages concerning hunger has the power to heal unhealthy relationships with food in consumerist societies. Authentic embodiment exposes the misconception that consuming the pain of other beings raised on factory farms is adequate nourishment for bodies we honor. It also tells us that the lives that sustain ours, embodied as we ourselves are, deserve reverence, care, and thanksgiving—as well as the honor of a full life lived according to the requirements of their species. Such awareness prompts us to demand an end to the industrialized farming that fails these ethical standards, even as it destroys soil fertility and displaces indigenous and localized agriculture in the global arena, substantially contributing to world hunger (Wright, 2005).

In order to heal life-destroying rifts between mind and body—and thus the rifts between self and others, humans, and nature—we must recognize and work with our physical vulnerability. To counter the flight from the body pervasive in Western philosophy, we must come home to our flesh in its depth and complexity—in its capacity for suffering as well as joy. It is our fragile skin, capable of being wounded, that allows us both to touch and be touched. Indeed, our physical vulnerability not only teaches us our interdependence within the circle of natural life, but provides the ground of human culture. In the extended period of childhood dependency is the occasion for passing on human culture, and the physical vulnerability of elders facilitates sharing their stories with succeeding generations.

There is distinct resilience to a community attentive to its most vulnerable members. Our presence in our bodies tells us we are, none of us, less vulnerable than those subject to patriarchal violence, and we must thus work to banish such violence. Physical compassion for those stricken by cancer should mobilize us to prevent the pollution centrally implicated in the current cancer epidemic. Likewise, responding to the vulnerability of our young impels us to work to ban toxic chemicals currently endemic in human embryonic fluid and mother's milk. Situated in these precious bodies, we cannot downplay our dependence on one another or on an intact natural world. We all rely together on the stable climate, for instance, which has nurtured global ecosystems for the past ten thousand years. Our embodied vulnerability situates us in compassionately connected bodies to those subject to drought and flooding seas resulting from climate disruption—and with future generations of all species that depend on our actions to remedy this crisis.

Authentic embodiment places us within the natural order that interweaves all life and our own turn at life with others. As *Godfather Death*, a folktale with versions told throughout the world, illustrates: healing entails accepting our place in the cycle of life and of death (Holden, 2002). Though it takes courage and wisdom—and personal generosity—to acknowledge our mortality, that same mortality is the potential ground of our transcendence, in which we become as large as memory behind us and legacy before us, knowing we are each, in the eloquent words of Chickasaw writer Linda Hogan (1995, p. 159), "the result of the love of thousands."

To use Armstrong's Okanagan insights, our bodies impart to us "the great gift of our existence," giving us to "understand we are everything that surrounds us" (Armstrong, 2005, pp. 36–7). Fully inhabiting our bodies allows us to connect in delight and responsibility not only with embodied lives of other times but other species. Taking place "when all the animals were people," many indigenous stories facilitate their audience's understanding that other species have families, homes, and feelings of their own. Whole-bodied attention to other lives establishes distinct knowledge of those lives as well as accountability for human actions that affect them. In translating the Okanagan word for "community" as "sharing one skin," Armstrong (p. 39) also tells us that that skin embraces all natural life, such that it is her own body that "is being torn, deforested, and poisoned by 'development'."

In sharing one skin we become acutely aware of our responsibility to shape our behavior in a way that avoids harm to other lives and of the deficits of grandiose schemes that would excuse us from having to do so—such as geoengineering the global climate to combat climate change, which Hamilton (2013) has aptly assessed for its destructive potential. Since it is our attempt to dominate the natural world that has brought about our ecological crises today, we do not need more of the same. Assuming responsibility for our own actions by decreasing our greenhouse gas emissions and changing our farming methods is the workable alternative. This will take resolve, but it is clearly within our power, and alleviates rather than expands the potential for harm to other lives.

We need sustaining visions of the kind that led indigenous Californians to increase the abundance and fertility of their landscapes over hundreds of years, that led indigenous peoples along the Columbia River to nurture the yield of that river's salmon at sevenfold the current catch, and that today leads the community of Gaviotas to facilitate the restoration of thousands of acres of rainforest in Colombia (Lichatowich, 1999; Anderson, 2005; Weisman, 2008). We need visions that draw fully on our creativity even as they enact our responsibility toward other lives and future generations—visions deriving from our full presence in the precious body in which each of us inhabits our place on earth. "Someday," Warm Springs elder Lizzie Pitt from Eastern Oregon declared, "The land will be our eyes and skin again" (Stowell, 1987, p. 104). When that day comes, we will be able celebrate together the gift of life in our bodies with their limits, their mystery, their distinction, their power—and the conscious responsibility this brings us for other lives and future generations on this earth we share.

Note

1 This is an abridged version of this essay: For a more complete version with a slightly different emphasis see Holden (forthcoming).

References

Anderson, M. K., 2005. *Tending the Wild.* Berkeley, CA: University of California Press.

Armstrong, J., 2005. Community: sharing one skin. In: J. Mander and V. Tauli-Corpuz, eds, *Paradigm Wars: Indigenous Peoples' Resistance to Economic Globalization*, San Francisco: Sierra Club. pp. 35–39.

Berger, J., 1972. *Ways of Seeing*, New York: Penguin.

Berry, T., 2006. *Evening Thoughts, Reflecting on Earth as Sacred Community.* San Francisco, CA: Sierra Club.

Berry, W., 1999. Back to the land. *The Amicus Journal*, 20(4), pp. 37–40.

Callahan, B., ed., 1979. *A Jaime de Angulo Reader.* Berkeley, CA: Turtle Island Press.

Debold, E., Wilson, M. and Malavé, I., 1993. *The Mother-Daughter Revolution*, New York: Bantam.

Ewen, S., 1988. *All Consuming Images.* New York: Basic Books.

Federici, S., 2004. *Caliban and the Witch.* New York: Autonomedia.

Fox, M., 1983. *Original Blessing.* Santa Fe, NM: Bear & Company.

Gilday, K., 1990. *The Famine Within*, Kandor productions.

Hamilton, C., 2013. *Earthmasters, The Dawn of the Age of Climate Engineering.* New Haven, CT: Yale University Press.

Hillman, J., 1972. *The Myth of Analysis.* Chicago, IL: Northwestern University Press.

Hillman, J., 1989. *A Blue Fire.* New York: Harper Collins.

Hillman, J. and Ventura, M., 1992. *We've Had a Hundred Years of Psychotherapy and the World's Getting Worse.* San Francisco, CA: HarperSanFrancisco.

Hogan, L., 1995. *Dwellings.* New York: Simon and Schuster.

Holden, M., 2002. Godfather death. *Parabola*, 27(2), pp. 22–6.

Holden, M., 2009. Re-storying the world: reviving the language of life. *Australian Humanities Review*, 47, pp. 141–57.

Holden, M., 2010. Indigenous peoples. In: D. Mulvaney and P. Robbins, eds, *Green Politics, An A to Z Guide*. Thousand Oaks, CA: Sage Press. [Online] Available at: http://knowledge.sagepub.com/view/greenpolitics/n70.xml, accessed 2 September 2015.

Holden, M., [Forthcoming] Reclaiming authenticity and ethics: an ecofeminist vision of re-embodiment. In R. Pellicer, V. DeLucia and S. Sullivan, eds, *Contributions to Law, Philosophy and Ethic: Exploring Re-embodiments*. London: Glasshouse Press.

James, D., 1980. *From Grand Mound to Scatter Creek*. Olympia: State Capitol Historical Association.

Kilbourne, J., 1999. *Deadly Persuasion*. New York: The Free Press.

Lichatowich, J., 1999. *Salmon without Rivers, A History of the Pacific Salmon Crisis*. Washington, D.C.: Island Press.

Lifton, R. J., 1986. *The Nazi Doctors: Medical Killing and the Psychology of Genocide*. New York: Basic Books.

McCloud, J., 1994. On the trail. In J. White, ed, *Talking on the Water*. San Francisco, CA: Sierra Club.

Mellor, M., 1997. *Feminism and Ecology*. New York: New York University Press.

Merchant, C., 1980. *The Death of Nature*. San Francisco, CA: Harper and Row.

Messner, M., 1992. *Power at Play*. Boston, MA: Beacon.

Mies, M. and Shiva, V., 1993. *Ecofeminism*. London: Zed.

Mueller, L., 1996. *Alive Together*. Baton Rouge, LA: Louisiana State University Press.

Nye, A., 1990. *Words of Power: Feminist Reading of the History of Logic*. London: Routledge.

Orenstein, P., 1994. *SchoolGirls*. New York: Doubleday.

Plumwood, V., 1993. *Feminism and the Mastery of Nature*. London: Routledge.

Roszak, T., 1999. *The Gendered Atom: Reflections on the Sexual Psychology of Science*. Berkeley, CA: Conari.

Shiva, V., 2005. *Earth Democracy*. Cambridge, MA: South End Press.

Somé, M., 1997. *The Healing Wisdom of Africa*. New York: Jeremy Tarcher.

Stoltenberg, J., 1994. *What Makes Pornography 'Sexy'?* Minneapolis, MN: Milkweed Editions.

Stowell, C., 1987. *Faces of a Reservation*. Portland, OR: Oregon Historical Society Press.

Van Gelder, S., 2013. Why Canada's indigenous uprising is about all of us. *YES Magazine*, 65, pp. 12–5.

Weisman, A., 2008. *Gaviotas: A Village to Re-invent the World*. White River Junction, VT: Chelsea Green.

White, C., 2014. *Grass, Soil, Hope: A Journey Through Carbon Country*. White River Junction, VT: Chelsea Green.

Wright, A., 2005. *The Death of Ramón Gonzales, The Modern Agricultural Dilemma*. Austin, TX: University of Texas Press.

18 Mending the Earth

Phoebe Godfrey

Figure 18.1 Mending the Earth. Black and white photograph, 2015. This image is part of a photo series that explores the regenerative power of the Earth to mend Her ecological damage as a result of industrialization by using, thereby transforming, the 'Master's Tools'. Bark (Earth), Metal (Fire), Paper (Air) are being sown together using a Crystal (Water), thereby uniting the Four Elements back together.

Part III
Fire

Conflict Mobilization Heat Resistance Transformation

Figure III.1 Fire

19 Before I was baptized

Noah Matthews

before I was baptized
I didn't know
I was born on a ship that had started to capsize a long time ago
now I know
I will breathe in
my last gasp of oxygen
and release
to the greens
my last gift of symbiotic symmetry
now I turn to you
the ability to turn the season
into new leaves
who'll yearn for reason

I can now submerge my flesh in water willfully
and bathe in a wealth of what I know skillfully
down in the depths
the difference between a breath and a sigh is undefined
the pressure rises
bubbling to the surface
my consciousness evaporates
my friends died restlessly that life
for they relied desperately
on matterless hope
that humanity could peer through time's scope
and feel the fraying destiny of our bloodline's rope
and from destructive caterpillars
chewing leaves
spewing smoke
evolve to sprout wings like butterflies
renewing this planet's ability to cope
pursuing methods to prosper other lives
to provide solid hope for generations yet to flutter by

in the midst of the revolutionary recognition
that this world in which we live in
has natural systems on the verge of disposition
due to the actions and decisions made by men and women with nearsighted vision
I fear the masses aren't capable of making adequate revisions
I feel the weight of our fate may remain an inescapable collision

it's the underlying wealth and greed
for which the mainstream seems to scheme
despite the consensus that society is relatively senseless
to the natural issues that stress the trees, the bees, the breeze, and the seas
it seems the evident disconnect between modernity and planet is due to sheer neglect
it doesn't take an environmentalist's intuition
to sense an immense fault with this delusional vision
that since the collapse of the natural systems is no one man's fault …
it's just a compilation of moral clashes and social divisions
yet our renewable resources are as finite as the light that passes through a prism
on the cusp of the trash ridden currents
the currencies that claim to float the seas
they burn in flames of instability

it's a sensational ability that resides within you and me
to be
to leave
behind any entities that represent entropy
to believe in what could potentially be achieved
when the worker bees work with intents to leave
the hive with a queen and more baby bees
with enough honey to endure winter's intensity
to continue to work beneath Mother Earth's expertise

20 The Eagle's Eye

Tina Shirshac

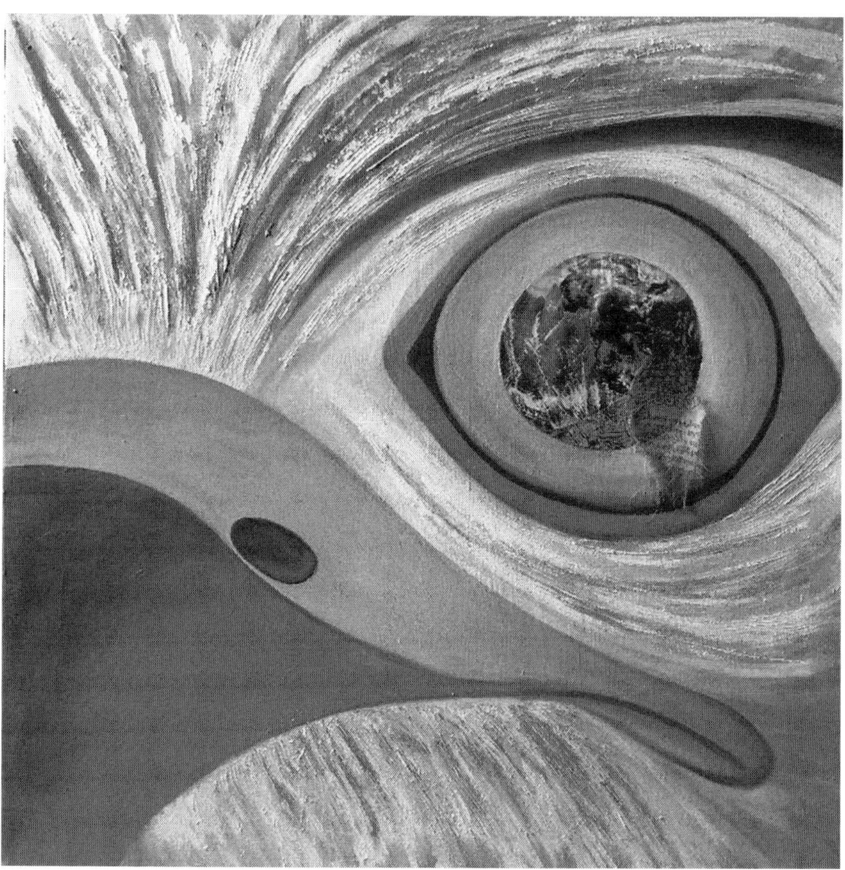

Figure 20.1 Eagle's Eye. Oil and mixed media on canvas, 32" × 32", 2012. A bird's eye view offering tearful prayers for all corners of the earth

21 Incorporating the arts is the key to building social movements

Cori Redstone

The age of magical thinking

Western culture has disconnected us from living in the present and created mass anxiety. Our collective ethical compass has been lost, and in its place we find an obsession with consumption as a means of personal fulfillment. Collective anxiety is a symptom of collective cognitive dissonance. Our choosing to remain on the sidelines and watch oppression only adds to our angst. Today's social norm regarding controversial topics is to avoid confrontation for fear of causing offense or to be forced to debate. Those who push against the status quo are shamed and sent back to their corners. The rebels then either conform by toning down their demands or they are subjected to societal alienation.

Ever-increasing economic pressures have added to cultural contempt for lengthy and messy democratic debate. Rather than address society's problems, people are distracted with job obligations that come from constant financial stress. Workers are kept so busy trying to get ahead, they rarely have time to stop and ponder the state of one's life or the situation of contemporary humankind. The arts open that space. Traditionally, the arts have provided room to wonder, to think and to contemplate something beyond ourselves and to act as a mirror for self-examination.

Few will ever achieve prosperity despite the threadbare sentiment that common people can become wealthy if they just work hard enough. The game has been rigged and today most wealth is inherited rather than earned (Picketty, 2014, p. 379–383). This "wealth" also refers to the advantages experienced by those coming from middle class families who assist young adults with basic financial needs. Those living at the social and economic margins are excluded from the shaping of popular culture and mass movements. The conformists, the children of the wealthy and the middle class reinforce their perceived norms of race, class and culture and society slowly becomes more conformist, less idealistic, less just and more likely to slide towards conservatism.

The ideology of endless economic expansion and wealth as a reward to those who deserve it is a symptom of magical thinking. Continuing to pollute without considering the consequences for our own and future generations is also a symptom of magical thinking. Because these narratives go largely unchallenged, people buy into the system with the slim hope they too will one day live their dreams. This narrative is reinforced by our media, education system, government policies, and the language of political pundits.

When artists encounter idealistic notions of what art can be and attempt to reconcile them with what art is in the present, they become politicized (Davis, 2013, p. 4). Artists shape culture. The contributions of artists concerned with ethics are now invaluable and will be essential in creating the resilient movements that take our economies beyond the exploitive methods of profit that are causing global climate change.

From fence-sitter to organizer

Following the news has been a lifelong obsession. I was caught up in the international Obama frenzy in 2007 and served on my state's steering committee for the campaign. My entry into politics came when I saw the ways, good and bad, that policy decisions affect real people every day. The Obama campaign introduced me to basic organizing strategies and political platforms. I found myself involved in state and national politics, and things went from there. I grew up as a Mormon in Utah. As I became more politically aware, I grew disenchanted with the undue influence the church played in local and state policy.

Feeling dejected and frustrated by a lack of response on the part of my elected officials, I started looking for other ways to affect change. About that time, a man I met at our Unitarian Church interrupted a federal Bureau of Land Management auction. His name was Tim DeChristopher. At the time, his nonviolent action of interrupting the auction was being considered for federal prosecution. Tim was forming a group to organize around issues of climate change. Inspired by his boldness, I took a trip to Powershift in Washington, DC in January of 2009 with Tim and the people who would also form Peaceful Uprising. Tim did end up serving almost two years in federal prison while we continued to organize. My life changed. It was my true political awakening and I found my calling as an artist and organizer during that time. I chose a path that was not only outside our dominant economic system, but living and working to destroy that system. As Chris Hedges said in his call for people who devote their lives to destroying injustice and modern capitalism, "There is nothing rational about rebellion. To rebel against insurmountable odds is an act of faith" (Hedges, 2014, n.p.).

For several years I organized with Tim and the passionate activists in Peaceful Uprising in Utah. Peaceful Uprising was and still is a climate justice group, not an environmental group. I served as the art director and as a strategist. We were known for our large, colorful banners, giant puppets, humor, pranks, celebratory gestures, street theater and singing activists. Lifelong friendships were developed and I finally learned how it felt to belong in a community. It was one of the happiest and most personally rewarding times in my life. It was also at times the most stressful as I struggled to balance earning a meager income and family obligations with what I felt were my obligations to the larger society. I held multiple part time jobs and struggled with being a parent while putting myself through college.

Creating art that connects to the heart

Painting and drawing were my passions from an early age. As soon as I had a home of my own I devoted space and time to daily art making. Sometimes that meant an easel in the corner of my cramped living room, other times I had the luxury of a small bedroom as an art studio. When it came to political organizing, I thought my art skills may come in handy to paint nice letters on signs. Early on in Peaceful Uprising, bay area organizer David Solnit devoted several days to teaching us how to create art in order to serve a movement. He showed us how to make puppets in the tradition of the Bread and Puppets Theater and how to mass-produce banners for big demonstrations. David taught us about messaging strategies and practical ways to apply art to organizing. Those few days with David completely changed my approach to politics and organizing. I will forever be grateful to him. He taught me to share my love of an issue with people through my love of art. It was natural that they merge.

Art and celebration are needed to balance activist burnout. Resistance becomes exhausting, whereas celebration, poetic expressions, and joy channel the human desire to build deeper connections. Art and activism can be synthesized through the mediums of drawing, painting, sculpture, teaching, song, dance, community building, consciousness raising, acts of nonviolent civil disobedience, performance, street theater, guerrilla art, puppetry, strategic protest, sign making, absurdist interventions, rallies, events, and storytelling. These forms of expression can be used to confront the institutionalized injustice within government, education, and industry in order to build a healthy and just world.

New forms of direct action and nonviolent civil disobedience

Aesthetics are at the core of successful social movements and therefore, artists belong at the core of social movements. This has never been more the case than since the beginning of the technological age. Our world is ruled by the digital. We have entered the age in which our economies are based on information much more than production. We are now more able to share knowledge and educate others utilizing high-tech platforms. Some of the most exciting activist work is being done by those on the margins of these new platforms. Hackers who steal and blackmail for their own gain give Hacktivists a black eye. As a result, too often the actions of hacktivists alienate the left, right, and center simultaneously. Governments condemn and hunt them. Much of the public despises and fears them. Organizer and theorist Saul Alinsky once spoke about the life of a true radical; "Life is an adventure of passion, risk, danger, laughter, beauty, love; a burning curiosity to go with the action to see what it is all about, to go search for a pattern of meaning, to burn one's bridges and because you're never going to go back anyway, and to live to the end" (Alinsky, 1969, p. viii).

We have new avenues for creative, nonviolent, and direct actions through digital activism beyond signing petitions. Hacking can be a peaceful and direct act of nonviolent civil disobedience, especially when done with artistry. In the case of a high school girl raped in Steubenville Ohio in 2012, hackers forced the hand of local prosecutors to pursue charges when the hacktivists found and made public proof the young woman was raped.

Pentagon papers leaker Daniel Ellsberg has admonished whistleblowers to continue their work but stay out of prison if the crime can be classified as one deserving of extraordinary prosecution. The United States Government has made examples of activists by sentencing them to inhumane prison terms. An alternative to risking arrest with nonviolent actions can be public artistic interventions.

Art steers social movements

In a public lecture, writer and radical Chris Hedges advocated art playing a key role in social movements. Hedges said grief, beauty, the struggle for mortality, meaning, and love can only come through art. It is not accidental the origins of religion were always fused with art. Culture is as important as the traditional infrastructures of resistance. Artists using the methodologies of more traditional and conceptual art can challenge the power of the state. Chris Hedges believes the transcendent forces of art are powerful because they remind us we are human and what life is about (Hedges, 2013).

Art has been used in political campaigns and special interest groups to shape public opinion. Think back on the ambiguous platitudes of the Obama Campaign: hope, change, etc., were all over posters, buttons, and t-shirts. The art became the branding of the campaign rather than articulated policy positions. What politicians and the state say they are doing too often has almost no relationship to reality. Strategically minded activists can counterbalance the visual and musical language of the agendas of the state with their own artistic expressions while prioritizing ethics above rhetoric. Art allows people to find their common interests, grow with a cause and fracture binary political systems.

Pranks, flash mobs, impersonations, and especially play are what reinvigorate activists and share core messages. These actions are strategically significant because creative, nonviolent actions are unexpected and more effective in changing minds by creating a sense of delight, a moment of clarity, or a rupture in societal expectations. Strange or theatrical actions containing a key political message have a better chance of capturing public imagination (Boyd, 2012).

Oppressive governments suppress not only labor but arts for strategic reasons. Time has borne this out. Social movement theorist Gene Sharpe has closely studied the alignments of arts and labor and spoken about the need for artistic interventions in social causes. Albert Camus said, "The first concern of any dictatorship is, consequently, to subjugate both labor and culture. In fact, both must be gagged or else, as tyrants are well aware, sooner or later one will speak up for the other" (Camus, 1961, p. 95). Many contemporary visual artists consider themselves laborers and closely align themselves with the platforms of labor rights organizations.

A matured climate justice movement

Enduring narratives that sustain public support are the infrastructure of a movement. In a November 2014 article on CommonDreams.org, Chris Hedges defined the ways in which organizations can be truly revolutionary in nature:

> There are environmental, economic and political grass-roots movements, largely unseen by the wider society that have severed themselves from the formal structures of power. They have formed collectives and nascent organizations dedicated to overthrowing the corporate state. They eschew the rigid hierarchical structures of past revolutionary movements—although this may change—for more amorphous collectives. Plato referred to professional revolutionists as his philosophers. John Calvin called them his saints. Machiavelli called them his Republican Conspirators. Lenin labeled them his Vanguard. All revolutionary upheavals are built by these entities … The revolutionists call on us to ignore the political charades and spectacles orchestrated by our oligarchic masters around electoral politics. They tell us to dismiss the liberals who look to a political system that is dead. They expose the press as an echo chamber for the elites".
>
> (Hedges, 2014, n.p.)

Stories succeed in penetrating our collective consciousness. The rhetoric of extremists, saying that climate change either doesn't exist, is natural, or that there will be scientific solutions we don't yet have reinforces not only ignorance, but also a deep-seated death drive. The death drive has compelled human action since our ancestors began making symbolic marks on cave walls. 'The sky is falling' is a more exciting mantra than 'our actions today will add up and cause really bad things to happen.' Sweeping biblical catastrophe has always made for better television. World religions have furthered this desire for mass catastrophe by tying a future apocalypse to religious salvation. The reality of climate change is that the evidence is cumulative, resulting in the rare, sudden tragedy. If we expect the sky to fall, we find ways to make it happen. The idea that the total annihilation of humanity is inevitable has captured our collective imagination. It is evidenced in popular shows and culture.

Contemporary society is actively finding ways to make the apocalypse a self-fulfilling prophecy. We have accepted the story that we are doomed to self-destruct. Apocalypse is much more exciting than the monotony of self-reflection, overwork and over consumption. We have accepted that humanity is not worth saving. Because catastrophe is accepted as something that must happen, the larger society refuses to intervene. To balance the angst of ever looming danger, we accept imaginary solutions to environmental problems, other social ills and the transcendence of death by financial means. Magic thrills more than human frailty, social responsibility and conservation. Ever-changing messages that portray calm and order juxtaposed to dire warnings of imminent doom cause a constant state of angst and rage. The resulting confusion leaves us feeling powerless. Sincerity is scorned in our political, intellectual and private institutions. The disconcerting state of cognitive dissonance gives us a preference for putting faith in perceived leaders rather than actively interrogating what we are told is the truth. As a result, our dominant and most engaging sociological narratives are based upon death and destruction.

Our inner turmoil is placated with what is sold as morally justified violence. Whether it's the overt violence of war or the passive violence of pollution or poverty, mass culture accepts the prime time promises of politicians and willfully ignores the dire warnings of those who study state policy. Pathological optimism stands in place of truth, which is too boring, complex or contradicts one's core beliefs. The result is a perennial oscillation between manic hope, despair, blood-lust and bewildered chaos.

This is why sincere and honest optimism plays a key role in driving social change. Wonder, surprise, nonviolent disruptions, and nonviolent confrontations best disarm systemically corrupt industries that rely on exploitation. Artistic interventions can rupture dominant narratives that obscure the truth of our past, present and future. Committing to nonviolence is a strategic and deliberate decision.

Many of the stories that ignite a movement come from what Western society assumes to be unlikely places. Activist, scientist, and educator Dr. Vandana Shiva regularly reminds us to erase the monoculture of the mind. We cannot comprehend the complexity of issues in the absence of dialogue. Ongoing discussions with those in frontline communities and even those working for industries we oppose are not only necessary strategically, but ethically called for. Environmental problems disproportionately affect the impoverished and people of color. Those affected should be steering movements. Activists protesting without listening or stepping back reinforces the social structures or 'isms' that are the blueprint to keep societies divided and prevent uprisings. The 'isms' are namely, sexism, racism, ageism, ableism, classism, etc. Our stories will compose visions of utopia and bring us glimpses of solutions as we listen.

As for my feelings of responsibility towards organizing, humorist Leo Rosten said it best: "The purpose of life is not to be happy. But to matter, to be productive, to be useful, to have it make some difference that you have lived at all" (Rosten, 1962 n.p.) Perhaps I can tie that attitude to deeply ingrained Mormon teaching and the religious emphasis on generosity, but I believe that nonetheless.

Writer Terry Tempest Williams has said that we can collectively transcend intimidation with expressions of joy and a commitment to resolve. Activists have a duty to bear witness to tragedy and instigate a dialogue of ethics. Tempest's call for joy and resolve applies to the coming battles that will be waged in the struggle to stop human caused climate change. Movements can find moments of beauty and celebration, and balance those with expressions of justified anger and sacred rage.

As Adrienne Rich said: "If you are trying to transform a brutalized society into one where people can live in dignity and hope, you begin with the empowering of the most powerless. You build from the ground up" (Rich, 1986, p. 158).

The most serious consequences of climate change are now inevitable but can still be mitigated. The motivations of climate justice work are as selfish as they are altruistic. We are protecting our own futures. Our destinies are bound together.

Life gets much more interesting when we are awake and actively engaged. I will do this work in many different forms for the rest of my life. I graduated from Cal Arts in 2015 with my MFA in fine art. It is my dream to open a center for arts and activism in Los Angeles. I am currently teaching in a public institution as well as organizing with Rising Tide Southern California and numerous other organizations to build resistance and alternative systems.

I hope to see you on the front lines. In joy and resolve.

References

Alinsky, S., 1946. *Reveille for Radicals*. Chicago, IL: University of Chicago Press.

Boyd, A., 2012, *Beautiful Trouble: A Toolbox for Revolution*. New York: OR Books.

Camus, A., 1961. *Resistance, Rebellion, and Death.* New York: Random House.

Davis, B., 2013. *9.5 Theses on Art and Class*. Chicago, IL: Haymarket Books.

Hedges, C., 2013. *The Role of Art in Rebellion* [video], Truthdig. Available at: www.truthdig.com/avbooth/item/chris_hedges_on_the_role_of_art_in_rebellion_20131127, accessed 26 July 2015.

Hedges, C., 2014. Why we need professional revolutionists [online]. *CommonDreams*. Available at: www.commondreams.org/views/2014/11/24/why-we-need-professional-revolutionists,accessed 26 July 2015.

Picketty, T., 2014. *Capital in the Twenty-first Century*. Cambridge, MA: Belknap Press.

Rich, A., 1986. *Blood, Bread, and Poetry: Selected Prose 1979-1985*. New York: W.W. Norton.

Rosten, L., 1962 Address delivered at The National Book Awards, NYC, published in the *Sunday Star*, Washington, DC.

22 Pathological and ineffective activism – what is to be done?

Marc Hudson and Arwa Aburawa

Despite clear and specific warnings, (see: The Change Agency, 2009) the 'climate movement' in the West decided to bet on the 2009 Copenhagen UNFCCC conference as a shortcut to galvanising the 'public' and expand its depth and breadth. The bet didn't pay off. We have spent the last five years in the shadow of this error. There's no immediate sign – in the United Kingdom at least – of the movement's corpse so much as twitching, anti-fracking efforts notwithstanding.

So, in a time of 'abeyance', it is important to take the time to do some reflection on what went wrong. Rather than critique this decision, or that NGO's opportunism, we want to try to make a broader critique of the sub-culture of activism. The critique hinges on a few terms that will be new to most/all readers – the smugosphere, ego-fodder, transruptive – and distinctions between affinity groups, friendship networks, and cliques, between demonstrations and protests and between mobilizing and movement-building.

Critique is easy, and too comfortable. So after that, we move on to specific things that groups—be they social movement organizations or NGO branches, or trades unions—could do to make themselves more dynamic, more welcoming to new members, and more likely to be useful nodes in a widening web of movement organisations, and the 'movement of movements.'[1]

The problem

The vast majority of social movement organizations active around climate change in the UK and Australia (at least, the ones the authors have witnessed and experienced) are part of the 'smugosphere'. What's that?

The smugosphere is not a place you'll find on a map. It's a state of mind: It is the protective fog of (unjustified self-congratulation and) complacency that surrounds any 'community' of people claiming to be making the world a better place, but which lacks commonly accepted and utilised measures of success and failure. It's the place where deeds are done not so much because they might actually have a positive effect on the world but because they will raise the status or self-esteem of the person/group doing them, or at least keep the group ticking over.

Sure, if the action succeeds in its stated goals (whatever they are), then even better, but success and failure are, in the smugosphere, bourgeois terms thrown around by people who don't understand how Important it is to Take A Stand. Man. (One good way to tell if you are in a smugosphere is to expose the 'leaders' to Jo Freeman's 'The Tyranny of Structurelessness', a classic early second-wave feminism text. If they don't like it, it's probably a smugosphere).

Because we are small in number, because we expect to lose, because we are doing this for friendship and self-validation as much as to win, we don't look for feedback. And because we are so few in number, and the people we would be criticizing are our close friends, we shy away from offering even constructive criticism.

Sometimes we do things not because they make it more likely our cause will win but because:

- we can
- it's the way we've always done things around here
- we want to impress ourselves and our social peers
- they get us attention from the media and the authorities
- we can't think of anything else to do

So we just keep on doing the same thing (marches, meetings, rallies, petitions).

And because feeling good about ourselves and 'our movement' is so important, we have no metrics for success or failure, no accountability mechanisms for dealing with people who consistently promise and then don't deliver, and other forms of destructive parasitism. And so, busy people, who would help if they thought that their efforts would contribute to a larger whole and would be recognised, leave or stay away.

This smugosphere is related to our misuse of three related concepts: 'friendship networks,' 'affinity groups,' and 'cliques.' Each is a useful tool, but who drives a nail with a saw?

Two more neologisms. The first is for public meetings and marches. 'Ego-fodder' is our term for the 'audience' at any event that has not been designed for genuine cross-clique mingling. Ego-fodder is the people sat in rows listening to the sage on the stage. The audience is sometimes collusive in this. There are members who quite enjoy the passivity, the opportunity to experience outrage without having to then do anything about that emotion. But that is, perhaps, something for a different essay.

The second is 'transruptive'. It's a portmanteau of 'transform' and 'disrupt.' Transform implies that things can change from the Bad Place we are now to the Good Place contained in the vision without pain, struggle and confusion. Disrupt is too negative—thus, transruptive.

Friendship networks, affinity groups and cliques

Cliques get a bad rep, but are useful for sustaining people over the long haul, helping them feel valued when mainstream society is sneering or indifferent, and defeat is being piled on defeat. The problems—aside from the obvious ones that cliques tend to be of people who are the same race, class, gender—start when the clique members are oblivious to how unwelcoming they are to potential new members of a social movement. Similarly, there is a problem when people will only work with activists whom they like, on principle. Affinity groups, in which people are accountable to one another, without having to necessarily like each other, or 'hang out' together are the way forward. Too often we allow our shining rhetoric of inclusivity to disguise the fact that we are not a swarm but a rabble.

Mobilization 'versus' movement-building

Activists talk about Movement-Building all the time but *do it* hardly ever. Instead, we do what is 'easier' and more visible, namely mobilizing.

One allows for:

- a fleeting sense of momentum and camaraderie
- photos to slap on website and in annual report
- it's easy to measure success and to boast of it and to use it for more of the same
- it's 'finite' (i.e. not an open-ended commitment)

The other is slower, harder, and invisible like the bulk of an iceberg. Guess which one gets done?

Mobilization often involves simplification and pacification. The repetition by those with the pre-knowledge (banner making, placards, booking coaches, selling tickets, etc.), that allows them to gain kudos while never stepping outside their comfort zone.

Movement building relies on finding out exactly what people want, can offer and what the movement needs. It doesn't always work; it's time-consuming and frustrating…

Solutions

There are other concepts, and other reasons for the movement's relative failure, not least the existential dread and horror of the topic itself—nobody wants to think about the end of civilization, after all. But space is limited, and we are focussing on 'internal' activist culture.

Get your fucking game face on

A couple of years ago, one of the authors had an email conversation with a good friend (both of us are white middle-class cis-gender men). In part it ran:

> I was saying to an academic today, that I can understand, if not forgive, the Council being so hopeless. It's the uselessness of the activist/campaigning lot that does my head in. I went to the first day one of the climate camp

gathering. Every bit as shit and deluded as I had expected of course. Couldn't bring myself to blog it … yet.

And along came this reply:

1. hmm, I think I'd differ a bit there. I share your frustration (though I should add, frustration with myself, as I am often among the massed ranks of the useless activists) but I'd say I'm more pissed off with council uselessness when it comes down to it: they are trained professionals with resources at their disposal, paid by the public to do a good job, and they have 40 hours a week to do it.
2. Without excusing the activist crapness, I always try to bear in mind that most activists are having to use stolen hours of their free time away from jobs or study (which most of their mates are spending relaxing) taking on a stressful and at times seemingly impossible task, having to learn on the spot about the impossibly complex questions like how to affect social change, with minimal or non-existent resources, and with the whole of the rest of society (including often their family and peers) calling them an idiot for doing so.
3. Also, despair, stress, the pressures of work, family and the high attrition rate caused by failure to achieve goals, mean the overwhelming proportion of activists are young (climate camp more so than most), and therefore more likely to lack knowledge, be prone to hare-brained ideas and incompetence.
4. And at a time when general apathy and disillusionment is at such an all time high, then activists don't even have the advantage of being many in number – which goes some way for compensating for a lot of the problems concerning competency, and also means you'll have a larger pool of talented people.
5. From my perspective then, without condoning or excusing it, I can understand activist incompetence more than I can the council …

Here are the replies:

1. They may be 'professionals' (in what though? In my experience most of them seem to know virtually nothing about climate science/policy/engagement etc.), but they are prey to laws of bureaucracy like the Peter Principle and Pournelle's law. And having to kiss the arse of everyone higher up the food chain than them. So they are too busy puckering up to actually get anything, you know, *done*. Urrgh.
2. Stolen hours? How many 'activists' spend endless hours watching shite television, sitting in the pub talking bollocks etc. How many have looked into 'time management?' They've never had to, because they never put themselves under proper unavoidable deadlines to get anything done. Because the costs of lunching something out are so low.

3. Despair, stress and pressures of work. Yep, activist burnout is a problem. Largely to do with choosing unachievable goals and ramping up expectations too high. But I remain to be convinced that we could retain loads of people, in a 'zone of legitimated peripheral participation' if we were *honest* about biographical availability, if we didn't run groups as cliques, if we didn't allow 'lunchoutism' to dominate. I see no acknowledgement of the identifiable and semi-preventable CAUSES of the problems you identify. I see lots of people saying 'but it's always like that, it's just how it is' when activism pathologies are mentioned. Curiously, if you say 'but capitalism's always like that, that's just how it is' they argue vigorously…
4. Our numbers are indeed way too small. But whose fault is that? Surely our own, for running such shite meetings, and shite groups and shite campaigns. Smug, irrelevant, directionless, poisonously reactive, boring and depressing. There was an excellent analysis published in July 2009 of the dangers of Copenhagen. And all the groups – liberal, state socialist and *soi-disant* radical non-hierarchical – ALL ignored it. We are merely reaping as we have sown.
5. Well, I can understand both Council and activist incompetence. But at least the Council doesn't go around pretending to have this wonderful analysis of how Wicked hierarchical power is and how Wonderful it is. You know where you are with the Council – they're shite and everyone knows it (except themselves, of course). What irks is the *smugosphere* nature of activists, always willing to let themselves off the hook because they Mean Well, or they are Under-resourced. That's true, but the human resources they let slip through their fingers all the time is staggering. It's a bit like a teenager convicted of the murder of his parents begging for mercy from the judge because he is an orphan. I have seen precisely NO tactical innovation from our city's climate 'activists' over the last four years. Just the same old shit, with the same old results. And if you ever call them out on it (using the same analytical tools they applaud you for when you are skewering the Council/Airport/local newspaper) then you become the anti-Christ!!

Basically it comes down to this. As you and I sit here reading/debating this, with electricity (by definition) readily available to us, clean water coming out of our taps, formal legal freedoms allowing us to write and assemble, there are tens of thousands of activists around the world struggling in situations far more dangerous, and without any of our advantages. And if I were one of them, listening to us whine about how difficult it is to get the Council to do anything, or about how we had no time and our friends and family were not supportive and this and that and the other, I would be very very tempted to quote the line Matt Damon's character in the movie Green Zone says to a reluctant colleague: 'Get your fucking game face on.' So, in that spirit, we offer the following ideas.

For organizers

Have some empathy with people who walk through the door of your meeting room for the first time. This below is a blog post written in 2010, which can be found at https://dwighttowers.wordpress.com/2010/08/31/adventures-in-the-liminal-zone-why-do-newbies-not-come-back/#comment-156 (last accessed September 10 2015).

'Went to a really good meeting tonight and in t'pub afterwards started quizzing people why they think "newbies", [sic.] who come to one or maybe two meetings, are then never seen again (almost all of them – a very small number stay involved, and replace old hands who drop out/go travelling, leaving groups with static or slowly declining numbers)'.

Below is a rough first list of why people don't stay with groups. There are *so many* different reasons that stop people sticking around. And the same person might have – at different stages in their 'career' of potential activism – different reasons for not getting involved. It isn't what they were expecting it would be.

Not welcomed to the meeting

- Everyone stand-offish
- Active/perceived hostility from members of the meeting (for whatever reason – they think you're a cop, or too liberal, or they're jealous of your looks, youth, intelligence, energy, naivete or whatever)

Meeting is confusing

- Jargon thrown around with no explanation
- Group doesn't have a clear objective
- Not clear how decisions get made. There seems to be a clique. Decisions are probably going to be made in the pub afterwards

Meeting is kind of scary

- Rhetoric of 'smash the state' 'all politicians war-mongers'
- Conflict between established people there
- Cynicism from the 'old hands'
- Don't think have the time to be involved at the level the group seems to expect/need (see 'legitimated peripheral participation')
- The gender dynamics are "normal" (i.e. Men/white people doing most of the talking and all of the interrupting)

Group is not credible

- Group has a clear objective but it seems unrealistic
- There seems to be a lot of stuff that was supposed to be done that hasn't been done, and no reports from people who aren't in the room about the work they were supposed to do (leeches credibility and morale away)

Nothing to connect you to group

- No potential friendships/contacts made
- Not given a job
- Given too big a job/not given support (buddy system)
- You don't have the time to get heavily involved in the group, but would have been willing to do a set piece of work. No-one suggested, so you're gonna take your energy elsewhere
- And if you come to the next meeting, having done your bit of the larger project to find that your time has been wasted because other people have not done THEIR bit, then that (IMHO) would piss you right off, especially if the problem wasn't acknowledge/dealt with
- OR the work you have done is not acknowledged, or is unfairly critiqued, or is critiqued in the wrong forum (e.g. on a shared email discussion)

The blog post drew a brilliant response from "Sarah," who says:

> Loads of useful points here. Keeping internal 'dirty linen' and tedious procedural wrangles (which often do need to be discussed) for core-group meetings rather than public ones is very important. The idea of re-framing the question as 'what would new people have been wanting/expecting which would have made them stick around' could be useful – it demands concrete proposals of better ways of doing things, rather than yet more depressing lists of how lousy meetings can be.
>
> To the many great suggestions here – especially 'buddying' for newcomers so they don't feel like they've walked in on someone else's party and have to stand around feeling isolated while everyone else catches up on gossip, I'd add: don't assume new people have nothing to offer except raw enthusiasm. There's an assumption sometimes that they must by definition know nothing, just because WE don't know them. But even if they're a fresh-faced teenager they'll probably have better social media skills than most 'adult' members of most groups, and if they're an adult they'll probably have a whole range of skills and assets – from childcare to marketing, event organising to a car that can be used to take stall kit around – so I think one of the most important things (without making them feel like they're going to have every spare job going dumped on them) is to make them feel like they'll be really valued and trusted and can really contribute.
>
> Yes, there are the tedious people who go to meetings for the sake of meetings, but most people go to meetings because they want to find out how to be useful, which means finding out what it is they can do, and finding out the best way of getting them doing it – get them on the email list or link them up with the person in charge of stalls/the website/fundraising or whatever so they can start talking about what can be done.
>
> And stop calling them 'newbies.' Being patronized is also deeply off-putting."[2]

'Is it possible to get/keep people involved in social movements without expecting them to come to soul-sucking meetings?'

We don't know, but *it had better be*, or we are stuffed. We've attended too many that were dispiriting, boring, formulaic, and unhelpful for people looking to actually DO something. We need to do two things at once:

1. Drastically improve the form and content of our meetings, thus
   a) making them more attractive to those who currently attend out of a sense of duty,
   b) making them LESS attractive to those who currently attend for the grandstanding opportunities/as a social activity rather than as a way of being useful,
   c) making them more attractive to those who want to know more about issues/perhaps see how they can help, without having to worm their way into the group's controlling clique, and
2. Create the infrastructure and expectations that people can gain public esteem as a pro-active part of a campaign/group without attending soul-sucking monthly/fortnightly/weekly meetings.

There are people who will never attend meetings or consider themselves 'activists,' but who might well be willing to offer a specific amount of their time and expertise if asked by the right person in the right manner.

The first thing is, the campaign group has to have some plausible medium-term goals, under a SMART rubric (the business world's Specific, Measurable, Achievable, Realistic Time frame). If the campaign is staggering blindly from next rally to next march to another meeting, none of the techniques suggested below will alter its final painful death (though they might, sadly, prolong it). If a campaigning group can't – or won't – 'orient', then within the 'OODA loop' (observe, orient, decide, and act) there can be no sensible decision or action. Anyone who cares about their long-term mental and political health will vote with their feet.

Before a meeting

Think about *the purpose of the meeting*. How will it help the broader growth of your group, and of a social movement on climate change in general? Is it mostly to 'give information to people'? Are there better ways of doing that (hint: almost invariably yes – newsletters, youtubes, podcasts).

Why will people be coming to the meeting? What will they want?

You must be able to answer the following questions:

- What could their ongoing participation in the group be?
- What will they get out of involvement in the group?
- What is the hoped-for change in the short to medium term (six months to two years)?

- How is success defined?
- How can they can decrease (or increase) their level of involvement with the group over these coming months, without losing touch or being 'forgotten'?

This technique is so that people who want to take on specific tasks can do so. It is NOT to force everyone who comes into the meeting space to leave with their 'marching orders.'

Phone/email out an agenda framed around 'questions' and 'tasks.' Ask people to say if they are coming or not! If not, is it because there is a one-off clash/weather/they are dog-tired, or is it because life is on top of them at the moment to such an extent that they can't commit? If the latter, do they have ANY time this month to be involved? Even an hour is helpful. Perhaps the person no longer wants to be part of the campaign/group. Do NOT try to convince them otherwise! If you bully/guilt them into staying, they will resent you and go soon anyway.

Say you understand, and then choose any or all of the following: *ask them if they would be willing to say why* (brutal honesty) – Have their time, energy and priorities have changed? Is there a personality clash with other member(s) of the group? Do they feel the group is not effective? Is there anything that would make them change their mind/reconsider at a later date?

This information should be considered confidential (for sharing only with other core people, under conditions of anonymity). If you start blurting this stuff out, you destroy your own credibility, and – even worse – you will not get honest feedback from other people who leave the group at a later date.

At the meeting

Orientation

After welcomes, housekeeping, intros, 'agenda agreed? etc., focus on the campaign in question – very quickly remind everyone of what the specific problem it hopes to address is, what change is hoped for, when the campaign started, successes so far, when it hoped to 'win' by, where things are up to right now, and what the next step (under discussion) is. In a perfect world, it will not be the facilitator or the same person doing this bit. If you can get someone else to do a recap of this, this will stop it seeming like a one-person/two-people show, and it will also spread valuable public speaking experience.

Recap briefly what has been done since the last meeting by people who are not at the meeting (this goes in the minutes too). This will give people a sense that a) there are others not at the meeting who are also pulling in the same direction (useful if the meeting numbers are low!) and b) that their efforts will be recognized if there are meetings they do not attend.

It *may* then make sense for a quick go-round of who has done what since the last meeting. It could be energizing and give a sense of momentum. BUT this can quickly create a hierarchy of who has done the most, and also rub the noses of those who've done little/nothing/are new in the fact that they are 'second class citizens'/not Heroes of the Soviet Union. A judgment call.

Alternatively, you can have a 'traffic light' system of a flip chart sheet with all the tasks that need to be done, and if they are completed/on track (green), behind but under control (yellow) or way behind and mission critical/have been lunched out (red).

ALSO consider having Gantt charts and Ishikawa diagrams to represent progress. Like these:

CHECKLIST FOR ORGANISATIONS

"Tail wagging the dog" syndrome

Have we made sure that we are not letting the prestige of our speakers and our policy wonks' need for validation/a captive audience sabotage the stated aims of this event? ☐

Have we got a plan to actively encourage and facilitate mingling, countering the tendency of people to clump together with other people whom they know if in a new setting? ☐

Have we got a plan to find out what people already know and are already doing? ☐

Have we got a plan to get people sharing their knowledge and activities without boring each other, and without us as a central gatekeeper? ☐

Easy as PIE

☐

Do we have a start to formal proceedings that is Positive, Inclusive and Energising? ☐

Have we got a plan so that we and our speakers do as much 'informing' about the issues as we can, *before* the event? ☐

Have we made sure that our workshops are really workshops and not just mini-plenaries in thin disguise? ☐

Have we got good plans for keeping people who wanted to come, but who couldn't, involved and valued? ☐

Figure 22.1 Checklist for organizations

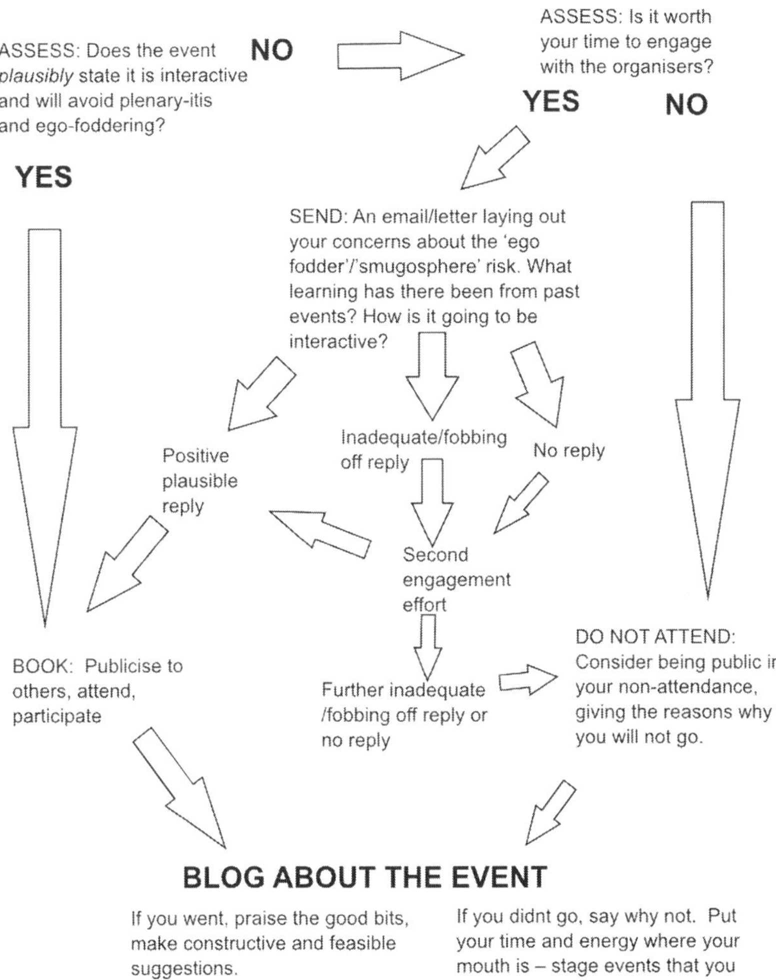

Figure 22.2 Our responsibility to our movement

Much snazzier than a list!

Allow time for questions from people either new or returning to the process. They may well come up with time and energy saving ideas, short-cuts, new perspectives. The facilitator must keep the ideas/suggestions/volunteering offers coming, without letting the attention focus onto a Shiny New Thing when there is some mundane 'who's going to get the leaflets printed and distroed' (distrubuted) decisions to be made…

Small groups focused on specific tasks will generate ideas/solutions more quickly than one big group. New people can more easily volunteer ideas etc. in small groups (though the quality of this process depends on reliable reporters-to-plenary, who capture ideas without ignoring those of new people and highlighting their own).

REMEMBER, people are going to be scared that if they stick up their hand for one thing, they'll be volunteered for others. Some people will come to multiple meetings before they make any decision to get involved.

DEBRIEF OF THE "BAD" STUFF TOO (with compassionate timing!). Look at the non-completed tasks.

- Did initially responsible people hand over tasks to others in a timely manner (in which case, they should be applauded.)?
- Were tasks then completed by a new team? At what cost to their other work/morale/energy?
- If the task wasn't done/handed over; why not? What assurance is there that this was a one-off?
- How did non-completion affect other people's work? Were their efforts rendered useless or less effective?
- If there weren't any consequences of the task not getting done, is there a lesson here – why do work that doesn't matter?

But be careful not to do this at the start or finish of the meeting. If at the start, you'll lower enthusiasm. If at the end, it's what people will remember the most! Maybe do it immediately before the break? Trouble is, sometimes the issues take time to thrash through, and are uncomfortable, and people will use the need for a break as a way of closing the discussion down.

Have a break in the meeting

People who have said they might be up for things are texted/phoned to see if they are willing to do a task. Mingling, flirting, eating a biscuit, browsing flip charts, etc.

Second half

Maybe debrief (see above)? There is then a report back after the break to say who has taken on what task, who has declined and who couldn't yet be contacted.

This will tell newbies that they don't HAVE to come to every meeting, and that they might get a phone call/text at some point, but they are free to say 'No, too busy.' 'Declined tasks' will then definitely have to be re-brainstormed: who else can do it? who will ask them? If no one comes forward, does the task need to be redesigned?

Have a standard form for jobs, on either A4 or A5 sheets (anything smaller will get lost):

- Task description: (in max of two sentences)
- Materials required
- Skills required
- Estimated time needed
- Completion date ('if cannot be met, please tell us')

Have two copies, with carbon paper in between (or take a digital photo of it), so that one copy goes with the person accepting the task and the other stays with the organizer.

Finish the meeting

Recap to everyone that:

- the group's chosen tasks are worthwhile and achievable
- what the next objective is and how tonight's decisions/input help towards that
- there are people not in the room tonight who are making valuable contributions and will continue to do so
- what we do matters. If we can't complete a promised task, it's crucial to communicate this to the organiser(s) in a timely fashion, and to get an acknowledgement that the message has been received (texts and emails can go astray)
- we are winning

After the meeting

Aha. Now the fun starts. If at all possible/appropriate, the organizer(s) should go on to the pub with people from the meeting. Let your hair down, live a little, etc. The next day the outcomes of the meeting must be communicated to the wider world. You can do this just by typed-up minutes. And you DO need to circulate minutes, that contain 'decisions taken', 'tasks taken' (does not have to include names etc., because those people AT the meeting have taken a task slip, and those who weren't at the meeting but said yes via text/phone message will be getting a separate email) and 'tasks outstanding, with description, deadline.'

BUT, since when were YOU ever inspired by minutes? Think of all the campaigns you'd help out with a few minutes of your time and expertise if they seemed like they wanted it and that your input would make a difference. Why not try: podcasting the 'decisions made' and 'jobs that need doing' of the meeting? People can then listen to the list while answering their emails etc. Why not slap together a quick YouTube video with some good music over the top of it? Here's what I mean:

Narration: "Hi, we're group x. (Logo) We're doing y to solve problem z. We had a meeting last night and divvied out loads of worthwhile jobs to ourselves to make y happen. BUT we still need people with [x amount] of spare time in the next [number] weeks to do the following tasks.

> Scroll list of outstanding tasks
>
> With music: perhaps "take this job and shove it" or "working for the weekend" or "9 to 5" or some other song – can be a different one each time …
>
> Narration: "If you can help with any of these, please get in touch via either this facebook account or this email." [and an image explaining you won't try to draw them in to other stuff against their will]. If you want to know more about us, but don't want to come to meetings, check out our website xxxx. If you want to come to our next meeting, you are very welcome. It's on xxx at xxxx.
>
> And ask people to share the video. It creates knowledge among the 'general public' of what it is your group is up to, why, what skills you are looking for. Put your 'jobs list' on the website, and then put a comedy 'taken' sign on any that do indeed get volunteered for. Maybe a cheesy way of showing how many tasks still outstanding after a week.

Inevitably, a small group of people are going to do MOST of the work, but with effort it is possible to push from 80:20 to say 65:35. And to nurture some new inner core people. We hope.

Between meetings

Once someone has said they'll do it, they ARE going to need to be chased, with a progress report/part of the work needed before the deadline, so that if the task isn't going to happen, there's time to get a Plan B. That's just how it is. Before you know it, it'll be time to think about the next meeting.

As potential attendee at another organization's event

Why not send an email/letter to the organizers of events that you are thinking of going to? See the algorithm below:

> Dear x, I've seen your event [details] advertised. Before I commit my limited time, energy and money, could you explain to me what steps you are taking to make sure it won't involve us sitting in rows listening to experts, followed

by Q and As dominated by the most knowledgeable, confident and/or vociferous? What lessons have been learnt from previous events, and what innovations are taking place to get us away from plenary-itis and workshops dominated by 'sage on the stage?' Is 'open space technology' going to be used, for example? If not, why not? Yours sincerely, Comrade y

1. Plausibility is the key. "Participation" is one of those motherhood and apple pie words that gets thrown in. What is the organization's ideology? What is its actual track record?
2. Is it even worth contacting them? Do they have 'form' in promising one thing and not delivering?
3. Two strikes and you're out. One is not enough – the movement matters. Three is more than a busy individual can afford.
4. Erm, we MUST get into habits of reflection, reflexiveness, public discussion of what has gone well and what hasn't. And if not us, who? If not now, when?

Conclusion

Even if the 'activist' culture were healthy, our species and its 'civilization' would in all probability still be accelerating towards a brick wall labelled Pending Ecological Debacle. We can't guarantee that any of the above ideas will help. But then, we don't really need to, because—as activists never tire of pointing out when denouncing the State and Capitalism—the status quo is leading towards catastrophe.

Notes

1 Most of the suggestions come from work written over the last three or four years, while we have been participants in, and observers of, social movement organisations in North West England. One of the authors also spent the best part of a year in Australia, where the activist culture seems no healthier.

2 Available at: www.dwighttowers.wordpress.com/2010/08/31/adventures-in-the-liminal-zone-why-do-newbies-not-come-back/#comment-156, accessed 10 September 2015.

Reference

The Change Agency, 2009. Anticipating and avoiding demobilisation (June/July 2009) De-mobilisation: Avoiding the post COP doldrums [online]. Available at: www.thechangeagency.org/01_cms/details.asp?ID=110, accessed 20 September 2014.

23 My life from the projects to the farm

Karen Washington

I grew up in NYC and have been growing food in the Bronx for over 28 years. People have asked me if I had any prior experience growing food or what advice could I give them? I would reply I had no prior experience growing food nor had family members who were farmers. My life has been a journey from a little girl growing up in the projects to an activist for social and food justice.

Lately I have seen the food movement make claims about working towards a just and equitable food system and slowly watch it fail along racial and socioeconomic divisions. I hear people use the term "green washed" when talking about the larger sustainability movement, I beg to differ because to me it remains "whitewashed".

My story begins in 1985. I bought a house in the Bronx that had a big backyard. I had three options;either cement it, put a lawn on it or grow food. I decided to grow food. I had no previous knowledge about how to grow food, but was brave enough to give it a try.

The first thing I wanted to grow was collard greens. I felt as part of my family's culture and tradition it was important to grow collards. Next, I wanted to grow tomatoes, peppers, eggplant, cucumbers and squash. I bought books on how to grow vegetables, and talked to my elders. For me it was important to hear their stories and get their advice. Today it is a missed opportunity not to capture the stories of our elders as our narrative around food continues to change.

So I planted seeds in the ground, added some compost. To my amazement, it worked. I grew vegetables, and a fruit. That fruit was a tomato and it changed my life (you see most folks don't know that a tomato is really a fruit and not a vegetable). Once I bit into that juicy ripe tomato I was hooked. The taste was remarkable, clean, crisp, sweet and full of the essence of summer. So hooked was I that, I wanted to grow everything in sight such as; mangoes, avocados and pineapples and quickly found out I could not and why.

In retrospect I never realized a connection between health and food. My body was trying to tell me, but I wasn't ready to listen … Not ready to connect the dots. You see living in a low-income neighborhood surrounded by fast food and junk food, your body learns to quickly adapt to the cravings of sugar and salt.

We are the prisoners of what I call the "three Food Groups in the Hood": processed food, fast food and junk food, all having one thing in common: no nutritional value.

As a Physical Therapist for 37.5 years, I saw many of my elder patients diagnosised with food and diet related diseases such as hypertension, heart disease, diabetes and obesity. Subsequently the treatment would be either medication, surgery or a machine such as dialysis. Statistics shows that the group with the most diet-related diseases are predominately low-income people of color. There is a strong a correlation between being poor and sick that falls along racial and social class boundaries.

I also noticed that in many poor neighborhoods in the late 1970s and early 1980s, there were many abandoned buildings and empty lots, again mostly in low income neighborhoods full of people of color.

Those that could move left; those unable to move were forced to endure the wrath of poverty and despair. With little to no help from the city due to the city's fiscal crisis, residents fought off the drug dealers, prostitutes and unscrupulous landlords to protect their neighborhoods. Many lived next to empty lots full of garbage. I can remember one incidence when I overheard a police officer call us a bunch of animals because we lived next to a garbage strewn lot. Can you imagine how we felt? The fact of the matter is that many times, outsiders would come in the middle of the night, bringing their trash to dump in our neighborhoods. We were all hard-working people but all this officer saw was garbage and not the plight of the people who lived there.

Many fought hard and long, to change those garbage-strewn lots into community gardens. I found myself doing the same thing on a property located right across the street from my home when the developer abandoned a project to build more homes … It took the community and the NYBG Bronx Green Up program to help turn it into a community garden. Today it is called the Garden of Happiness. Do you ever wonder why they are called community gardens? Because it takes the efforts of many and not one person alone.

As I look back now, that might have been the beginning of my activism. Those garbage-strewn lots became garden oaseis and community resource centers throughout the city. People gathered together to beautify their neighborhood and invest in its people I can truly say that the food movement we see today came from resistance. I learned that gardening is much more than just planting seeds. I learned first-hand about the social and economic issues plaguing my community, looking through the lens of environmental and structural racism.

Someone once said that "Power lies in numbers and a community that votes has power". The most important aspect of power is that it can change once those in power realize you have something they want: in this case, the power was our votes.

When you vote, you must remember that you have power of electing a politician to office. He or she now works for you. It is up to the people to make sure that political leaders are doing their job in the best interest of the community and to hold them accountable for their actions.

As a result of community activism, the New York City Community Garden Coaltion and La Familia Verde Community Garden Coalition were formed. NYC community gardeners had an agreement with the city that lasted 10 years.

The agreement expired 2010, as the New York City Parks and Recreation Dept. took over and started offering 2 year renewal garden leases. To this day, the battle over community garden preservation and development continues.

It has been a long struggle; many gardens were lost along the way, but we never gave up nor will give up, on garden protection and preservation. Looking back, we have gained some victories. Garden leases are now 4 years instead of two, we no longer have to pay for liability insurance, and we have access to free water. The elephant in the room remains access to permanent land.

In the interim, I became President of the New York City Community Garden Coalition to continue the work of garden protection and preservation. During my tenure we reopened the state office of community gardens in Albany and met with Livia Marques who was then the director of the federal government's People Garden program in D.C.

As a group it was important to have both state and federal officials know our committment to the protection and preservation of community gardens and that we were a force to be reckoned with.

Eventually our community lens started to expand as we took a closer look at more impending social issues, such as housing, labor and health. Our annual health fairs addressed the health concerns our community had on such pressing issues as AIDS, HIV, diabetus, hypertension-, infant mortality, and obesity to name a few. Education and awareness were the key to getting people who were most at risk to be screened and checked. Each year we did this in partnership with the Bronx Health Dept and the Mary Mitchell Family and Youth Center.

Still, what was constantly coming up in our garden meetings were issues surrounding food. The fact that we were growing healthy food in our gardens, yet unable to find the same food in our grocery stores was a mystery to us all. When took a closerlook at our food system we realized that many healthy food options did not exist.

When people call poor neighborhoods 'food deserts', I take issue with that. No one in our neighborhood uses that term;it is an outsider's term used for statistical purposes as well as to minimize the fact that the real issue is hunger and poverty. I rather call it food apartheid.

To answer the call of a broken food system in 2002, we decided to start a farmer's market. At first people laughed at us, told us that it would never work. We felt, as community gardeners, since we already knew how to grow food that was fresh, local and culturally appropriate, so why not start our own farmer's market?

You would think such an idea of starting a farmer's market in a low-income neighborhood would be receptive, but in fact we faced many obstacles. For one, we were told that farmers would not travel to the Bronx because it was too far, that it was unsafe, and that poor people cannot afford to buy healthy produce.

So let's dispel this myth; first off when you say it's too far, you have to travel through the Bronx and Harlem in order to get to 14th Street Union Square market located in Manhattan. That argument is a joke. Secondly when you state that it's unsafe, I have to wonder why. If you're afraid for your safety around poor people of color, just because they are poor and not white, then you have other problems. It's a fact that most of the neighborhoods of color where people once were afraid to live are being gentrified, so the argument holds little merit anyway. Lastly, when you state poor people can't afford farm fresh produce, poor people as a whole spend more money on food per capita than any other group.

Such negativity was just the impetus that we needed propel us forward in this endeavor. You see, for so long people have been telling us what we can't do and never asking us what we can do. This stems from a superiority complex that people feel they have over poor folks. Frankly, it is 'power over' instead of 'power within.' So we looked at the power within and asked our community residents what they wanted; overwhelming, they wanted a farmer's market. With the help of Just Food, the Mary Mitchell Family and Youth Center, Community board 6, the Bronx Health Department, our city councilman Joel Rivera, and our community, we launched the La Familia Verde Farmer's Market.

The La Familia Verde Farmer's Market is going into its thirteenth year. We have four rural farms and five community gardens working together to bring fresh, local, and affordable produce to people who need it the most. We take pride in saying, "no one is turned away because of hunger".

This journey has taken me to challenge the status quo and speak up against the injustices I have seen first-hand in my community and communities that look like mine.

In 2008, I had the opportunity to be an apprentice at the Center for Agroecology and Sustainable Food Systems – CASFS in Santa Cruz California. There, I learned about growing food organically, but came away with a growing need to challenge a food system that was unjust and racist. I did this by first looking at the landscape of farmers and saw no one who looked looked like me. I saw that all of the farmers were white and male, and the farmworkers were people of color. Even though many of the farmers no longer worked in the fields, they were still given the title of farmer. Those that were actually farming were given the status of 'farm workers.' Such a distinction made me take a closer look at the food system. You see, everyone had been talking about a sustainable food system that was supposed to be fair and just; I found neither. The food system was built along racial and economic lines. If you look at the history of the food system in this country, you'll see it was built on the backs of the poor and disenfranchised. Slavery, sharecropping, and immigrant labor have built the food industry in this country. Today, most of the people who are working in the fields, receiving low paying jobs, eating poor quality food, and having an array of health issues are the poor and people of color. They are at the bottom of the food chain, whereas those with power and wealth, who have access to healthy fresh food and high-paying jobs are overwhelmingly white.

When I came back home after a 6 month stint in Santa Cruz, there were many questions that needed to be answered. For instance, looking at the agricultural census in New York State, did you know that in there are 55,000 white farmers, and only 115 Black farmers? If you take a closer look at the census across the United States, there are some states that have no black farmers at all. How can we talk about a fair and just food system if the control of our food is predominantly white.

Compelled to do something and uncover this deep dark secret of institutional and structural racism, I met with my friend Lorrie Clevenger to talk about doing a conference around black farmers and black agriculture. We would amass black farmers and leaders throughout the country to discuss issues that were affecting us. The title would be *What's for Dinner?* In 2010, we had the first Black Farmers and Urban Gardeners conference in NYC.

Again, the negativity and obstacles were there, but this was good because we knew we struck a nerve, a need to see black voices and leadership in action.

We triumphed! It was standing-room only; what was so inspiring about the conference was that the majority of the people there were our youth. Since then, the conference has been an annual occurrence. In 2014 it was held in Detroit, and in 2015 it was held in Oakland. The satisfaction that I get out of the conference is when people – especially the youth – come up to thank me for providing a platform to be seen and heard. Goosebumps!

By becoming an activist and advocate for food and social justice, I have travelled around the country, speaking at colleges, universities, conferences, and local neighborhoods. My message is loud and clear, people who have been marginalized must be part of the solution. The lens through which I look at my community has changed; I no longer look with despair but with hope, resiliency, and admiration. There are assets and leaders in our communities that go unnoticed and untapped. Given the chance and resources, we can fix our own problems.

We don't need outsiders or problem solvers to save our communities. Many times they do more harm than good because they never stay nor seek out the true leaders in the community. Problems are not solved, just delayed.

My advice for all you outsiders: if you want to come into my community, ask first, just don't assume you know our problems and have all the answers. Then, if you really feel you can help or make a difference, share your knowledge and resources. But most importantly, seek out the leadership that is already there. Believe it or not, even in poor neighborhoods there are people with great minds and leadership capabilities. Lastly, please work alongside the community and leave the savior mentality at home.

In framing and reclaiming our narrative around food, another thing that has been left out in our conversation around a sustainable food system is food sovereignty. (The ability for one to grow their own food, on their own land, that is healthy and culturally appropriate). I was told once there is no agriculture without culture.

It is important for African Americans that our history an agrarian people is preserved and that farming or agriculture are not synonymous with slavery, which we have been brainwashed to believe. Farming and agriculture are in our DNA. Even George Washington Carver, one of the greatest minds and scientists of agriculture, is barely mentioned among the champions of the food movement.

Finally, I have been blessed to embark on a new chapter in my life. In 2014, I retired from Physical therapy, leaving a 6 figure salary to become a full-time farmer.

That's right: with my friends, we started a farm in Chester, New York called Rise&Root farm. We are all women, some are people of color and LGBTQ. We want to create a farm that not only grows food, but a place where people feel welcome. We have been so supported by our family and friends as we make this transition from urban to rural farming. It is exciting, hard work but the love of the land is what keeps us going.

Through it all, I have been blessed to have loving family and friends that keep me grounded. I have remained humbled in all that I do, knowing that I am doing God's work and paying homage to my ancestors. I am not finished yet; I will continue to be outspoken against an unjust food system, and I hope to be an inspiration to the next generation of farmers and food activists.

24 Triumphant

Melica Bloom

Figure 24.1 Triumphant

Source: Digital media, 2016

25 Family farmers can feed the world and cool the planet!

The food sovereignty struggle in the climate justice movement

John E. Peck

Having grown up in the Midwest amid family farmers, it is hard to escape the topic of weather. Those who live close to the land realize everyday just how much their livelihood depends upon nature. Whether or not there is enough moisture in the topsoil for seeds to germinate, whether an early warm up will stall the maple syrup season or a late spring cold snap will frustrate fruit pollination, whether a tornado takes out a barn or a hail storm flattens a crop, whether a flash flood or heat wave snatches away prize vegetables or animals, whether rising average temperature means a longer growing season or worse pest outbreaks – these are serious factors that all farmers – and small family farmers in particular – can't ignore. And with global climate change (GCC) comes impacts around the world that threaten people's survival even further.

Family Farm Defenders (FFD), the grassroots organization I have worked with for close to two decades, has long seen GCC as the worst threat to our global food/farm system. Originally founded by dairy farmers opposed to mandatory commodity check-off programs and federal approval of the first ever patented biotech food product, Monsanto's recombinant bovine growth hormone (rBGH), FFD has since grown to include anyone who cares about restoring health, justice, safety, integrity, and democracy to agri-culture. Our work is increasingly inspired by the seven principles of food sovereignty first elaborated back in 1996 by La Via Campesina (LVC), the largest umbrella organization for family farmers, farm workers, fishers, foresters, herders, hunters, gatherers, and indigenous peoples in the world (Wittman, Desmarais and Nettie, 2010). FFD is also a founding member of the United States (US) Food Sovereignty Alliance (USFSA), which is working hard to popularize and implement these principles closer to home so as to bridge the north – south divide that often exists between global food/farm activists.

Natural disaster or state terrorism?

The signs of a sick planet and the threats to our food supply are now almost an incessant drumbeat in headlines. Yet, even my own awareness about how closely intertwined the survival of family farming is with the struggle for climate justice took decades to develop. At first I found myself stuck in a reactive mode, responding to one natural disaster after another, dealing with calls from bankrupt suicidal farmers pushed to the brink by another 'Act of God.' Eventually, though, it became apparent that these disasters were not so 'natural' and certainly not 'divine' in origin. Rather, they were often predictable and avoidable consequences of misguided public policies, distorted economic incentives, and – to be honest – institutionalized discrimination and neocolonial exploitation.

When it comes to GCC many people forget that agriculture is itself responsible for an estimated 15–20 percent percent of overall greenhouse gas (GHG) emissions – not just in primary production but also food transport, processing, packaging, refrigeration, and retail distribution (Grain, 2012). A farm rich state like Wisconsin still has the lion's share of its food trucked in from far flung parts of the US or even overseas. On average, food in the US now travels 1500 miles from field to plate, and takes 7–10 times more energy to produce than it contains (Heller *et al.*, 2000, pp. 40, 42). And increasing industrialization and consolidation has created new pollution on a scale one will never see on an Amish farm. For example, methane emissions in the US increased roughly 34-49 percent between 1990 and 2006 as more livestock were confined in factory farms and their manure placed in anaerobic lagoons (Environmental Protection Agency [EPA], 2006).

In the US 'who' raises 'what' 'where' – and thus has access to taxpayer subsidies, school lunches, and disaster bailouts – is largely a function of pork barrel politics under the Farm Bill. It is not just 'free trade' that destroyed Native American agriculture and created modern forms of malnutrition on reservations (LaDuke, 1999). It is not just 'productive efficiency' that drove 98 percent of all black farmers in the US off their land over the last century (Cowan and Feder, 2013).

This corporate driven agenda is replicated and even amplified in US foreign policy. It is not just 'humanitarian assistance' that allowed US grain giants to hijack relief efforts after a tsunami or earthquake in order to dump commodities, penetrate markets, and bankrupt farmers in Indonesia or Haiti (Kripke, 2005; Bell, 2010) It is not just 'national security' that spawns death squads to murder peasants who stand in the way of U.S. Agency for International Development (USAID) funded palm oil biodiesel plantations in Colombia (Bacon, 2007).

Such decisions reflect deep-seated systems of oppression going back centuries. As Eric Holt-Gimenez notes, reflecting on the legacy of Ferguson, "all the organic carrots and farmers' markets in the world are not going to end hunger unless we also end racism" (Holt-Gimenez, 2014). We will still end up eating bitter fruit unless we tackle the roots of historic injustice and restore dignity and fairness across our entire food/farm system (Holt-Gimenez, 2014).

Hurricane Katrina – a wake up call for climate justice

In February 2007 I was one of over 500 delegates from about 80 countries invited to the village of Selengue in Mali for the Nyeleni Food Sovereignty Forum. As part of a miniscule US delegation, it was humbling to be amongst so many women of color who do most of the heavy lifting when it comes to worldwide agriculture. Contrary to popular belief, the US is not feeding the world, and the typical farmer is not an old white guy on a John Deere tractor in the Dakotas.

One of the most difficult Nyeleni workshops I attended examined how food sovereignty is often the first victims of catastrophe, war, or occupation. A Sami herder lamented losing his reindeer due to radioactive contamination after Chernobyl. A Palestinian farmer told of how ancient olive groves are bulldozed under Israeli occupation. A Sri Lanka fisherman related his village's eviction by a World Bank post-tsunami tourism scheme.

Being the only U.S. citizen on hand, I was called out to explain how Katrina became such a racist nightmare for the earth's superpower (Squires and Hartman, 2006). FFD was among the first grassroots groups to respond to Katrina's landfall on August 29th, 2005. Watching the horror unfold on television, I thought if there was ever a time for mutual aid this was it. By Sept. 15th FFD had dispatched a veggie oil powered bus with eight Wisconsin volunteers and 15,000 pounds of food, medicine, and other emergency supplies. Some of the communities in Alabama, Mississippi, and Louisiana that the relief bus first reached had seen no other help since the disaster two weeks earlier. Ironically enough, when the FFD bus arrived in New Orleans and set up a Food Not Bombs kitchen, Blackwater mercenaries hired by the White House to maintain 'law and order' were among the first to enjoy the free food. Meanwhile, desperate residents were being attacked by vigilantes and arrested as 'looters' simply for salvaging what they could to survive.

In April 2007 FFD delivered nine tractors and other equipment donated by Midwest farmers to the Mississippi Association of Cooperatives as part of our annual meeting held amidst the wreckage of Katrina. Most of the recovery efforts we saw underway had been organized by local people themselves. Six months after the disaster New Orleans' Ninth Ward was still a devastated landscape – fishing boats perched on rooftops, children's toys rotting in former living rooms, abandoned dogs roaming the streets, corpses still being found in now condemned homes, drive-by speculators sizing up what was left for "clean-up" gentrification. Toxic trailers provided by the Federal Emergency Management Agency (FEMA) had even been 'conveniently' parked atop once thriving community gardens. Those victimized by systemic discrimination and climate injustice had been reduced to the status of throw-away second class citizens in their own country. In October 2010, when FFD went back to New Orleans to receive the Food Sovereignty Prize from the Community Food Security Coalition, many local people cam e up afterwards to thank us for our solidarity – not charity – rapid response. Given state neglect, the do it yourself (DIY) power of food sovereignty came through.

Climate chaos strikes America's heartland

Soon enough, the Midwest faced its own GCC induced disasters as torrential rains ravaged the region in 2007 and 2008. A sad joke was that corn planted at one end of the Mississippi River found itself at the other. Many small farms were especially hard hit, losing entire fields of transplants; greenhouses, tractors, and packing sheds were swept away in mere minutes. One Wisconsin sheep farmer had thirty animals drowned in their paddock before they could be pulled from the rising water. Sand bags lined the main street of Reedsburg to hold back the swollen Baraboo River. Near Spring Green, cropland that had been compacted by too much heavy machinery turned into a stagnant 'lake' for months on end. Some organic farmers even lost their contracts as faraway corporate executives claimed their cropland was now 'contaminated' by floodwater and would have to be 'recertified.' As with Katrina, solidarity sprang into action, and FFD quickly raised close to $25,000 in donations from across the country to assist 40 family farmers and farm workers.

GCC cuts both ways: In 2011 and 2012 the Central Plains were suffocating from the worst droughts in a century. At the Farm Aid Concert held in August 2011 in Kansas City we heard firsthand from ranchers about desperation sales of starving livestock. Predictions of depressed crop harvests were already wreaking havoc in global markets and lining the pockets of commodity speculators. With support from Farm Aid and the Teamsters, FFD organized an emergency hay lift of 17 semi-loads from Wisconsin to Oklahoma and Texas. When the drought crept north the next year, FFD raised another $6,000 to assist nine Wisconsin farm families in their recovery. Scientists soon issued a disturbing report outlining potential adverse impacts of GCC on Wisconsin and urged prompt adaptive strategies (Wisconsin Initiative on Climate Change Impacts [WICCI], 2011) Still there were powerful voices in the mass media and elected office, including Wisconsin's own Republican governor, Scott Walker, who self-servingly denied that climate change even existed, while others waited in the wings to feed off the human misery that came with such ignorant paralysis.

Family farmers – unwitting pawns in the game of disaster capitalism

Family farmers are not just victims of climate injustice, they are also being used – and abused – to justify 'false solutions' to the global crisis. In her earlier 2007 book, *The Shock Doctrine*, Naomi Klein describes the rise of disaster capitalism – predatory behavior by multinational corporations eager to take advantage of any crisis for their own gain. Appropriately enough, I was in the midst of reading this book on my way to the Fifth International LVC Conference being held in Maputo, Mozambique in October 2008. Like the Nyeleni Forum a year before, this event drew hundreds of delegates from all across the globe, including activists that would not even be allowed into the US for various post-9/11 geopolitical reasons. Yet, even at this convergence, the snake oil peddlers were present.

In his welcome address to the conference, Mozambican President, Armando Emilio Guebuza, lauded the crop jatropha as a wonderful opportunity for African farmers. Jatropha is but the latest 'green fix' for GCC being promoted by the global agrofuel industrial complex. Such crops, increasingly genetically engineered and grown in vast monocultures destined for overseas markets, hardly deserve to be called 'biofuels' since they have no life affirming qualities about them and undermine all the basic principles of food sovereignty (Peck, 2009a). In a subsequent report, LVC warned:

> Leaving aside the insanity of producing food to feed cars while so many people are starving, industrial agrofuel production will actually increase global warming instead of reducing it. Agrofuel production will revive colonial plantation systems, bring back slave work and seriously increase the use of agrochemicals, as well as contribute to deforestation and biodiversity destruction.
>
> (LVC, 2009, n.p.)

To put it mildly, Guebuza's remarks were not well received by the crowd, and people wondered: who had bent his ear? The forces of corporate agribusiness never sleep, and chief among these is the Nairobi-based Alliance for a Green Revolution in Africa (AGRA). Launched in 2006 with a $150 million grant from the Rockefeller and Gates Foundations, AGRA's mission is to create a beachhead for biotech crops and their kindred agrochemicals on the continent. Left conveniently unsaid is why Africa needs or wants another 'Green Revolution' given that the last one failed so miserably (Shiva, 1989). The heavy handed imposition of expensive high input industrial agriculture for export is now known to have simply displaced more sustainable subsistence oriented agroecological farming, increasing hunger and poverty in the process (Holt-Gimenez, Altieri, and Rosset, 2006). As one African critic of AGRA, Mukoma Wa Ngugi, noted "Once the mask of philanthropy is removed, we find profit hungry corporations" (Mittal and Moore, 2009, p. 30). For many food sovereignty activists this was again old wine served up in a new bottle.

While some powerful figures – such as US President George H.W. Bush at the 1992 Rio Earth Summit – have argued that the US life style is not negotiable (Deen, 2012), the ongoing food versus fuel debate has forced many to think otherwise. Staple food prices skyrocketed 88 percent between March 2007 and March 2008 (Food and Agricultural Organization [FAO], 2010), triggering deadly riots in dozens of countries. Behind the food shortage was a perfect storm – runaway speculation in commodity markets (Kaufman, 2010), crop failures induced by climate change, and – as even the World Bank had to admit (Mitchell, 2008) – the agrofuel boom. Some US farmers had even plowed up highly erodible marginal land, forsaking soil conservation and wildlife habitat, in order to get on the ethanol bandwagon. Since 1990 it is estimated that acreage worldwide devoted to soy, sugar, oil palm, corn, and canola – much of it for agrofuels – increased by 38 percent, while the land area devoted to human staples like wheat and rice declined (GRAIN, 2012).

US farmers who invested their life savings to pioneer ethanol co-ops back in the early 1990s soon lost their shirts, muscled out by agribusiness. Back then half of US ethanol output was farmer controlled, but today over 90 percent is in corporate hands (Smith, 2010). The same energy giants with rather sordid track records from their fossil fuel extraction activities, such as British Petroleum, Chevron, and Royal Dutch Shell, now dominate the agrofuel sector, along with other familiar grain, timber, biotech, and finance giants: Cargill, Weyerhauser, Syngenta, and JP Morgan Chase. This corporate takeover would not have been possible without legislated agrofuel requirements and massive taxpayer handouts. Even the distillers' waste, a byproduct of ethanol once touted as a 'valuable' livestock feed supplement, is now being found to be unhealthy for animals that are designed by nature to eat grass – not grain – especially when it is chock-full of illegal antibiotic residues (USDA, 2010). By the time the ethanol subsidy program was mercifully euthanized in 2012, it had cost taxpayers over $20 billion and siphoned off 40 percent of the US corn crop into gas tanks (Pear, 2012).

Where will all this agrofuel come from if not from subsidized US suppliers? Well, Brazil already has six million hectares devoted to agrofuels and is poised to increase sugarcane acreage fivefold to meet global demand. Indonesia wants to establish the largest oil palm plantation in the world – 1.8 million hectares in Borneo. Dubbed 'deforestation diesel,' this palm oil bonanza has already destroyed vast tracts of pristine rainforest, jeopardizing biodiversity and indigenous peoples alike. And, then there is the aforementioned jatropha. India has earmarked 14 million hectares of 'wasteland' for jatropha plantations (once the residents are evicted), while a German consortium is negotiating to purchase 13,000 hectares in Ethiopia, including portions of an elephant sanctuary, for the same purpose (Rice, 2009).

Corporate agribusiness helps scuttle climate justice in Copenhagen

Climate justice advocates were out in force in Copenhagen for the 2009 United Nations (UN) Climate Change Conference (commonly referred to as COP 15). Over 100,000 people participated in a march on the eve of the official meeting – the largest protest in Danish history. I was part of the LVC contingent, highly visible with our green "Food Sovereignty Now!" flags. Sadly, the phony accord which President Obama hatched through secret side negotiations at COP15 proved to be a disastrous step backward from the Kyoto Protocol which the US under President George W. Bush had refused to ratify anyway. "We have nowhere to run," warned Apisai Ielemia, prime minister of Tuvalu, one of the Pacific island nations doomed to disappear with rising sea levels, but his plea fell on deaf ears. The fact that reducing energy demand remains the best cure for GCC was lost in the official rhetoric of Copenhagen – and that is because such a simple solution does not make money for the capitalist elite. In fact, the largest accredited non-governmental organization (NGO) at COP15 was the International Emissions Trading Association, a thinly veiled front group representing 170 companies that hosted 66 of their own side events (Peck, 2009b).

One Ponzi scheme that continues to dupe many is known as REDD (Reducing Emissions from Deforestation and forest Degradation). Under REDD wealthy polluters will be able to buy 'clean air' credits from less developed countries and indigenous communities with intact forests, mostly in the global south – and, in exchange, these areas are supposed to be preserved to offset emissions elsewhere. The trouble with this commodity variant of green capitalism is that it once again elevates profit over people. As Gustavo Castro Soto, an activist with the group Otros Mundos in Chiapas, Mexico notes: "When a natural function like forest respiration becomes a product with a price, it's easy to see who is going to end up with control of the forests" (Field and Bell, 2013, p. 112). REDD has even been described as a 'you pay or I cut' modern day protection racket, whereby pristine forests and indigenous peoples will end up being held hostage by foreign investors demanding carbon credits (Guttal and Monsalve, 2011).

As mentioned, US style industrial agriculture is among the least energy efficient in the world. For instance, it takes thirty-three times more energy to grow corn in the US as compared to Mexico and eighty times more energy to grow rice in the US compared to the Philippines (FAO, 2000). In the era of peak oil, yield per acre is no longer the proper goal. The myth of 'productivity' was still being peddled in Copenhagen, though, as USDA Secretary Tom Vilsack reaffirmed President Obama's support for biotech crops and agrofuels as a 'green fix' for the climate crisis. To make matters worse, he also announced a new Memorandum of Understanding (MOU) between the USDA and Dairy Management Inc. to push methane digesters. As the USDA's own COP15 press release noted, just two percent of all US dairy farmers would be "candidates for a profitable digester" (USDA, 2009, n.p.). Small-scale grass-based livestock operations, which don't have a manure 'problem' worth digesting and account for a third of all operations in a dairy rich state like Wisconsin, stand to gain nothing for their responsible environmental stewardship as filthy industrial operations get more taxpayer handouts.

Of course, agribusiness giants who are partly responsible for the pollution problem are ready to offer their own market solutions to stricken farmers for a steep price. Monsanto is now moving to privatize drought resistant genes for their next generation of patented 'climate ready' crops. In 2007 Monsanto and BASF unveiled a $1.5 billion collaboration to develop crops more tolerant to adverse environmental conditions and promptly filed patents for over two dozen 'climate ready' genes (Smith, 2010). The Union of Concerned Scientists, though, warned farmers to not believe the hype about "more crop per drop" since Monsanto's DroughtGard corn underperformed many other non-biotech drought resistant corn varieties (Gurian-Sherman, 2012). Indian activist, Vandana Shiva's critique went even further,

> Say there are 1,500 climate resistant genes and we go to the gene bank to map drought resistant genes and make a bet on 100 varieties that have the highest potential. We still don't really know what's contributing to drought resistance. It is not a reliable way of finding drought resistant varieties … Diversity has to be our partner in adaptation and resilience".
>
> (Tran, 2013, n.p.)

Land grabbing – the final frontier for climate change green washing

The latest insidious twist to the GCC crisis is the speculative trend of turning land itself into a market derivative through land grabbing. It is estimated that 50 million hectares have been sold off in the last decade in the Global South – an area equal to the size of one football field each second (*New Internationalist*, 2013, n.p.). Those familiar with European history will recall the Enclosure Movement which transferred the commons to feudal landlords and basically paved the road for the Industrial Revolution by forcing small farmers off their homesteads and into factories as wage labor. Others who have heard about the Diggers, Black Hawk or the Zapatistas will also know there has been a proud history of grassroots resistance to such land expropriation and resource privatization. Access to land is one of the founding principles of food sovereignty and – as Thomas Jefferson argued – insures the economic autonomy that underlies modern democracy. The flipside is that a state will offer food security to its loyal citizens, while it denies enemies their land and uses hunger as a weapon. The experience of my own ancestors, who fled as undocumented immigrants from Ireland into indentured servitude in the US thanks to a famine triggered by genetic monoculture and colonial subjugation a century and a half ago, has been replicated countless times since.

Even in the US – a supposed bastion of private property and personal freedom – less than half of farmers now own the land they cultivate, and this concentration of ownership is only aggravated as land becomes another 'hot' commodity for speculation (Tett, 2011). For example, UBS AgriVest, a unit of the Swiss banking giant, recently paid $67.5 million – or nearly $7,000 per acre – for 9,800 acres in southwest Wisconsin, well above the going market price that existing farmers can afford (Ivey, 2013). Current Wisconsin law limits foreign ownership to 640 acres, with some exceptions – such as when a foreign hedge fund has a pliant US subsidiary – but Governor Walker wants to eliminate this cap altogether to grease the skids for footloose capital. It is assumed that the mutual fund pension giant TIAA-CREF is behind many of these modern day land rush schemes. In 2011 TIAA-CREF launched its own $2 billion Global Agricultural unit and recently received another $1.4 billion to expand its landholdings, making it one of the largest private landowners in the U.S. Real Estate Investment Trusts (REITs) which used to focus on apartment complexes, shopping malls, and commercial buildings are now also expanding to allow even more absentee investors to make money off renting farmland. In many cases the goal behind these holdings may not be to grow any food, but to extract other benefits – such as carbon credits or export markets. In Cameroun US based Heracles Capital recently claimed a 99 year lease on 73,000 hectares of rainforest for conversion to palm oil for biodiesel production, while Chinese, Saudi, and Indian investors have gobbled up thousands of hectares in Ethiopia to grow cash crops, forcibly evicting the current residents (*New Internationalist*, 2013, n.p.).

When it comes to climate justice, black lives really do matter

Ten years after Katrina many in the Black Lives Matter movement are now drawing the parallels linking systemic racism and environmental injustice (Zaitchik, 2015). Those most victimized by unmitigated GCC cut across other lines of identity, as well as class, gender, and even species. While wealthy elites ('the 1 percent') can often 'escape' the consequences of the climate chaos they create – at least temporarily – the rest of the 99 percent cannot. Naomi Klein's 2014 book, *This Changes Everything*, exposes just how lucrative GCC denial can be and goes on to argue that it will take direct action – not polite lobbying – to stop this greed driven capitalist juggernaut in its tracks. A November 24, 2014 blogpost on the Movement Generation website also reveals the potential for solidarity: "Social inequities are a key form of ecological erosion. Ferguson & St. Louis … Palestine … Detroit … Chiapas … The Gulf Coast. These communities are all at the frontlines of the ecological crisis. And they are also at the forefront of change" (Movement Generation, 2014).

That is why it is so critical at this historic tipping point to bring together the food sovereignty and climate justice movements. When the Cowboy and Indian Alliance (CIA) rode their horses into Washington DC on Earth Day 2014 to express their opposition to the Keystone Pipeline their action was not just a media cliché, but a grassroots harbinger of what is already underway. The Enbridge 61 Pipeline, which will have double the capacity of the Keystone if allowed to proceed, has been repeatedly blocked by native activists at Red Lake, MN (Ball, 2013). Idle No More is forging solidarity between indigenous peoples and their allies against corporate resource grabs on both sides of the US-Canada border (Jarvis, 2013). Defending the integrity of wild rice (manoomin) from genetic biopiracy, toxic mine runoff, and global warming has become a rallying point for the indigenous and many other food sovereignty activists across the Midwest. Anishinabe activist, Winona LaDuke, argues that one of the most powerful strategies in fighting GCC is for people to grow and enjoy their own local foods. A similar perspective is held by Malik Yakini, director of the Detroit Black Community Food Security Network, which has been reclaiming abandoned urban land to grow traditional African American food. Yakini asserts that farming is not only honorable work, it is also a powerful form of community self-determination (Jackman, 2013). Other projects such as the Open Source Seed Initiative (OSSI) are working to expand community seed libraries and reaffirm that seeds are not 'private property' but the common heritage of all (Shemkus, 2014).

On June 6, 2015 I was able to join the more than 5000-strong Tar Sands Resistance March in St. Paul, MN. Labor activists walked hand in hand with family farmers, while native elders shared the stage with church leaders – all to oppose fossil fuel pollution and demand more earth friendly alternatives. If the majority of those hurt by GCC are poor, people of color, and indigenous, and if those best situated to solve the climate crisis are those who already living sustainably, close to the land and nature, then it is time we all recognized and empowered this frontline collaboration for social transformation. To paraphrase one popular La Via Campesina slogan, small farmers are not only feeding the world, they are also cooling down the planet! And they are ready to do both, while also kicking agribusiness out of our agriculture.

References

Bacon, D., 2007. Blood on the palms – Afro-Colombians fight new plantations. *Dollars and Sense*, 271, July/Aug.

Ball, D., 2013. Red Lake Band of Chippewa determined in blockade atop four Enbridge pipelines in Minnesota. *Indian Country Today*, 4 April.

Bell, B., 2010. Miami Rice: the business of disaster in Haiti. *Other Worlds*. [online] Available at: www.otherworldsarepossible.org/another-haiti-possible/miami-rice-business-disaster-haiti [Accessed 9 December 2014].

Cowan, T. and Feder, J., 2013. The Pigford cases: USDA settlement of discrimination suits by black farmers. *Congressional Research Service*, 13 March.

Deen, T., 2012. US Lifestyle is not up for negotiation. *IPS News*, 1 May.

Environmental Protection Agency (EPA), 2008. *Inventory of U.S. Greenhouse Gas Emissions and Sinks: 1990–2006*. [pdf] Available at: www3.epa.gov/climatechange/Downloads/ghgemissions/08_CR.pdf [Accessed 14 September 2015].

Field, T. and Bell, B., 2013. *Harvesting Justice – Transforming Food, Land, and Agricultural Systems in the Americas*. New Orleans, LA: Other Worlds.

Food and Agricultural Organization (FAO), 2000. *The Energy and Agriculture Nexus*. Rome: FAO [pdf]. Available at: www.fao.org/uploads/media/EAN%20-%20final%20web-version.pdf, [Accessed 14 September 2015].

Food and Agricultural Organization (FAO), 2010. *Price Surges in Food Markets*. Rome: FAO [pdf]. Available at: www.fao.org/docs/up/easypol/822/price-surges_food_markets_264en.pdf, [Accessed 14 September 2015].

GRAIN, 2012. *The Great Food Robbery – How Corporations Control Food, Grab Land, and Destroy the Climate*. Capetown: Pambazuku Press.

Gurian-Sherman, D., 2012. *High and Dry: Why Genetic Engineering is Not Solving Agriculture's Drought Problem in a Thirsty World*. Cambridge: Union of Concerned Scientists.

Guttal, S. and Monsalve, S., 2011. Climate Crises: Defending the Land. *Development*, 54 (1), pp. 70–76.

Heller, M. and Keoleian, G., 2000. *Life Cycle-Based Sustainability Indicators for Assessment of the U.S. Food System*. Ann Arbor, MI: Center for Sustainable Systems – University of Michigan.

Holt-Gimenez, E., Altieri, M., and Rosset, P., 2006. *Ten Reasons Why the Rockefeller and Melinda and Bill Gates Foundations' Alliance for Another Green Revolution Will Not Solve the Problem of Poverty and Hunger in Africa*. Oakland, CA: Food First.

Holt-Gimenez, E., 2014. Tangled roots and bitter fruit: what Ferguson and New York can teach the food movement. *Common Dreams* [online] 4 December. Available at: www.commondreams.org/views/2014/12/04/tangled-roots-and-bitter-fruit-what-ferguson-and-new-york-can-teach-food-movement [Accessed 14 September 2015].

Ivey, M., 2013. Huge Grant County farmland sale linked to state law on foreign ownership. *Cap Times* [online], 12 April. Available at: http://host.madison.com/news/local/writers/mike_ivey/huge-grant-county-farmland-sale-linked-to-state-law-on/article_f1259d94-a087-11e2-baf7-001a4bcf887a.html [Accessed 14 September 2015].

Jackman, M., 2013. Detroit black community food security network's growing determination. *Detroit Metro Times*, 2 October.

Jarvis, B., 2013. Idle no more: native-led protest movement takes on Canadian government. *Rolling Stone*, 4 February.

Kaufman, F., 2010. The food bubble – how Wall Street starved millions and got away with it. *Harpers Magazine*, July.

Klein, N., 2007. *The Shock Doctrine: The Rise of Disaster Capitalism*. New York: Metropolitan Books.

Klein, N., 2014a. *This Changes Everything*. New York: Simon & Schuster.

Klein, N., 2014b. Why #BlackLivesMatter should transform the climate debate. *The Nation* [online] 12 December. Available from: www.thenation.com/article/what-does-blacklivesmatter-have-do-climate-change/ [Accessed 14 September 2015].

Kripke, G., 2005. *Food Aid or Hidden Dumping?* [online] Available from: www.oxfam.org/sites/www.oxfam.org/files/bp71_food_aid.pdf

La Via Campesina, 2009a. *Industrial Agrofuels Fuel Hunger and Poverty*. [pdf] Available at: http://viacampesina.net/downloads/AGROFUELS/EN/LVC-AGROFUEL7.pdf [Accessed 14 September 2015].

La Via Campesina, 2009b. *Small Scale Sustainable Farmers are Cooling Down the Planet*. [pdf] Available at: http://viacampesina.org/downloads/pdf/en/EN-paper5.pdf [Accessed 14 September 2015].

LaDuke, W., 1999. *All Our Relations: Native Struggles for Land and Life*. Boston, MA: South End Press.

Mitchell, D., 2008. *A Note on Rising Food Prices*. DC: World Bank. [online] Available at: http://elibrary.worldbank.org/doi/abs/10.1596/1813-9450-4682 [Accessed 14 September 2015].

Mittal, A. and Moore, M., 2009. *Voices from Africa – African Farmers and Environmentalists Speak Out Against a New Green Revolution in Africa*. Oakland, CA: Oakland Institute.

Movement Generation, 2104. Ecology & Racial Justice: On Ferguson, State Violence, & the Proper Management of Home. [online] 24 November. Available at: http://movementgeneration.org/ecology-racial-justice-on-ferguson-state-violence-the-proper-management-of-home/ [Accessed 14 September 2015].

New Internationalist, 2013. *Land Grabs Issue*, May.

Pear, R., 2012. After three decades, tax credit for ethanol expires. *New York Times*, 2 January.

Peck, J., 2009a. Via Campesina confronts the global agrofuel industrial complex. *Z Magazine*, January.

Peck, J., 2009b. Corporate agribusiness helps scuttle climate justice. *Cap Times* [online] 29 December. Available from: http://host.madison.com/news/opinion/column/john-e-peck-corporate-agribusiness-helps-scuttle-climate-justice/article_aa6b0714-e6d0-5aea-b26b-5a4b4cccd6d2.html [Accessed 14 September 2015].

Rice, A., 2009. Agro-imperialism. *New York Times Magazine*, 22 November.

Shemkus, S., 2014. Fighting the seed monopoly: 'We want to make free seed a sort of meme.' *The Guardian*, 2 May.

Shiva, V., 1989. *The Violence of the Green Revolution*. London: Zed.

Smith, G., 2010. A harvest of heat: agribusiness and climate change – how six food industry giants are warming the planet. *Organic Consumers Association* [online]. Available at: www.organicconsumers.org/news/agribusiness-and-climate-change-how-six-food-industry-giants-are-warming-planet [Accessed 14 September 2015].

Squires, G. and Hartman, C., eds, 2006. *There is No Such Thing as a Natural Disaster: Race, Class, and Katrina*. New York: Routledge.

Tett, G., 2011. The real bull market – why Americans are making pensions out of prairies. *The Financial Times*, 14 October.

Tran, M., 2013. Vandana Shiva: 'Seeds must be in the hands of farmers.' *The Guardian*, 25 February.

US Dept. of Agriculture (USDA), 2009. Agricultural Secretary Vilsack, dairy producers sign historic agreement to cut greenhouse gas emissions by 25 percent by 2020. [press

release] 15 December.
US Dept. of Agriculture (USDA), 2010. *FSIS National Residue Program for Cattle* [pdf]. Available at: www.usda.gov/oig/webdocs/24601-08-KC.pdf [Accessed 14 September 2015].
Wittman, H., Desmarais A. and Nettie, W., 2010. *Food Sovereignty: Reconnecting Food, Nature, and Community*. Oakland, CA: Food First Books.
Wisconsin Initiative on Climate Change Impacts, 2011. *Wisconsin's Changing Climate: Impacts and Adaptation*. Available at: www.wicci.wisc.edu/publications.php [Accessed 14 September 2015].
Zaitchik, A., 2015. After the Deluge – Building Climate Justice From the Wreckage of Hurricane Katrina. *New Republic*, 25 June.

26 Environment of the margins

Reconsidering environmental racism for sustainable action

Laurens G. Van Sluytman and Phoebe Sheppard

There is an old Guyanese proverb, '*lil bit dutty mek dam.*' This translates into, '*a little bit of dirt makes a dam.*' In other words, small actions over time can make a difference; perhaps preventing a deluge, much less rolling back the tide. Metaphorically, the tide is persistent; each drop regulates our innate, environmental and structurally derived selves. For example, in casting my vote along with many others, Marilyn Mosley emerged as the State's Attorney general for Baltimore, Maryland. Several months later she would play in critical role as Baltimore burst into flames – members of the impoverished community, Sandtown/Winchester, retaliated against police brutality. Under the umbrella of Black Lives Matter, the community struggles to address the actions of law enforcement. However, police brutality is all but one of multiple spines to the umbrella designed to contest the multiple streams of structural violence this community experiences.

This is a community plagued by poor housing stock and public transportation, limited access to healthy food, and schools that fail to meet the goals of educating the next generation. The stark landscape, absent of any trees, laid bare the largely untold history and impact of deindustrialization, community revitalizations, community destabilization. The uprising struggles to sustain a conversation concerning the community's vulnerability to disparate health outcomes such as high rates of HIV, diabetes, respiratory and cardiovascular disease. Further economic marginalization and fewer resources to negotiate fair treatment and counter susceptibility to further predation and decay – fruit left unattended in a parched environment.

The environment, despite our tendency to see it as 'out there,' as in a place we 'go visit,' is of course fundamental to all life. It shapes and is shaped by our activities. It embodies specific socio-cultural values and actions. Often these actions are driven by the desire for capital accumulation on the part of the elite and determine both the built and lived environments for the majority. These actions, in turn, determine how business is conducted in communities with few resources and political clout. They determine the environmental context within which these communities live. In contrast, the residents who perform the labor required to run, literally and figuratively, industries' operations in their communities face the challenge of saying 'No' to poverty reducing opportunities. Thus, the intersection of race and socioeconomic status deeply informs both the elites' interests in land and labor exploitations, as well as the communities' susceptibility and reaction to this exploitation and the resulting environmental degradation.

Central to this interactive relationship debate is the disparities in health of communities burdened by disproportionate environmental degradation. Black and other poor communities, in particular, shoulder a disproportionate burden of environmental pollution and degradation despite research on health disparities demonstrating an association between negative health outcomes due to environmental degradation among communities of color. Over the course of years, there have been many efforts to enact legislation in the services of reducing the disproportionate impact of environmental degradation on disadvantaged communities. Persistent and recalcitrant abuses by industry, and community outcries demonstrate a clear association between the need for effective policy and continued community organizing. To examine this association, this paper first employs a public health lens in reviewing the literature on the contribution of pollutants to the health of Americans. Next, we discuss the history of national and local efforts to address environmental degradation and its impact on marginalized communities. We argue that, despite cases of galvanized and august movement activities spanning multiple decades, many communities remain vulnerable. We posit that deindustrialization within the USA and the surge of economic and political policies focused in wealth accumulation and reduced government spending have reduced essential cultural and community capital within economically depressed communities (Van Sluytman, in press). Rendered marginal, these communities are susceptible to pressure from government and corporate forces that deploy a rhetorical frame emphasizing personal responsibility. That is, nothing is impossible, including resisting oppression if one takes charge of one's life – pulling oneself up by the bootstraps. This rhetoric often neglects the absence of bootstraps and at times boots, themselves. Thus, we propose that sustained action to counter environmental racism requires participatory community engagement in contesting environmental degradation, through problem definition and examining and reframing prevailing ideologies that reinforce subordination to the economic and political elites, locally and globally. Correspondingly, we introduce a model for conceptualizing community action through mindful engagement, critical consciousness, and critical mass. Study of the community actions to improve lived environments yields valuable information not only for safety and wellbeing of vulnerable populations, but for informing us of new ways to develop community driven actions designed to confront environmental degradation and intersecting vectors such as racism and poverty, and to sustain progressive social change.

Public health – public pollutants

The literature demonstrates a relationship between race and exposure to environmental pollutant such as air pollution (Just *et al.*, 2012; Nachman and Parker,

2012). What is more, focus on health disparities have inaugurated an extensive literature demonstrating that African American bear a disproportionate burden of poor health outcomes (CDC 2011; Cooper *et al.*, 2013, Klinnert, Price, and Robinson, 2002) across a number of diseases such as asthma (Moorman, Person and Zahran, 2013), coronary heart disease and stroke (Gillespie, Wigington and Hong, 2013), colorectal cancer (CRC) (Steele *et al.*, 2013), diabetes (Becklesand Chou 2013), and hypertension (CDC, 2011; Gillespie and Hurvitz, 2013). For each disease, there is an association with environmental degradation giving rise to what is known as environmental racism (Bullard, 1993, 2000, 2005). Chavis (1994) asserted that environmental racism codifies the collection of laws, policies, and regulations thereby permitting the targeting of communities of color as sites for facilities that store or produce toxic pollutants.

The United State accounts for 25 percent of global emissions in the form of CO_2 emissions from fossil fuel burning, cement production, and gas flaring (Maryland *et al.*, 2003) despite comprising only 5 percent of the world's population (U.S. Census Bureau, 2005). The Environmental Protection Agency (EPA) (2013) outlined the impact of air pollution in general. Among harmful elements in the air we breathe are Ozone (O_3), Nitrogen Dioxide (NO_2), Sulfur Dioxide (SO_2), Particulate Matter (PM-10 and PM-2.5), Lead (Pb), and other nitrogen oxides. While contributing to reduced air quality for all Americans, they pose substantial threat to communities with disproportional disease loads such as asthma, cardiovascular disease, and diabetes (Type II). Similarly, exposure to Lead (Pb), often found in older homes, can result in kidney and neurological impairment and disturb reproductive health. Other classes of air pollution such as Ozone (O_3), Nitrogen Dioxide (NO_2), and Sulfur Dioxide (SO_2) contribute to poor health outcomes as well as environmental degradation. Ozone (O_3) or Ground level ozone is associated with gasoline vapors, chemical solvents, large industrial and small industries, gas stations, and small businesses. Ozone can result in high ozone concentrations over heavily populated areas such as Los Angeles and the Northeast. Among the health effects are respiratory inflammation and increases in hospital admissions for respiratory problems. Ozone also contributes to loss of crops and correlate cost of getting food to market for populations living with depressed wages, underemployment, and chronic unemployment. Likewise, Nitrogen Dioxide (NO_2), belonging to a class of gases called nitrogen oxides (NO_x), derives principally from motor vehicles, electric utilities and industrial boilers. NO_2 impacts lung functioning and reduces resistance to respiratory infections. By itself, SO_2 may compromise those who live with asthma, chronic lung disease or cardiovascular disease. Together, SO_2 and NO_x further contribute to environmental degradation through acid rain, penetrate indoors, and result in premature death related to cardiac and respiratory functioning. Often, communities at highest risk and who have the highest levels of exposure are poor and/or minority and are extensively socially, economically, and politically marginalized.

Historic efforts to address environmental degradation

Title VI of the Civil Rights Act of 1964 and National Environmental Policy Act of 1969 (NEPA) intended to promote nondiscrimination in Federal programs affecting health and environments. To achieve these aims, President Lyndon B. Johnson's administration suggested improving community access to public information, meetings, notices and other documents, public participation, and requiring that all analysis of environmental effects and measures to mitigate them include the economic, health, and social impacts on minority and low-income communities. Almost 20 years later, in 1994 with the issuance of Executive Order #12898 'Executive Order on Federal Actions to Address Environmental Justice in Minority Populations and Low-Income Populations,' the Clinton administration sought to achieve environmental justice by focusing Federal attention on the environmental and human health conditions in minority communities and low-income communities. (EPA, 1994). And more recently, in 2011, the Executive Office of the President Council on Environmental Quality declare that several Federal Agencies – the Department of Health and Human Services, Department of the Interior, Department of Energy, Department of Labor, Department of Housing and Urban Development – had signed an Environmental Justice Memorandum of Understanding to develop environmental justice strategies to Protect Health of U.S. Communities Overburdened by Pollution.

Over the course of fifty years, many (Sexton, 1997; Pastor, Morello-Frosch, and Sadd, 2005) have argued that these legislations and pursuant recommendations have done little to reduce the disparate rates of environmental degradation. For example, two incidents demonstrate particular disregard for the impact of environmental pollutant on poor, in general and specifically black communities.

In 1982, protests against environmental degradation in Warren County, North Carolina, a predominantly black and poor community, may have inaugurated the efforts to confront environmental racism within Black communities. Those rallies and demonstrations resulted in the first national study, Toxic Wastes and Race, in 1987, which posited that there is a relationship between waste site locations and population characteristics. Still, Black communities and poor communities continue to face disproportionate burden. For example, in 2010, the Alabama Department of Environmental Management, driven by potential profits, agreed to allow the Tennessee Valley Authority (TVA) to move 4 million cubic yards of poisonous ash to the Arrowhead Landfill near Uniontown from Kingston, Tennessee. Kingston is a predominantly white, middle-class community; Uniontown is predominantly African-American and poor. The actions of the Alabama Department of Environmental Management effectively subjected the residents of Uniondale to an undue burden of environmental degradation (Rector, 2014; Taylor, 2014).

In addition to disproportionate exposure to environmental toxins in rural areas, it is also important to consider the social and physical built environment of Black and marginalized communities. Like other migrants before them,

African Americans migrated to and began to establish black communities in urban environments, seeking greater opportunities than those offered in rural communities. Subsequent waves of people from across the African diaspora followed in like manner, often settling in communities occupied by other African Americans and building diverse black communities in many of the nation's largest metropolises. In all but one of the top 10 US cities, the percentage of black residents outnumbers their overall state percentage (Census Bureau, 2011).

Given the history of injustices, there should be little surprise that when mapping Superfund Priorities List 2011 and Toxics Release Inventory (TRI) Sites 2011 (EPA, 2014) the highest concentrations of are in communities that are highly populated by Black or African American residents (US Census Bureau, 2011).

From industry to vulnerability

For many years, either by the forces of segregation or desire for community, the residents of these communities exchanged valuable resources. Over time, physical transformations in the built environment allowed and constrained access to resources and at times posed the threat of environmental degradation. Industrialization and ensuing deindustrialization pave the way for changes in the landscape. For instance, while industrialization led to the building of bridges, roads, and factories, it stripped away trees and absorptive land. Increases in exposed urban surfaces result in urban regions that are warmer than their surroundings – heat islands. These areas are at times 50–90°F hotter than the air and can be more than 1.8–5.4°F (1–3°C) warmer than their surroundings. On a clear, calm night, the temperature difference can be as much as 22°F (12°C) (EPA, 2013). While impacting health – respiratory and cardiac function, heat islands also increase energy consumption, subsequent air pollutants, and greenhouse gas emissions and ground-level ozone as well as storm water temperatures that impact surrounding ecosystems.

Community organizing: from ideology to not just rhetoric

But what or who are known polluters? Such questions often devolve into ideological bluster. For example, in June 2014, the Environment Protection Agency (EPA) unveiled its planned new regulations intended to reduce carbon emission by 30 percent by 2030. Some known environmentalists such Former Vice President Al Gore supported the measure, calling it "the most important step taken to combat the climate crisis in our country's history." However, others criticized its failure to go far enough.

> While a step forward, this rule simply doesn't go far enough to put us on the right path. The science on climate change has become clearer and more dire, requiring more aggressive action from the president.
>
> (Erich Pica, Friends of the Earth)

And no sooner had President Obama introduced elements of the plan than it was mired in ideological rhetoric:

> If these rules are allowed to go into effect, the administration for all intents and purposes is creating America's next energy crisis.
>
> (Mike Duncan, president and CEO of the American Council for Clean Coal Electricity)

Perhaps most germane to this chapter was Republican Senator Mitch McConnell's response, "The impact on individuals and families and entire regions of the country will be catastrophic, as a proud domestic industry is decimated – and many of its jobs shipped overseas" (Associated Press, 2014). Raising the specter of further deindustrialization. He positions international communities as harmful interlopers – stealers of our jobs. The threat of job loss serves a distinct political purpose, not only the threat of lost resources but a distinct divide between vulnerable peoples. In exposing the key underpinning of this argument we must turn to community actions.

Taking personal responsibility from the margins

The prevailing neo-liberal ideological position states that under ideal circumstances and in line with current austerity messages, each citizen is equally responsible for protecting the environment (Lavelle, 1994). Such a proposition raises an important paradox concerning the position of marginalized and disadvantaged communities. First, how does one take responsibility for the production of environmental degradation due to waste, when one has no say in their production and one is reliant on the processes leading to environmental degradation for both employment and consumption?

Herein lies a dilemma in setting the policy agenda and actions designed to achieve desired sustainable change. This dilemma encompasses an intersection of marginalized identities – comprising race, gender, class, and the ideological arguments that govern action. In the case of the policy agenda, truly setting the agenda requires defining the problem and identifying those who are negatively affected by a given social problem. Ideology then determines action. Few would deny that American identity is undergirded by the belief in rugged self-determination. Neither would they deny that individualism, egalitarianism, liberty, populism, and laissez-faire – hand off (Lipset, 1996) are crucial ideological features of American exceptionalism. From this perspective many would say that all citizens have the opportunity to advance or at least to maintain their economic position. Others would argue that this is clearly historically and currently untrue, citing evidence of the unequal playing field and that many are doing less well than previous generations. Still, pragmatism concedes that prevailing ideological positions and the policies associated with them chafe at the notion of government intervention, regulation, or 'over-reach.' Therefore, efforts to confront environmental racism – collective in nature – must engage the prevailing narrative in shaping it actions to defeat attempts to exploit community deficits and threaten efforts to build coalitions.

Strategies to go forward

Confronting environmental racism and achieving sustainable change must deploy consideration of the collective economic, environmental, and human right decisions and actions required to protect communities. Communities must be able to protect themselves from both present and further degradation of environments and associated health outcomes, have access to resources, and hold responsible parties accountable for the their actions (Benford and Snow, 2000). Agyeman and colleagues (2003a, 2003b, 2004) define sustainability as a processes that recognizes the limits of existing ecosystems while ensuring ongoing, equitable improvements in quality of life. They argued that sustainability:

> cannot be simply a 'green', or 'environmental' concern, important though 'environmental' aspects of sustainability are. A truly sustainable society is one where wider questions of social needs and welfare, and economic opportunity are integrally related to environmental limits imposed by supporting ecosystems.
>
> (Agyeman *et al.*, 2002, p. 78)

Thus, addressing environmental racism also requires deep participation by members of impacted communities. Movement forward for communities requires methods devised through the participatory engagement of community members in reframing strategies that move beyond immediate problems of pollution and diminished resources to long term goals that build coalitions on both local and international levels.

Participatory engagement acknowledges that all citizens do not have the same capacities to enact change nor do they all share the same concerns. But all citizens represent potentials to engage communities who despite material poverty may possess rich social ties and capital. César Chávez once warned against forsaking the community, "We cannot seek achievement for ourselves and forget about progress and prosperity for our community ... Our ambitions must be broad enough to include the aspirations and needs of others, for their sakes and for our own." Still another apt adage would be, "there's no good sense like common sense." These maxims acknowledge that there are ways of being that are interactional and require investment in the communal. The individual and communities' progress and prosperity are inextricably intertwined. Chávez's words remind us that we must recognize that knowledge is both produced and augmented within the community's processes. These processes are inherently ecological – representing a form of social capital making use of networks within abutting and encompassing communities to disseminate information and produce knowledge that is context driven and relational. They make use of knowledge of the individual, the community, and the social institutions, within the community and at large, to develop strategies for addressing, coping with, and adapting to perceived risks and emerging possibilities. These community processes are assets as calls for austerity, decentralization, smaller government, and place-based services prevail within the national and global dialogue. In fact, The World Bank (2010) asserts that with time-, place-, and event-specific knowledge, communities have greater capacity to manage the impact of local climate hazards on local social and ecological relationships, rules, and institutions. The context (i.e., time and place) rich decision-making process cuts cost, increases decision making autonomy, and stimulates adaptations, sustainability, and cost-effectiveness.

Therefore we must continue to ask, "how do communities fortify themselves to take conscious, critical steps towards confronting environmental degradation, among other disparities?" (McDonald, 2004) For when we do it is clear that confronting environmental racism that concerns black and other economically marginalized communities must be expanded to include the basic needs of the community as well as the oppressive actions that maintain the margins. McDonald continues "at its core, environmental justice is about incorporating environmental issues into the broader intellectual and institutional framework of human rights and democratic accountability." To this end, Glass and McAttee (Glass and McAtee 2006) provide a meaningful framework. Glass and McAtee's 'Axis of Nested Hierarchies' posits that, though we are initially shaped by our genomic composition and associated risks, our actions and behaviors over the course of a lifetime are constrained and permitted by the ways in which we engage with the boundaries of larger social structures – immediate groups such as families and networks, work, access to health care, and local, national, and international processes.

Embodiment represents the expression of experiences with the larger social environment. For example, positioned within certain marginalized communities, residents may have few if any options concerning education opportunities for themselves or their children. They may face work conditions resulting in income insecurity. The embodiment may manifest in psychological distress. Confined to economically depressed communities and vulnerable to pollution, children may express embodiment in the form of asthma and other respiratory issues. Johnson-Hanks and colleagues (2011) offered further elaboration of the Glass and McAttee model in asserting that the person flowing downstream interacts with others comprising the flow and together they are guided by the banks of the conceptual river that have been hewn by generations before. Thus numerous individuals and their networks manifest the stress together. Although generations before have transformed the river banks, it is incumbent upon present and coming generations to disrupt the patterns that maintain marginalization. Disrupting these patterns to promote greater social / economic and political equity requires building a critical mass. Critical mass requires critical consciousness; it is essential.

Critical consciousness recognizes that personal and social concerns and individual thoughts about these concerns do not occur in isolation. They are inextricably tied to others, both within and outside of the community. Critical consciousness affirms the experience of experiencing, preforming, and opposing various forms of oppression and powerlessness, thereby removing polarizing arguments, which are inherent to ideologically driven agenda setting. By these means, members of the communities identify individual and intersecting identities, increasing coalition-building opportunities. The movement efforts heed the tenets of intersectionality, now so popular in academic circles. It is one that is concerned with gender, race, class, and ethnicity among other marginal identities. It is one that incorporates ideological position and associated arguments into a critical analysis of community organizing.

Conclusion

Marginalized communities have historically taken action to achieve justice. Taking action requires building critical mass. Critical mass is achieved through engaging multiple actors in defining the social problem and action steps. In the service of efforts to actively engaging communities in confronting environmental racism, efforts locally and in larger processes of the state, partnership with worried individuals must be community-based and participatory. Engagement must move beyond victim status claiming to declare a stake in environmental degradation and subsequent efforts to marshal practices that deplete resources and generate waste.

Finally, both Weber (1946) and Michels (1949) suggested that organizations – in this case, organizations designed to confront environmental racism – run the risk of stagnating ineffectiveness through a general set of processes. They assert that, as organizations garner support, they increasingly adopt rigid bureaucratic characteristics designed to preserve the organization. These developments also threaten sustainability. To avoid stagnation or rigid bureaucracies that plagued previous movements and ultimately led to their demise, practices must be designed by the community to increase accountability and afford means for informed decision making. Thus, movement activities must employ critical consciousness to engage in recursive processes within an inclusive and diverse body that includes researchers culled from the communities themselves. In 2002, Adger stated that inequality, in its multiple manifestations, represents the greatest threat to sustainable development. Inequalities render collective action and implementation impossible, as they interact with the individual position in society. Anti-environmental racist efforts raise our gaze to include all. Age, race, gender and gender identity, (dis)ability, and sexual orientation stand on equal footing as relevant categories intersecting disadvantaged communities' efforts to challenge disparities.

Recognition of intersectionality is a necessary step towards building alliances. However, deconstructing the constituting elements of communities exposed to environmental injustice offers a valuable opportunity to recognize specific risks and needs for continued support of existing actions. Therefore, this discussion of environment justice within black communities recognizes the intersection of multiple identities and discriminations in communities. It permits an analysis of the impact of environmental injustices on the individual, community, and larger social/global level. Within these recursive processes members have the opportunity to redefine the problem (Zald and Ash, 1966), and themselves as the definition transforms or needs contract, expand and supplant passive acceptance of environmental inequities locally and globally.

References

Adger, N., 2002. Inequality environment and planning, *Environment and Planning*, A 34, 1716–19.

Agyeman, J., Bullard, R. D., and Evans, B., 2002. Exploring the nexus: bringing together sustainability, environmental justice and equity. *Space and Polity*, 6 (1), 77–90.

Agyeman, J., Bullard, R., and Evans, B., 2003 *Just Sustainabilities: Development in an Unequal World*, London: Earthscan/MIT Press.

Agyeman, J., and Evans, T., 2003. Toward just sustainability in urban communities: building equity rights with sustainable solutions. *The ANNALS of the American Academy of Political and Social Science*, 590(1), 35–53.

Agyeman, J., and Evans, B., 2004. 'Just sustainability': the emerging discourse of environmental justice in Britain? *The Geographical Journal*, 170(2), 155–164.

Associated Press, 2014. Reaction to Obama's global warming plan. Available from: www.washingtonexaminer.com/reaction-to-obamas-global-warming-plan/article/feed/2138134 [Accessed 8 July 2015].

Beckles, G.L., and Chou C., 2013. Diabetes – United States, 2006 and 2010. *Morbidity and Mortality Weekly Report* (MMWR), 62 (03), 99–104.

Benford, R.D., and Snow, D.A., 2000. Framing processes and social movements: an overview and assessment. *Annual review of sociology*, 611–639.

Benz, J., Tompson, T., and Agiesta, J., 2014. The People's Agenda: America's Priorities and Outlook for 2014. [pdf] Available from: www.apnorc.org/PDFs/Peoples%20Agenda/AP-NORC-The%20Public%20Agenda_FINAL.pdf

Bullard, R.D., 1993. *Confronting Environmental Racism: Voices from the Grassroots*. Cambridge, MA: South End Press.

Bullard, R.D., 2000. *Dumping in Dixie: Race, Class, and Environmental Quality* Vol. 3. Boulder, CO: Westview Press.

Bullard, R.D., 2005. *Quest for Environmental Justice*. San Francisco, CA: Sierra Club Books.

CDC, 2011. Prevalence of hypertension and controlled hypertension – United States, 2005–2008. MMWR 2011:60;94–97. In: CDC. CDC Health Disparities and Inequalities Report – United States. *Morbidity and Mortality Weekly Report* (MMWR), 60 (supplement, January 14, 2011).

CDC, 2011. Surveillance of health status in minority communities – racial and ethnic approaches to community health across the US (REACH US) Risk Factor Survey, United States, 2009. *Morbidity and Mortality Weekly Report* MMWR, 60 (no. SS-6).

Chavis, B.F. Jr, 1994. Preface. In Bullard, R.D., ed, *Unequal Protection: Environmental Justice and Communities Of Color*. San Francisco, CA: Sierra Club Books, xi–xii.

Chesapeake Bay Foundation, 2008. Bad Water and the decline of blue crabs in the Chesapeake Bay. Available from: cbf.org/badwaters

Cooper, L., Boulware, L., Miller, E., Golden, S., Carson, K., Noronha, G., Brancati, F.,

2013. Creating a transdisciplinary research center to reduce cardiovascular health disparities in Baltimore, Maryland: lessons learned. *American Journal Of Public Health*, *103*(11), e26–e38. doi:10.2105/AJPH.2013.301297.

Council on Environmental Quality, 2011. Obama administration advances efforts to protect health of US communities overburdened by pollution. Available from: www.whitehouse.gov/administration/eop/ceq/Press_Releases/August_04_2011

Environmental Protection Agency, 1994. EPA insight policy paper: executive order #12898 on environmental justice. Available from: www.epa.gov/fedfac/documents/executive_order_12898.htm

Environmental Protection Agency, 2013.What is an urban heat island? Available from: www.epa.gov/heatisland/about/index.htm

EPA, 2012. Effects of acid rain – forests. Available from: www.epa.gov/acidrain/effects/forests.html

EPA, 2013. Health effects of air pollution. Available from: www.epa.gov/region7/air/quality/health.htm

Fodha, M., and Zaghdoud, O., 2010. Economic growth and pollutant emissions in Tunisia: an empirical analysis of the environmental Kuznets curve. *Energy Policy*, 38(2), 1150–1156.

Gilens, M., 2009. *Why Americans Hate Welfare: Race, Media, and the Politics of Antipoverty Policy*. Chicago, IL: University of Chicago Press.

Gillespie, C.D. and Hurvitz, K.A., 2013. Prevalence of hypertension and controlled hypertension – United States, 2007–2010. *Morbidity and Mortality Weekly Report* (MMWR) 62(03), 144–148.

Gillespie, C.D., Wigington, C. and Hong, Y., 2013. Coronary heart disease and stroke deaths – United States, 2009, November 22. *Morbidity and Mortality Weekly Report* (MMWR) 62(03), 157–160.

Glass, T.A. and McAtee, M.J., 2006. Behavioral science at the crossroads in public health: extending horizons, envisioning the future. *Social Science & Medicine*, 62: 1650–1671

Glass, T.A., Goodman, S.N., Hernán, M.A., and Samet, J.M. 2013. Causal inference in public health. *Annual Review of Public Health*, 34, 61–75.

Grossman, G.M., and Krueger, A.B. 1995. Economic growth and the environment. *The Quarterly Journal of Economics*, 110(2), 353–377.

Johnson-Hanks, J., Bachrach, C., Morgan, S.P., Kohler, H.P., Hoelter, L., King, R., and Smock, P., 2011. Understanding family change and variation: structure, conjuncture, and action. [Manuscript submitted for publication].

Just, A., Whyatt, R., Perzanowski, M., Calafat, A., Perera, F., Goldstein, I., and Miller, R., 2012. Prenatal exposure to butylbenzyl phthalate and early eczema in an urban cohort. *Environmental Health Perspectives*, 120(10), 1475–1480. doi:10.1289/ehp.1104544.

Klinnert, M., Price, M., Liu, A., and Robinson, J., 2002. Unraveling the ecology of risks for early childhood asthma among ethnically diverse families in the southwest. *American Journal Of Public Health*, 92(5), 792–798.

Lavelle, M., 1994: Environmental justice. In *The 1994 Information Please Environmental Almanac*, World Resources Institute, Boston MA: Houghton-Mifflin, 183–92.

Lipset, S.M., 1996. *American Exceptionalism: A Double-Edged Sword*; W. W. Norton: New York, p. 98

Maryland Department of the Environment (n.d.) Stormwater Fee FAQ. Available from: www.mde.state.md.us/programs/Marylander/Pages/StormwaterFeeFAQ.aspx

McDonald, D.A., 2004. *Environmental Justice in South Africa*. Juta and Company Ltd.

Michels, R., 1949. *Political Parties*. Glencoe, IL: The Free Press.

Moorman, J.E., Person, C.J., and Zahran, H.S., 2013. Asthma attacks among persons with current asthma – United States, 2001–2010. *Morbidity and Mortality Weekly Report* (MMWR), 62(03), 93–98.

Nachman, K., and Parker, J., 2012. Exposures to fine particulate air pollution and respiratory outcomes in adults using two national datasets: a cross-sectional study. *Environmental Health: A Global Access Science Source, 1125*. doi:10.1186/1476-069X-11-25.

Pastor Jr., M., Morello-Frosch, R., and Sadd, J.L., 2005. The air is always cleaner on the other side: race, space, and ambient air toxics exposures in California. *Journal Of Urban Affairs*, 27(2), 127–148. doi:10.1111/j.0735-2166.2005.00228.x.

Rector, J., 2014. Environmental Justice at Work: The uaw, the war on cancer, and the right to equal protection from toxic hazards in postwar America. *Journal of American History*, *101*(2), 480–502.

Sexton, K., 1997. Sociodemographic aspects of human susceptibility to toxic chemicals: do class and race matter for realistic risk assessment? *Environmental Toxicology and Pharmacology*, 4(3), 261–269.

South African Environmental Justice Network, 1997. *South African Environmental Justice Networker*, Autumn.

Steele, C.B., Rim, S.H., Joseph, D.A., King, J.B., and Seeff, L.C., 2013. colorectal cancer incidence and screening – United States, 2008 and 2010. *Morbidity and Mortality Weekly Report* (MMWR), 62(03), 53–60.

Taylor, D.E., 2014. *Toxic Communities: Environmental Racism, Industrial Pollution, and Residential Mobility*. New York: NYU Press.

The World Bank, 2010. *World Development Report: Development and Climate Change*. Available from: http://siteresources.worldbank.org/INTWDR2010/Resources/5287678-1226014527953/WDR10-Full-Text.pdf [Accessed 23 June 2014].

Tsosie, R., 1996. Tribal environmental policy in an era of self-determination: the role of ethics, economics, and traditional ecological knowledge. *Vermont Law Review*, 21, 225.

US Census Bureau, 2011. The Black Population: 2010 Census Briefs. Available from: www.census.gov/prod/cen2010/briefs/c2010br-06.pdf. [Accessed 8 July 2015].

US Census Bureau, 2011. 2010 Census Shows Black Population has Highest Concentration in the South People Who Reported as Both Black and White More than Doubled. Available from: www.census.gov/newsroom/releases/archives/2010_census/cb11-cn185.html

Weber, M., 1946. *Essays in Sociology*, eds. H. J. Gerth, and C. W. Mills. New York: Oxford University Press, 297–301.

Williams, D., 2015. Transgressive Design, retrieved from http://etd.auburn.edu/bitstream/handle/10415/4543/package2.pdf?sequence=2&ts=1433755764229

Wing, S., Horton, R., Muhammad, N., Grant, G., Tajik, M., and Thu, K., 2008. Integrating epidemiology, education, and organizing for environmental justice: community health effects of industrial hog operations. *American Journal of Public Health*, 98(8), 1390–1397. doi:10.2105/AJPH.2007.110486.

Zald, M. N., and Ash, R., 1966. Social movement organizations: growth, decay and change. *Social forces*, 44(3), 327–341.

27 Ubuntu

Lindsy Floyd

Ubuntu

It's a concept that I had heard of before. It had floated around, in and out of conversations I had been in, but it didn't stick until a dear friend of mine got the word tattooed on her arm. Lizzy and I were instant friends our freshman year of college. She and I had both arrived at college ready to change the world and without a clue as to how to do it. But it didn't matter. That first year, I created Political Awareness Campaign, a student club that would eventually help contribute to getting every college student in New Mexico registered to vote during the 2004 Presidential Election. Lizzy was my Vice President. And, when she started a chapter of Amnesty International at our school, I was her Vice President. Our sophomore year, we began our own journeys. I transferred to a school in Hawai'i, and Lizzy spent semesters abroad.

Years after Lizzy came back from South Africa, she had a tattoo designed for her right arm, a sleeve as it's called. At the bottom of the sleeve, near her wrist, is the word 'ubuntu.' She described it to me as the concept that a person's well-being is inextricably linked to the well-being of others.

It's an African concept that, when translated into English, roughly means, 'I am what I am because of who we all are.' In 2008, Desmond Tutu defined Ubuntu:

> One of the sayings in our country is Ubuntu – the essence of being human. Ubuntu speaks particularly about the fact that you can't exist as a human being in isolation. It speaks about our interconnectedness. You can't be human all by yourself, and when you have this quality – Ubuntu – you are known for your generosity. We think of ourselves far too frequently as just individuals, separated from one another, whereas you are connected and what you do affects the whole world. When you do well, it spreads out; it is for the whole of humanity.
>
> (See Ubuntu Women's Institute, 2012)

The concept finally stuck with me when I saw it on her arm in ink. *I am what I am because of who we all are*. Would it matter if I told you my gender? Would it matter if I told you how much money I had in my bank account? Or if I told you what color my skin is? What if I only disclosed facts about me that pertained to when I was oppressed or discriminated against? I'd like to think that those facts alone cannot tell much. If I believe in the concept of ubuntu, which I do, then it shouldn't matter what I myself have endured. It matters, rather, what we have endured collectively. After all, an injustice to one is an injustice to all.

Within my daughter's lifespan, which has only been six short years, my country has fought wars and killed over oil, or was it weapons of mass destruction? A friend of mine, who was twenty-three at the time, died of a heroin overdose. I had a miscarriage. The BP Oil Spill in the Gulf of Mexico destroyed an ecosystem. A giant cluster of trash, the Great Pacific Garbage Patch, floats around the Pacific Ocean, slowly being digested through the food chain. Corporate executives walked away with billions while the American middle class and poor paid for it. The list goes on. We've got a whole world of injustices. And this is the world I had to offer my daughter when she was born. It's a far cry from the stories in the fairytales that she loves. The enemies in real life are far more abstract and complicated than a fire-breathing dragon or a curse broken by true love.

When I think of my life as an activist, I wonder what that really means and how I can define it. What qualifies me to identify as an activist? At the very basic level, I think that everyone has the potential to be an activist. All that it requires is for anyone to look around them and observe. I look around me in my new Austin home and see a lot of problems with education.

Most Hispanic neighborhoods are poorer than the White neighborhoods, and the best schools happen to be in the White neighborhoods. Most of the Hispanic neighborhoods have Title I schools that struggle with truancy, crime, etc. My daughter is being taught to be allegiant to not only the American but also the Texas flag. Creationism seems to be brought up on a regular basis as something that must replace evolution in science classes.

Water, like most places in the West, is a concern, both in quantity and quality. Rural towns in West Texas have adopted cloud seeding to increase rain. Like most places in the world, invasive species are a problem. Population growth, while great for the booming and expanding local economy, encroaches on wildlife and promotes more industry and importing and resource use. That's just the tip of the iceberg. The point is: if you can notice problems and injustices around you, then you can be an activist.

My mom tells a story from when I was four years old. We were leaving a grocery store when I noticed a woman holding a cardboard sign. Although I don't remember, my mom recalls that I asked what the sign said. 'Homeless with three kids, starving, please help, god bless.'

I wanted to give all of our groceries to the woman. When my mom pointed out that doing that would leave us without food, I protested. She agreed to let me pick some things out to give to the woman. I don't remember what I gave her, but looking back, I think that it must have felt huge, pressing a large box of something against my small body. I was still upset that we couldn't give her more and refused to eat for three days.

I am an activist because I am terrified to think of what our world will look like when my daughter is in her twenties, her thirties, her forties, or beyond. I live in the American West. One unalterable fact about living in this arid landscape is a very limited and very finite amount of water. Water is a non-negotiable resource that every living thing simply must have. The most recent studies project that, by 2030, half of the world population will not have access to drinkable water.

The Colorado River, which is a huge supply of water for the West, including the busy and growing cities of Las Vegas and Los Angeles, no longer flows all the way to the Sea of Cortez as it used to. Reservoir levels are already low. In 2030, my daughter will be twenty-two years old. Will she have access to drinking water? I am an activist because I am terrified to think of the alternative of sitting around, doing nothing, just waiting for something to happen. I imagine what I would do to get drinking water. Right now, I walk to my kitchen, turn the faucet on, and fill up my glass with ease. Will my daughter be able to do that? If not, how will she get water? If she has the water, what will other people be willing to do to get that water?

I am an activist because I am literally scared for my daughter's future. I sometimes worry that there will be water wars. The idea of such reality isn't that far-fetched. Severe water shortages have been glamorized in cultures throughout the world and through time. Many of the stories, whether in fiction or film, begin the same: civilizations lose their access to essential resources and, if they are to survive, must find a solution. In some of these stories, societies change their geographical location, some kill, some die, and some use modern technology to create a solution. What will our solution be?

I am an activist because a recent study found that children who grow up in Salt Lake Valley will never develop full lung capacity because of the air pollution.

I am an activist because my grandfather nearly died because of a cancer that is completely preventable. Hürthle thyroid cancer, in my grandfather's case, was caused by exposure to arsenic. The doctors say that this cancer will be the thing that kills him. It may be in one year, it may be in five or ten years, or maybe even twenty. But it will, ultimately, kill my grandfather.

Becoming an activist was a journey. It did not happen overnight, and there was certainly a delay between when I became aware and when I began to act.

In 2009, I entered into the Environmental Humanities graduate program at the University of Utah. In my first semester, I took a class with award-winning author, Terry Tempest Williams. Toward the end of the class, I asked her, 'How can I be a good writer if I have no grief in my life?' I had just finished *Refuge*, Terry's book that parallels the rise of the Great Salt Lake's waters in the 1980s with the death of her mother. As I saw myself then, I had little to grieve about. I was the mother of an adorable little girl, my family was healthy, I had a lot of friends, and my marriage was strong.

I was well aware of the problems in the world. I knew that there were people starving and dying daily, some more slowly as in the USA, others more quickly as in some developing countries. I knew about epidemics and pandemics and war and greed and hate and consumerism and racism and rape and environmental degradation. I had spent four years in my undergraduate studies learning about all of the environmental problems in the world. Climate change was an increasing concern of mine.

In hindsight, I know that everything I knew then was abstract. I wrote research papers on invasive species. I created a club at my alma mater, Political Awareness Campaign. With the help of volunteers, I succeeded in getting every college student in Santa Fe, New Mexico registered to vote in 2004. I signed petitions. I donated money to local non-profit organizations. I switched my light bulbs to more energy-efficient ones. I never idled my car.

My brother calls this kind of environmentalist the 'L.L. Bean Edition' of hippies. There are plenty of them here in Salt Lake City. The L.L. Bean Hippie typically drives a Subaru, has a few dogs (usually Labradors), goes to the Patagonia sale every fall, shops at the farmer's market, doesn't idle their car, and has energy-efficient light bulbs throughout their home. Their Subaru is usually covered in bumper stickers: 'Obama/Biden 2008,' 'Make Love Not War,' 'Coexist,' or a sticker from a local ski resort. These L.L. Bean hippies are usually on the upper end of the socioeconomic scale, not in the very top 1% in the U.S., but certainly in the upper class worldwide. They're usually white, too. Now, I myself fit into this category of an L.L. Bean hippie, though I no longer drive a Subaru. I am not wealthy by American standards, but I am certainly wealthy compared to most of the world. Rarely have these L.L. Bean hippies been at the frontline of social and environmental problems. Their (or 'our,' since I should include myself) land has never been forcibly taken, they will always have a roof over their head, a meal to eat, and, most likely, have a smart phone and enough cash in their bank to make a $10 donation to their cause of choice to get that free bumper sticker to show the world that they care.

I say this not to judge my fellow Utahns, only to describe who I used to be. At that point in my life, in 2009, I was under the impression that, if I signed enough petitions, donated enough money, and made small changes in my own life, I could make a difference in the world.

The only problem is that I was lying to myself. It wouldn't be until the summer of 2010 that my perception would shift. That summer, I took another class from Terry. It was a ten-day course. In those ten days, I saw images and heard stories that are seared into my memory as much as the day my daughter was born.

Author Rick Bass told the class about his fight to save the Yaak region in Montana. Author and grizzly bear advocate Doug Peacock (a.k.a. Hayduke in Ed Abbey's, *The Monkey Wrench Gang*) told us about the stress of trying to protect the predator from humans. Activist Louisa Wilcox talked about how ranchers would unscrew the bolts on the tires of her car because they were so fed up with her ambition to protect wild wolves.

Jason Baldes, a Shoshone, explained the stress of finally getting approval to reintroduce bison onto a Reservation. Teresa Cohn talked about another Reservation. The Reservation can't have water rights unless they irrigate, but their culture never adopted irrigation practices, so they have no water.

What was important to me was the way that these individuals talked about their passion. Teresa paced back and forth when telling her story. Jason shook his head in sorrow. Louisa tugged at her hair with a nervous twitch. Doug had a twitch in his eye, though I can't be certain if the twitch was from his days in the Vietnam War or from his work with grizzlies. Rick summed it up quite accurately. He said, 'It's all shit-stew out there.'

On the final day of the class, I awoke to a beautiful sunrise over the Red Rocks Wildlife Refuge in Centennial Valley, Montana. I called my grandfather. He had just undergone a biopsy for Hurthle Anaplastic thyroid cancer. It was at stage four. His particular form of cancer is largely attributed to exposure to arsenic. He worked for a gold mine in the 1970s, pouring arsenic over mine tailings to extract more gold.

As I walked to class, I noticed a pile of feathers. A bird had been eaten that morning, and its predator had left behind this pile of white and gray feathers. I picked them up and brought them to class. I gave them to Terry.

Terry shared with us a quote from the environmentalist David Rockefeller Jr. that, 'Even a feather can tip the scales.' With that, Terry asked us to sit on the floor. She then silently shared photographs with us by photographer Chris Jordon. We saw dead albatross after dead albatross. Their carcasses had rotted on the island of Midway, revealing guts filled with plastic. Within the bellies of these beautiful birds, Chris had documented bottle caps, lighters, and even a dildo. The birds died because the plastic had blocked their digestive tract. It's called the Great Pacific Garbage Patch. Depending on who gives the numbers, it's either larger than Texas, larger than the continental US, or somewhere in between. The birds think that the tiny, floating pieces of plastic are food.

I came home to Salt Lake City a different person. I no longer had to ask Terry how I could be a good writer without grief. I had found my grief. My grandfather had a cancer that was rooted in human behavior. Birds were choking themselves to death on our rubbish. Wolves were being shot to save cattle. Bison couldn't be released to the wild over concerns that they would expose cattle to disease. You could add any number of disasters to the list: overpopulation, lack of clean water, disease or sickness (including cancer, whether genetic or environmental), famine, war, pollution, politics, etc., etc.

I was overwhelmed. I felt as though all of the things that I had done—all of the badges that I had put on my L.L. Bean Hippie sash—had been a waste of time. It was as though the class had ripped off my metaphorical Band-Aid and forced me to see the wound: our world is broken. It is, of course, necessary to mention the proverbial notion that this planet is not really ours. We merely attribute ownership. What is ours, though, is our actions and the way in which we interact with this planet.

I remained that way for months, soaking in the shit-stew. I was hopeless. I was depressed. I found no use in signing petitions or calling my elected officials. What was the point?

I was worried about seemingly every issue: reproductive health and rights, finite water in the desert with a growing population, over-population, food supplies, natural resources, ecological diversity…and the list went on. I was overwhelmed by the complexity of our problems. Each issue is linked to other issues. Race, class, gender, access to health care, gay rights, discrimination, pollution, climate change, ecological degradation—all of it—is a complicated web. One issue can't be fixed in isolation, because it tugs and pulls and stretches as the other issues tug and pull and stretch.

Ubuntu. I am who I am because of who we all are collectively. I recently found out that a high school friend was a victim of a hate crime in Minneapolis. He was riding the bus home, and two men followed him off the bus, yelling derogatory terms about his sexuality. They beat him so badly that he was in the hospital for days. I am a part of a society that condones that sort of behavior. That is a reflection on me. I wear that shame.

The crime is not isolated. It's tied to education and ethics and politics and class and gender and sexuality and public safety. I am a reflection of our collective selves.

Terry talks about sacred rage. She told me that I had to turn my anger into sacred rage. As I understand it, sacred rage reminds me of giving birth to my daughter. I had a natural home birth. Just before she was born, I remember feeling so frustrated that laboring hurt so badly. I remember wondering why it had to hurt so badly. I reminded myself then, back in 2008, that it hurt so badly because it was the evolution of the human species. Labor pains hurt so that the mother-to-be knows to drop what she's doing and find a safe place to give birth.

For me, being an activist is a responsibility. So long as my fellow citizens suffer, I too suffer. So long as I am aware of these issues, I cannot stand idly by. It is my obligation, as a citizen and as a parent, to stand my moral ground and to make our society as rich and wonderful as possible. After all, as Ed Abbey said, "Sentiment, without action, is the ruin of the soul."

Without action. To act. Action defined: 'the fact or process of doing something, typically to achieve an aim,' or, 'A thing done; an act.' I acted.

Tied together with zip ties, we walked into the nearest intersection: 400 South and Main Street in downtown Salt Lake City. A Judge had just issued a verdict that had enraged us; it had been unjust. It was rush hour. Within moments, cars were honking and a news helicopter hovered above. We continued to sing. My heart was racing, yet I was strangely confident. After some amount of time (it could have been minutes or hours for all I recall), a Trax light rail train approached. It honked its horn and the media ate it up, taking photos of my friends who were directly blocking the train with their bodies.

There was some concern. Observers were wondering why environmentalists would block public transportation. Of the twenty-six of us who were locked together, we brainstormed. Was it the right thing to do? I had so many thoughts running through my head. I thought of civil rights movements in recent history: the suffragettes and women's rights activists, racial equality activists, LGBT activists, and fellow environmentalists. We all decided as a group that blocking public transportation was morally justified, albeit illegal. My moral grounds superseded the law.

When my daughter comes to me, sometime in the future, and wants to know how we got to where we are (wherever that is) as a society, I need to have answers. More importantly, I need to be able to tell her that I did and still do everything in my power to create a bright future for her. The challenging thing about activism, though, is that it has almost become an obsession. I must have my voice heard. Activism, at least for me, takes on many forms. My teaching remains as a form of activism. I want younger generations to have a voice and to use it, and teaching is the perfect opportunity to give them that. I do my best to not indoctrinate, but rather, to question. Activism is not narrowly defined. Activism is teaching, it is civil disobedience, it is learning, and it is singing.

I think to myself about my daughter's future. I think of how I can explain to her about water scarcity or climate change or lack of resources. I wonder about how I can ethically apologize to her and give her a genuine apology. How can one really apologize for those things? I am not directly responsible for many of these injustices in the world. I am not directly responsible for imprisoning the innocent, for gender and racial inequality, for environmental problems such as deforestation or pollution, or even for climate change. And yet, I am indirectly responsible.

I cannot tolerate inaction. I do not wish to apologize for those in power or to those directly responsible for such ills. They can make their own apologies when they see fit. For now, I am content being a climate activist, in all of the shapes and forms in which it is presented. It is the only option. That, in essence, is why I am an activist. I am an activist because it is my responsibility as a human, as a parent, and as a citizen.

Reference

Ubuntu Women's Institute [UWI] USA, Inc. 2012. Brief meaning of Ubuntu'. Available at http://uwi-usa.blogspot.be/2012/01/ubuntu-brief-meaning-of-african-word.html

Part IV

Water

Flooding Drought Emotions Spirituality Turmoil

Figure IV.1 Water

28 Yemaya Madre de Agua

Isis Rakia Mattei

Her womb gushes amniotic fluids
from which creature thing emerged.
Her breasts flow rivers, streams, brooks
pooling into crystalline lakes and dark silent wells.
Her breath transforms the world into a dreamy misty realm.

She is life itself.
The great Mother who births new expressions ad infinitum.
Who nurtures her young with the bounty of her being.
Who cares after creation in a way only she can.

The sweat of our brow
tears of joy and pain
every cell in our body connects us to her,
to every other living creature
and to the celestial waters stirring deep within our imaginations.

She cleanses and purifies.
Refreshes us.
Cures us.
Grieves us.

For the destruction we sow
for our lack of sense
failure of heart
and refusal to see.

Why wound the Mother so?
Since she and we are one.

29 Yemaya

Imna Arroyo

Figure 29.1 Yemaya. Linoleum cut 30" × 22". The Madre de Agua print and poem were inspired by the biggest US environmental catastrophe of our time, the BP oil spill/ DeepwaterHorizon drilling disaster, which killed scores of birds, marine mammals and sea turtles in the Gulf of Mexico. Yemaya, the Yoruba Goddess of the Sea, is holding a sea mammal in her arms while she and the animals try to make sense of mankind's greed and destructiveness. The print and poem are a lament, a prayer and an apology, all at once.

30 Catholics, socio-ecological ethics and global climate change

Incarnations of green praxis

Christopher Hrynkow and Dennis O'Hara

Socio-ecological connectivity and Catholicism

Thomas Berry (1988) emphasized that humans are dependent on the Earth and Universe communities for any substantively rich form of existence. Greening the cosmological consciousness of Catholic predecessors, Berry combined his own training for the priesthood, his studies of cultures and Eastern religions, and a sense of the multi-faceted tragedy accompanying a widening anthropogenic ecological crisis in order to develop a functional cosmology and spirituality. For example, Berry sourced the ecological crisis in the dysfunctional commercial, industrial, and military understandings of progress, which he further characterized as no longer able to sustain humanity due to their major fault of failing to adequately recognize humanity's deep interconnectivity with and dependence upon the rest of Earth's community. In their place, he upheld the potential of reintegrating humanity within the creative life systems of Earth through a renewing functional cosmology, recognizing humanity's essential role as one of a multitude of players in a vast, evolving, and diverse universe story. Berry's social, economic, and ecological analysis, and his call for a profound reinvention of the human, has inspired many forms of Catholic green praxis (McFarland Taylor, 2007).

For Berry, this worldview, integrated into action, has the power to transform religion, law, politics, and commerce so that they allow all people to be present to the planetary community in a mutually enhancing manner. Noting that it is not possible to have healthy people on a sick planet, Berry also declares: "There is no way that the human project can succeed if the Earth project fails" (Berry, 2000, p. 127). Berry's sense of shared fate with the natural world has informed the work of a number of Catholic intellectual and eco-justice activists. For example, ecofeminist theologian and Berry scholar, Heather Eaton (2013), now situates her reflections on human-Earth and intra-human relationships within a framework of "socio-ecological crisis" (p. 109). This notion of interrelated crises, demanding an ecojustice response that couples social justice and ecological health, is a central feature of many Catholic initiatives to address climate change (e.g., CCODP, 2014).

During the second half of the 20th century, Catholic intellectuals and activists paralleled, contextualized, and contributed to a growing awareness of various intersections between human activities that damaged equality among people and the ecological crises threatening the future of humanity (e.g., Boff, 2006). In this period, official teachings of the Roman Catholic Church also reflected this emerging worldview, albeit with more hesitation and less clarity (e.g., Paul VI, 1971). Nevertheless, near the turn of the century and into the new millennium, several popes identified the deep interrelatedness between human flourishing and the heath and integrity of the rest of creation (John Paul II, 1989; Benedict XVI, 2009; Francis, 2013). Difficult ecological challenges, including global climate change (GCC), were presented as the product of human practices that neither respected the integrity of creation nor human dignity and, as such, generated both ecological and ethical issues. For instance, citing this legacy at the conclusion of a workshop on "Sustainable Humanity, Sustainable Nature: Our Responsibility," forty-five world experts and Vatican delegates concluded that:

> If current trends continue, this century will witness unprecedented climate changes and ecosystem destruction that will severely impact us all. Human action which is not respectful of nature becomes a boomerang for human beings that creates inequality and extends what Pope Francis [2013] has termed "the globalization of indifference" and the "economy of exclusion," which themselves endanger solidarity with present and future generations.
> (PAS-PASS, 2014)

The Vatican has repeatedly bemoaned the loss of healthy, just, and sustainable relations among peoples and between humanity and the rest of the Earth community. It notes that the distorted values and attitudes that have caused this loss of deep connectivity have also contributed to the advance of GCC. This point is representative of a general orientation on the part of popes and other Catholic bishops toward sourcing the ethical causes of various social and ecological crises in (currently) reparable breakdowns in proper relationships.

In summary, various Catholic intellectuals, ecojustice activists, and official Roman Catholic Church teachings hold that efforts both to more fully realize, analyze, restore, and revitalize integral connectivity between human and Earth communities, and to adopt values that address ecojustice concerns, represent a cogent path for responding to GCC. This emerging socio-ecological ethic can be further informed and understood by bringing the work of Catholic intellectuals, activists, popes, and other bishops into conversation with selected insights from intersectionality theory. Moreover, this analytical approach will help situate Catholic green praxis' actual and potential contributions to socio-ecological flourishing.

Intersectionality, Catholic social thought and climate change

Intersectionality theory can be understood as studying the melding points among various forms or systems of oppression and domination using such social-cultural categories as: class, gender, race and ethnicity, and indigeneity (cf. Yuval-Davis, 2006; McCall, 2005; Woodhams and Lupton, 2014). In the following subsections, we will employ these five analytical categories as a means to map, and therefore highlight, the nature of the intricate ethical problems that accompany GCC in our contemporary context. Specifically, we will triangulate each of the five analytical categories with both Catholic social thought and GCC research. In some ways this is an unnatural parsing, breaking apart an inter-connected whole. However, this methodological approach is justified since it provides a deeper and more particular understanding of the various ethical issues precipitated by GCC. Moreover, it should be emphasized that our mapping, in parallel with the cartography of natural terrain (as opposed to imaginary political borders) deals with contours that penetrate and weave along the landscape; therefore, they do not accord with the type of fixed boundaries that divide and demarcate between nation-states. As such, the five categories unfolded below ought to be considered as fluid and overlapping, and when they overlap, co-constructing a new analytical focus. Taking this approach, which is inspired by intersectionality theory, helps to lay bare the inter-locking ethical problems that are manifest with shifts in the Earth's climatic functioning. Yet, even as we parse for the sake of deeper understanding, it is crucial to remain cognizant of the totality of the ethical issues associated with the multilayered phenomenon of GCC.

Class and poverty

Overcoming social inequities and suffering exposed by the analysis of class and the realities of poverty have been central concerns of Catholicism from earliest Christianity (Acts 4: 32–35). In contemporary Catholic social thought, these concerns are reflected in the concept of the preferential option for the poor, which is based on the notion that those living in poverty or otherwise marginalized ought to be fundamentally favored by social and political processes. As the US Bishops (2011) explain: "While the common good embraces all, those who are weak, vulnerable, and most in need deserve preferential concern. A basic moral test for our society is how we treat the most vulnerable in our midst" (USCCB, 2011, para. 50). Further, the larger life community, understood by Catholics to be God's creation, can be considered a 'new poor' as not only climate change but ecological crises threaten the integrity of the functioning of life on this planet. Hence, the Canadian Bishops also highlight the need for a "preferential option for the earth, made poorer by human abuse" (SCACC, 2003, para. 10). The call for ecojustice by Catholic social thought recognizes that both people of lower socio-economic status and great swathes of other-than-human creation experience suffering sourced in oppressive economic and political structures. Accordingly, "the cry of the earth and the cry of the poor are one" (SCACC, 2003, para. 17); the oppression of the former is invariably linked to oppression of the latter.

Studies of class, poverty, and ecojustice reveal that those who are least responsible for the ecological crises are the ones who are bearing the largest burdens as a result of the same, particularly in the case of GCC. This can no longer be considered an abstract ethical problem. For example, DARA and the Climate Vulnerable Forum (2012) emphasize that "5 million lives are lost each year today as a result of climate change" (p. 14), with 98 percent of those deaths occurring in the least economically developed countries, which are also least responsible for generating the causes of climate change. The principle of a fundamental option for the poor points to a need to address disproportionate suffering as a result of climate change that is experienced by those who are marginalized by their socio-economic status (USCCB, 2009). Calls for distributive justice by these marginalized persons require fair responses that minimally provide: economic compensation; the means to adapt to the effects of climate change through technical support and the transfer of expertise and goods; just immigration, migration and environmental refugee policies; and redressing the equitable use of the atmospheric commons.

Gender

Despite the availability of alternative visions like a discipleship of equals, much of the focus on gender equality in the Catholic Church has centered on priestly ordination (Schüssler Fiorenza, 1993). Addressing the exclusion of women from the institutional priesthood, Pope Francis notes that the "demands that the legitimate rights of women be respected, based on the firm conviction that men and women are equal in dignity, present the Church with profound and challenging questions which cannot be lightly evaded" (2013, para. 104). While Catholic social teaching has presented a vision of gender equality dating from Pope John XXIII's (1963) encyclical, *Pacem in Terris*, the Church's applications of those insights have remained uneven. The Vatican practice of using exclusive language in its English translations, most notably employing "man" and its derivatives to encompass all humans, has long remained problematic for post-Vatican II reformers within the Church as well as its critical observers (Moore, 1985). In the 1970s, Pope Paul VI (1971) called for "a charter for women which would put an end to an actual discrimination and would establish relationships of equality in rights and of respect for their dignity … Developments in legislation should … be directed to … recognizing her independence as a person, and her equal rights to participate in cultural, economic, social and political life" (para. 13).

In practice, calls for women's equality by Church or state do not assuage the reality that women and girls continue to be denied equal rights and access to education, decision-making, and economic opportunities (UN, 2013). Furthermore, "women are more vulnerable to the effects of climate change than men—primarily as they constitute the majority of the world's poor and are more dependent for their livelihood on natural resources that are threatened by climate change" (UNWW, 2010, n.p.). Women are less well-positioned than men to cope with the adverse effects of climate change since they are hampered by social, economic, and political structures that limit their mobility, access to resources, and participation in decision-making processes (UNWW, 2010). As a result of such marginalization, Catholic ecofeminist theologians connect oppression of women and the planet, citing dually-alienating structures that are buttressed by the forces of patriarchy (Ruether, 2003). Catholic ecofeminists also tend to offer positive solutions to this malaise. For example, the Brazilian nun, liberation theologian and advocate for reproductive justice, Ivone Gebara (2003), shades the entire religious project "in light of making justice, of right relationships with women, men, and all living beings" (p. 103).

Race and ethnicity

As part of their struggle to overcome a Eurocentric orientation that too often resulted in racist actions and programming, Catholics have been able to draw on visions of equality with a long genealogy related to the premise that following Jesus Christ ought to be an identity marker that overcomes all racial and ethnic divisions (cf. Galatians 3:28). The bishops, assembled at the Second Vatican Council (1965), highlighted this point as essential to the Catholic Church's mission in the cotemporary world. The torch in this regard has been carried forward and contextualized by Catholic theologians who apply liberationist methodologies to issues of equity. Gustavo Gutiérrez (1988) notes, for liberationists "the use of a variety of tools does not mean sacrificing depth of analysis; the point is … to insist on getting at the deepest causes of the situation" (p. xxv). As a result, today there are Catholic liberationist thinkers who analyze the fracture of just relationships inherent in racism and ethnocentrism. For example, Catholic womanist liberation theologians such as Diana Hayes (2010) generate particular and embodied insights concerning "triple and often quadruple oppression in today's world" (p. 6) at intersections of race, gender, faith, and class.

Racial analysis applied to the prevalence of climate change-related effects offer some telling revelations regarding the manifestations of multilayered oppression. In terms of ethics, for example, a basic principle of climate justice is that the benefits and burdens of shifts in the functioning of the Earth's weather systems as a result of human activity should be shared equally (Hrynkow and O'Hara, 2014). Too often, racialized communities have born a disproportionate amount of the suffering associated with ecological ill-health as represented by the concept of 'environmental racism.' Though the concept traditionally deals with issues such as the placement of toxic sites close to racialized communities (Cole and Foster, 2001), extreme weather events like Hurricanes Katrina and Sandy preview how marginalization also reduces capacity to flourish under shifting climatic conditions. In the case of Hurricane Katrina, not only were Black neighborhoods more likely to be vulnerable to the impact of severe weather, they were less likely to receive financial support for their rebuilding (Bullard and Wright, 2009). Further, on a global scale, a perspective of intersectionality can serve to highlight how constructions of race reveal that those least responsible for GCC are suffering ill health and reduced life expectancy as a result of shifting weather patterns (Hrynkow and O'Hara, 2014).

Indigeneity

Despite the resistance of a creative minority represented by figures such as Bartholomé de Las Casas (Gutiérrez, 1983), the stories of enslavement and violent evangelization that accompanied the expansion of Catholicism into Latin America during the early modern period can be taken as clear evidence of the Catholic Church's failure to respect indigenous peoples and their worldviews. Healing such legacies has been a major challenge for contemporary Catholicism. However, such reconciliation processes (cf. CCCB, 1995), important for intra-human community healing, can have the added advantage of addressing root causes of GCC, which coalesce with the colonization and domination of peoples holding integral worldviews.

Indigenous cultures are intimately connected to place, and value traditional ecological knowledge (TEK) into the dynamics of the ecosystems in which they dwell (Berkes, 2009). While it has been argued that the blending of TEK with environmental sciences would provide a richer understanding of ecological communities and their co-management, this mutually respectful integration generally remains a distant goal (Sherry, 2002) and TEK is not adequately recognized as an equal dialogue partner in climate science and politics (Smith, 2012). Furthermore, indigenous peoples' relationships with the land have been shifted by commercial, industrial, and military processes that have generated socio-ecological crises. The direct effects of climate change serve to further degrade and unduly modify the ecological communities upon which indigeneity rests. A shifting climate depleting sea ice in the North would, for example, necessarily represent an end to seal hunting that has been a key feature of Inuit culture from time immemorial. And while TEK has extended the ability of Inuit communities to adapt to these changes, continuing climate warming is exhausting those adaptations (Leduc, 2010). Similarly, the cultures of Pacific Island peoples are losing their key existential reference point as the islands on which those cultural systems developed are becoming submerged below a rising ocean resulting from GCC. Further, as the climate shifts, TEK about ecosystems and human activity can also lose its contextual cogency. The end result can leave entire peoples as climate refugees with their time-honored sustainable ways of living in relationship with the larger life community marooned with diminishing chances for rescue.

GCC and Catholic green praxis

Catholics are involved in cultivating such a heartfelt global conscience through a remarkable number of green initiatives along with ecumenical, inter-religious, and secular coalitions of action and advocacy that seek to mitigate the negative effects of GCC. While there are many such initiatives, we can only focus on two examples, describing programming initiatives undertaken by an intentional community of Catholic nuns, and a coalition for responding to the challenges of a shifting climate from a broadly Catholic perspective. Nonetheless, even this large-scale mapping will serve to demonstrate something of the current and potential value of Catholic green praxis for the creative functioning of an Earth community under threat by climate change.

Genesis Farm: previewing a good life after peak oil

Founded in 1980 by the Sisters of St. Dominic of Caldwell, NJ after they received a donation of land, Genesis Farm (2014) has been a prominent intentional community working to increase ecological literacy and model sustainable living. The site is home to a 51-acre community-supported agricultural initiative, several residential courses on eco-spirituality and a transition culture, bioregional celebrations, spiritual events tied to the changing seasons, a meditative walk designed to stimulate cosmological consciousness, and a section of land that is reserved for other-than-human creatures. The farm's director, Mariam McGillis, is well known for building upon and applying Thomas Berry's eco-ethical insights. As Phoebe Godfrey (2008) argues, one key result of this synthesis is the combination of Berry's reflection on a functional cosmology with ecofeminist ways of knowing and being-in-the-world. In light of the discussion of intersectionality presented above, this expands the remit of the farm's initiatives to include socio-ecological content and action, making it a location of Catholic green praxis in a fuller sense.

This relationship has been mutually enhancing, offering a model of alternative sustainable living, growing from the Catholic tradition, reaching out into the surrounding bioregion and, because of the ethical force of an integrated approach to living, beyond to the wider world. A principle value of this work has been the incarnation of a low-carbon lifestyle buttressed by a re-contextualized Catholic spirituality, which has, in turn, inspired others to address the ecological crisis on the levels of insight and action. Moreover, a number of the sisters and their supporters, McGillis among them (Godfrey, 2008), came to the project with a background in peace and social justice work and these concerns have remained active in Genesis Farm (2014) programing.

Drawing on the sisters' own framing, Genesis Farm's contribution can be read as an effort to demonstrate that quality living in proper relationship with the Earth community is possible after peak oil. This effort is multidimensional in nature, extending out from the model community through education and action initiatives. Hence, it is not surprising that the community participated in the People's Climate March held in New York City (September, 2014) with the hope of generating increased momentum for an effective, binding global agreement to mitigate against GCC (Genesis Farm, 2014).

Bringing the example of this faith-inspired, intentional community into conversation with the five components of intersectionality we have mapped above helps to situate the tensions and promises in relation to Genesis Farm's contributions to Catholic green praxis. Genesis Farm's location and constituency tends towards an Anglophone, white, and middle class demographic, which is a feature of much eco-theology work in Canada and the USA. A corrective in this regard is the global ecofeminist witness for justice that includes, for example, indigenous, Latina, and Black/African women's voices, which is taught on the level of insight as part of Genesis Farm programing. This latter feature of their educational outreach is indicative of Genesis Farm's expansive ecofeminist gender consciousness. For example, in explicating her ecofeminist witness for social justice, Miriam McGillis (2014, n.p.) comments that "many of the roots of our racism, bigotry, aggression, sexism and arrogance [are] sustained by our separateness from the earth and aversion to anyone who is different." In line with McGillis' comments, intersectionality suggests that Genesis Farm actively seeks to form more prominent links with diverse communities. This process can be moved forward through participation in such events as the People's Climate March (September 2014), which allowed for encounters among people who hold strong visions for a just and sustainable future, but come to that commitment through diverse responses to marginalization related to class, gender, racial and ethnic, linguistic and cultural, or indigenous identities.

Catholic climate covenant: responding to a global challenge from a broadly Catholic perspective

Drawing on many of the discourses we have mapped with broad brush strokes above, Catholic Climate Covenant (CCC) centers its programming on ecojustice principles, as is suggested by its banner slogan "Care for Creation, Care for the Poor." The organization was founded in 2006 in partnership of the US Bishops to act as an expression of their environmental justice programming (CCC, 2014). As such, it is not surprising that the coalition's campaigns seek to connect ecologically responsible behavior with just outcomes for those on the margins of global society. For example, a present campaign highlights global justice issues linked to GCC precipitated by overconsumption in the USA, and then asks viewers of the CCC website: "Who's under your carbon footprint?"

Visitors to the webpage are given free access to a number of resources to support educational and practical ecojustice initiatives in a variety of Catholic contexts. Moreover, they are encouraged to take the "St. Francis Pledge to Care for Creation and the Poor." This pledge invokes the example of a medieval reform movement in the Church, given a large measure of momentum by Francis and Clare of Assisi, who were known for their closeness to nature, simple living, care for those on the margins of society, and taking on the mantle of the "poor Christ" (Boff, 2006). Today, there are a number of orders and movements of priests, religious brothers and sisters, and lay people within the Roman Catholic tradition and beyond who identify as Franciscan. As Dawn Nothwehr (2012) highlights, the inviolable nature of human dignity is at the heart of Franciscan views of personhood. Thus, many within the Franciscan tradition hold that structures or actions that do not simultaneously enhance relationships with creation and respect human dignity will not foster substantive peace and justice. Moreover, the pledge's nomenclature has the additional advantage of resonating with the excitement surrounding Pope Francis, who chose his papal name to reference the medieval saint and, at the time of writing, is set to release a social encyclical on ecology. With such resonances never far from the surface, people and organizations taking the St. Francis Pledge commit to pray, reflect, and learn about "the duty to care for God's Creation and protect the poor and vulnerable" (CCC, 2014). They also commit to assess and act to change personal and associational complicity in GCC. Moreover, they pledge to advocate "for Catholic principles and priorities in climate change discussions and decisions, especially as they impact those who are poor and vulnerable" (CCC, 2014). This initiative is part of the CCC's general goal of "creating a climate of solidarity" and, at the time of writing, over 11, 000 individuals and organizations had taken the pledge (CCC, 2014). A relevant question then becomes how the solidarity developed through CCC's outreach materials might also encompass class, gender, racial and ethnic, linguistic and cultural, and indigenous identities.

A media analysis of CCC's materials reveals a preponderance of references to statements by bishops and popes. This can be explained by the common practice for Catholic NGOs, activists, and academics (including the authors of this chapter) to buttress their work with references to Catholic institutional voices of authority in order to calm potential concerns about the Catholicity of their work. Frequently, these same 'grassroots' Catholics, activists and academics, while working within an atmosphere of scrutiny, were the first to develop insights on socio-ecological justice that bishops and popes later espouse. The practice of appealing to the episcopal class of Catholics in order to demonstrate the orthodoxy of one's work perpetuates the misconception that bishops and popes are the sole definers of Catholicism and at its forefront. As a result, this approach supports a potentially oppressive structure of class privilege. Despite a long history of Catholic Social Teaching championing a preferential option for the poor who live at the social, economic, and political margins of society, intersectionality theory exposes an often overlooked need to address structural marginalization within the Catholic community itself.

Furthermore, intersectionality theory raises another potential concern with the CCC's website. Their "Who's under your carbon footprint?" campaign uses images of those most affected by climate injustice such that their representation of victimhood could become a form of disempowering, racialized suffering. Again, this is a tension not unique to the CCC, yet it certainly merits vigilance in order to avoid any possible diminishment of the valuable work done by this group.

Conclusion: working toward socio-ecological flourishing as a *telos* for green Catholic praxis

As early as the 1970s, Pope Paul VI (1971) began to link nascent concerns about ecological health with previously stated concerns about social justice. As the inseparability of these two issues became increasingly clear during subsequent pontificates, the analysis became stronger and clearer (Benedict, 2009), and a socio-ecological ethic became theoretically articulated by the Vatican and

Catholic academics and incarnated by Catholic activists. More recently, this socio-ecologic ethic has been applied to analyses of GCC.

To unpack the genealogy of this ethic, we have employed five overlapping analytical categories inspired by intersectionality theory. Each of these analytical categories was triangulated with Catholic social teaching and GCC research. This approach not only revealed the dynamic, multiple, and overlapping layers of critical analysis and insight within a socio-ecological ethic, but it also showed the wide application of the ethic to the complex issues of GCC. This triangulation both showed the promise of Catholic teaching with respect to each of the analytical categories of intersectionality theory as they were linked with GCC, and exposed significant failings and gaps in both Catholic teaching and praxis. McFarland Taylor (2007) has noted that theologians and religious communities are generally more progressive in their responses to socio-ecological issues than the Vatican, and this is especially true for women religious orders. That is, they are more likely than the Vatican to engage more completely and comprehensively the various categories of intersectionality theory, especially as these relate to GCC. However, as noted above, gaps and failings have been revealed at all levels within the Church through the application of a socio-ecological ethic, not only in theoretical musings but in practical applications. These regrettable gaps are shown to be even wider when intersectionality theory is applied.

In light of our mapping of these insights and actions, we want to end by suggesting that, in seeking to overcome the unjust burdens of shifting weather patterns and fostering a climate of solidarity, the concept of socio-ecological flourishing provides a contextually appropriate *telos* for green Catholic praxis. Significantly, the concept indicates a positive set of goals that purposely respond to the multi-dimensional challenges indicated by the term 'socio-ecological crisis.' For example, in seeking to heal the present crisis brought about by GCC, socio-ecological flourishing points to how solutions to the ecological degradation can be situated within Catholic cultures. Or, put another way, socio-ecological solutions will have more resonance for the Catholic tradition when they keep concerns for social justice at the forefront of their practical and intellectual responses to ecological challenges, like GCC, which effect the creative functioning of the entire Earth community. As this chapter has begun to demonstrate, concepts of intersectionality can help map, deepen, and otherwise further inform Catholic green praxis when concerns associated with analytical concepts such as class, gender, race and ethnicity, and indigeneity are harvested from the tradition and re-contextualized because of their cogency in the service of an integrated socio-ecological ethic.

There are many promises inherent in such contextual integration. For example, it offers many points of entry for those among the world's 1.2 billion Catholics seeking explicitly Catholic foundations to heal the current damage and prevent future undesirable consequences from GCC. In this manner, working towards a state of socio-ecological flourishing becomes representative of a contextual recasting of the normative Catholic goals of right relationship with God and neighbor. Here, the social justice component will be necessarily intersectional, including overcoming marginalization and alienation attached to identity markers such as class, gender, race and ethnicity, and indigeneity. Such a formulation of right and socially just relationships views 'neighbor' as a category denoting both humanity and the rest of the ecological world so that deep equity among people is understood to be buttressed by the health of the other-than-human members of the Earth community, and vice versa, in a positive feedback loop. The end result is the promise that Catholic green praxis can contribute to denominational, ecumenical, inter-religious, and secular projects seeking to both mitigate the negative effects of GCC and transform the contemporary situation into one that is not only sustainable but also fosters socio-ecological flourishing. In this manner, Catholic green praxis can more unequivocally embrace a multitude of programing to help ensure that the world keeps turning as home to a diverse and vibrant Earth community. The extent to which it does so will depend on its ability both to embrace the promise of its commitment to a socio-ecological ethic and to address the gaps and failings in its responses to date.

References

Benedict XVI, 2010. If you want to cultivate peace, protect creation. *2010 World Day of Peace Message*. [Online] Available at: www.vatican.va/holy_father/benedict_xvi/messages/peace/documents/hf_ben-xvi_mes_20091208_xliii-world-day-peace_en.html [Accessed 20 December 2010].

Berkes, F., 2009. Indigenous ways of knowing and the study of environmental change. *Journal of the Royal Society of New Zealand*, 39(4), pp. 151–156.

Berry, T., 1988. *The Dream of the Earth*. San Francisco, CA: Sierra Club Books.

Berry, T., 1999. *The Great Work: Our Way Into the Future*. New York: Bell Tower.

Berry, T., 2000. Christianity's role in the earth project. In: D.T. Hessel and R.R. Ruether, eds, *Christianity and Ecology*. Cambridge, MA: Harvard University Press. pp. 127–134.

Boff, L., 2006. *Francis of Assisi: A Model for Human Liberation*, Maryknoll, NY: Orbis.

Bullard, R. and Wright, B. eds, 2009. *Race, Place, and Environmental Justice After Hurricane Katrina: Struggles to Reclaim, Rebuild, and Revitalize New Orleans and the Gulf Coast*. Boulder, CO: Westview Press.

CCODP (Canadian Catholic Organization for Development and Peace). 2014. Rio +20: a time to rethink the green economy. [Online] Available at: www.devp.org/sites/www.devp.org/files/IMCE/files/rio/devpeace_rio_summit.pdf [Accessed 17 August 2014].

CCC (Catholic Climate Covenant). 2014. Catholic Climate Covenant: care for creation. Care for the poor. Web.

CCCB (Canadian Conference of Catholic Bishops). 1995. Let justice flow like a river. *Brief to the Royal Commission on Aboriginal Peoples*. [Online] Available at: www.cccb.ca/site/images/stories/pdf/justice_flow_residential_schools.pdf [Accessed 20 August 2014].

Cole, L.W. and Foster, R., 2001. *From the Ground Up: Environmental Racism and the Rise of the Environmental Justice Movement*. New York: New York University Press.

DARA and the Climate Vulnerable Forum, 2012. Climate vulnerability monitor: a guide to the cold calculus of a hot planet, 2nd edition.

Eaton, H., 2013. Forces of nature: aesthetics and ethics. In S. Bergmann, I. Blindow, and K. Ott, eds, *Aesth/Ethics in Environmental Change: Hiking through the Arts, Ecology, Religion and Ethics of the Environment*. Berlin: LIT Verlag, pp. 109–126.

Francis, 2013. *Evangelii gaudium*. Apostolic Exhortation. Web.

Gebara, I., 2003. Ecofeminism: a Latin American perspective. *Cross Currents* 53(1), pp. 93–103.

Genesis Farm, 2015. Genesis farm: since 1980. Web.

Godfrey, P.C., 2008. Ecofeminist cosmology in practice: Genesis Farm and the embodiment of sustainable solutions. *Capitalism Nature Socialism* 19(2), pp. 96–114.

Gutiérrez, G., 1988. *A Theology of Liberation: History, Politics and Salvation*. Maryknoll, NY: Orbis Books.

Gutiérrez, G., 1993. *Las Casas: In Search of the Poor of Jesus Christ*. Maryknoll, NY: Orbis Books.

Hayes, D., 2010. *Standing in the Shoes My Mother Made: A Womanist Theology*. Minneapolis, MN: Fortress.

Hrynkow, C. and O'Hara, D.P., 2014. Catholic social teaching and climate justice from a peace studies perspective: current practice, tensions, and promise. *New Theology Review*, 26 (2), pp. 23–32.

John XXIII, 1963. *Pacem in Terris*. Papal Encyclical. [Online] Available at: www.vatican.va/holy_father/john_xxiii/encyclicals/documents/hf_j-xxiii_enc_11041963_pacem_en.html [Accessed 16 August 2014].

John Paul II, 1989. Peace with God the creator, peace with all of creation. 1990 World Day for Peace Message. [Online] Available at w2.vatican.va/content/john-paul-ii/en/messages/peace/documents/hf_jp-ii_mes_19891208_xxiii-world-day-for-peace.html

Leduc, T., 2010. *Climate, Culture, Change: Inuit and Western Dialogues with a Warming North*. Ottawa: University of Ottawa Press.

McCall, L., 2005. The complexity of intersectionality. *Signs: Journal of Women in Culture and Society*, 30(3), pp. 1771–1800.

McGillis, M., 2014. Quoted in Genesis Farm: A Community for Our Times. Web.

McFarland Taylor, S., 2007. *Green Sisters: A Spiritual Ecology*. Cambridge, MA: Harvard University Press.

Moore, M.E., 1985. Inclusive language and power: a response. *Religious Education*, 80(4), pp. 603–614.

Nothwehr, D.M., 2012. *The Franciscan View of the Human Person: Some Central Elements*. St. Bonaventure, NY: The Franciscan Institute.

Paul VI. 1971. *Octogesima adveniens*. Papal Encyclical. Web.

PAS and PASS (Pontifical Academy of Science and Pontifical Academy of Social Science), 2014. Statement of Vatican workshop on Sustainable Humanity, Sustainable Nature: Our Responsibility. Web.

Ruether, R.R., 2003. Ecological theology: roots in tradition, liturgical and ethical practice for today. *Dialog: A Journal of Theology*, 42(3), pp. 226 – 234.

Schüssler Fiorenza, E., 1993. *Discipleship of Equals: A Critical Ekklesialogy of Liberation*. New York: Crossroads.

Second Vatican Council, 1965. *Gaudium et spes*. Pastoral Constitution. Web.

Smith, H. and Sharp, K., 2012. Indigenous climate knowledges. *Wiley Interdisciplinary Reviews: Climate Change*, 3(5), pp. 467–476.

Sherry, E., and Myers, H., 2002. Traditional environmental knowledge in practice. *Society & Natural Resources*, 15(4), pp. 345–358.

SACCCB (Social Affairs Commission, Canadian Conference of Catholic Bishops), 2003. 'You Love All That Exists … All Things Are Yours, God, Lover of Life …'. Web.

UN (United Nations), 2013. Millennium development goals report 2013. Web.

UNWW (United Nations Women Watch), 2010. Women, gender equality and climate change. Web.

USCCB (United States Conference of Catholic Bishops), 2009. Why does the church care about global climate change. Web.

USCCB, 2011. *Forming Consciences for Faithful Citizenship: A Call to Political Responsibility from the Catholic Bishops of the United States*. Web.

Woodhams, C. and Lupton, B., 2014. Transformative and emancipatory potential of intersectionality research. *Gender in Management: An International Journal*, 29(5), pp. 301–307.

31 Our climate, our change

Using visual and interactive practices to expand participation and leadership in climate action

Jennifer L. Hirsch, Abigail Derby Lewis, Ryan Lugalia-Hollon, Lisa See Kim, Sarah Sommers and Alexis Winter

Since 2008, anthropologists, ecologists, artists, and communications specialists from the applied science division of The Field Museum of Natural History have worked, in partnership with the City of Chicago, to engage diverse Chicago communities in city and regional climate action efforts. This work began when the Chicago Department of Environment commissioned the museum to help them engage communities throughout the city in the Chicago Climate Action Plan, launched in October 2008. The Plan aims to reduce carbon emissions to 25 percent below 1990 levels by 2020 and 80 percent by 2050 by implementing five strategies focused on energy efficiency in buildings, clean and renewable energy, improved transportation options, waste reduction, and adaptation (City of Chicago, 2008).

The museum was seen as a strong intermediary between the City and communities for two reasons. First, as a scientific institution, it was viewed as apolitical. Second, it had been conducting action research and implementing urban ecology projects in communities for over fifteen years and was regarded by many as a trusted partner, bridging grassroots and institutional stakeholders and simultaneously focused on nature and people.

From 2008 to 2012, Museum staff conducted rapid research studies in nine communities and developed and led or co-led three major community action programs that built on research findings. The six authors of this chapter participated in this work. Communities were chosen to achieve geographic, ethnic, racial, and socioeconomic diversity. The research resulted in two energy efficiency programs initiated by the City in low-income communities: the Energy Action Network and the South Chicago Retrofit Project. Additionally, The Field Museum worked with community partners to develop the Chicago Community Climate Action Toolkit, discussed in detail later in this chapter.

The community action programs were designed to engage communities in both the Chicago Climate Action Plan and the region's other major climate action plan, the Climate Action Plan for Nature, which was launched by the

Chicago Wilderness alliance in 2010 with significant leadership from The Field Museum. The Climate Action Plan for Nature addresses the broad 'Chicago Wilderness' region that surrounds southern Lake Michigan. While the Chicago Climate Action Plan focuses primarily on people and the built environment, the Climate Action Plan for Nature specifically addresses climate change impacts and adaptation strategies for the region's plants and wildlife.

Our approach to working with communities includes four stages and goals, each described in Table 31.1. Stage 1, the first and most critical stage, entails research specifically targeted to identify *community assets*, or those strengths and resources that can be mobilized to effect change (see Box 31.1), as well as concerns. Stages 2, 3, and 4 build on the research conducted in Stage 1.

In all of this work, creative visual and interactive practices have been key to engaging urban residents in an issue that often seems distant and less than urgent. These practices include photo/object elicitation (using photos or objects to start a conversation), collecting and telling stories, participatory mapping/photography, drawing, art, data visualization, and more. Read more about these practices in the section below, 'Creative visual and interactive practices: making climate change relevant, tangible, and fun.'

This chapter introduces some of our most successful creative practices, drawing particularly from our research and our most recent project, the Chicago Community Climate Action Toolkit. We have found that these practices can help leaders and residents from racially, ethnically, and socioeconomically

Table 31.1 The four stages and goals of our approach to working with communities

Stage	*Goals*
1. Conduct rapid, collaborative community research	Work with community partners to identify local assets and concerns related to climate change and the region's climate action plans (see Box 31.1)
2. Report on findings, through written reports and presentations	Help municipalities; community leaders; other organizational partners; and broader audiences see the connections between Chicago community life; climate change; and climate action as a basis for increased community engagement
3. Create public education and engagement materials on climate change and climate action; and train community organizations, municipalities, and educators to use them	Increase understanding of climate change and how it is relevant to the Chicago region and people's everyday lives Equip community leaders to take the lead on climate action.
4. Develop and carry out climate action projects	Help form diverse, community-based partnerships that advance climate action and improve local quality of life

Box 31.1 What are community assets and concerns related to climate change?

Our research identified local assets related to climate change and climate action that can serve as springboards for increasing engagement in climate action, such as energy efficiency, climate-friendly gardening, recycling, etc. We looked for assets that are tangible, such as organizations (nonprofit, for profit, municipal), infrastructure (e.g. boulevards, rivers, parks), and people with valuable skills (e.g. community leaders, informal networkers, gardeners, fundraisers, artists). We also looked for assets that are intangible, such as values (e.g. frugality, sense of global citizenship), traditions (e.g. seasonal festivals celebrating nature, conserving water), practices (e.g. growing your own food, sharing with neighbors), ideas, and dreams.

Additionally, our research identified community concerns that might be able to be addressed by climate action work, such as a desire for additional after school programs for youth. The research also highlighted barriers to increased participation in climate action, including tangible barriers like gang activity in parks and intangible barriers such as perceptions of climate action or "going green" as a luxury.

diverse communities: understand climate change as something that has to do with their particular histories and lives; recognize the varied ways in which they are already taking climate action, even if they do not realize it; begin to envision themselves as a key part of climate action solutions; and then develop community projects based on these visions. The chapter concludes with a brief discussion on the potential of using visual and interactive practices to engage more groups of people in climate action and lift up diverse cultural perspectives within a field that is otherwise largely focused on technological solutions.

Deep engagement in climate action: a brief look at the literature

Our approach to engaging diverse communities in climate action can be seen as a case study of deep engagement. Anthropologists who study contemporary global climate change (a small but growing group) as well as scholars and practitioners of climate change communication have recently begun to call for expanded efforts to dialogue with the public on this issue, beyond just conveying scientific facts and information. Anthropologists explain the importance of exploring climate change as a cultural, rather than simply environmental, phenomenon. They argue that people experience climate change, both its discourse and its impacts, through the 'mediating layers' of their lives, such as social networks, cultural identities, and other issues that they care about (Roncoli, Crane, and Orlove, 2009; see also Crate and Nuttall, 2009; Crate, 2011; Barnes *et al.*, 2013).

Scholars and practitioners of climate change communication make similar arguments, although they often talk in terms of 'values' rather than 'culture.' They suggest that climate action efforts should engage people by:

- Working through groups or networks, rather than focusing on individuals (e.g., Michaelis, 2007)—including and perhaps especially groups that have not been engaged in climate- or environment-related issues and thus 'perform a critical role in spreading change through society' (Corner and Randall, 2011, p. 1011);
- Highlighting the *human* (in addition to the planetary) benefits of a low-carbon lifestyle, including but also going beyond the benefit of saving money (often referred to as 'co-benefits;' see Westphal and Hirsch, 2010);
- Constructing narratives, or stories, to help people situate climate change in relation to local conditions and their deeper or intrinsic values, 'such as duty, stewardship, self-reliance, and prudence' (Pike, 2012; see also, e.g., Agyeman, 2007; Marshall, 2012).

In a particularly strong call to move beyond social marketing—the application of marketing principles to social issues—Corner and Randall (2011) argue that the response to climate change must be cultural transformation, not just isolated behavior changes. This, they suggest, means that climate action efforts must work through social networks to create long-lasting 'pro-environmental social identities.'

Meanwhile, anthropologist Sarah Pink, who writes about the emerging field of applied visual anthropology (2007; 2011), explains that visual practices—including visual research methods, analysis of visual culture, and visual representation—provide 'routes through which other people's understandings, experiences and ways of doing things can become 'visible,' and therefore comprehended, explained to others' (2011, p.446). As such, these practices nurture empathy and connections and provide especially strong tools for helping lift up the voices of marginalized populations (2011). In this chapter, we build upon Pink's claims, demonstrating the key role that visual and other creative practices can play in moving climate change work beyond surface-level communication, particularly among people who have had little previous involvement in climate change-related efforts.

Creative visual and interactive practices: making climate change relevant, tangible, and fun

In our research and subsequent community climate action programs, we employed a wide range of ethnographic practices. These included the traditional methods used by anthropologists (interviews, focus groups, and participant-observation) as well as creative visual and interactive practices such as visual collages of climate-friendly practices, data visualization (e.g. word clouds), participatory photography, home tour interviews, and exercises using prompt questions such as, "What three words come to mind when you hear 'climate change?'" These latter practices are the focus of this chapter.

Climate change is an issue that people often do not relate to personally, and the creative practices we used were particularly effective during all four stages of our work in creating a place-based model for climate action: turning climate change from a technical, scientific, and distant topic requiring particular expertise into conversations and actions focused on Chicago and people's lives. These practices often helped people see that they are already taking many actions that can be considered climate action. This framing helped people make their own connections to the issue and to climate action practices, which they were able to share with us during the research stage and then incorporate into their local projects. These connections would likely not have come forth using traditional practices alone.

Our overall methods—collaborative research and story collecting—comprise two of these creative practices. Leaders from one to two community organizations worked with us on each of the nine research studies. In all cases, they connected us to residents and other organizations and also commented on report drafts. In three communities, Field Museum anthropologists worked with professional storytellers to train our community partners to use some of our tools to collect stories about residents' climate-friendly practices, which contributed key data to the research (see Figure 31.1). The stories that our researchers and partners collected were highlighted in our reports; used by partners in their community work; and performed by our storyteller partners in an online video, 'Telling Our Stories: Creating Green Communities' (http://vimeo.com/35764542).

The collaborative research process that we used, wedding the museum's scientific expertise with partners' story collecting, laid the groundwork for community-initiated action later on. By training community leaders in some of our research methods, we helped them realize that they could indeed take ownership over the climate science required for larger-scale climate action. Their participation in the research helped expand their knowledge of a pressing contemporary issue as well as the resources their communities have to address it. Additionally, since community partners conducted the research, they were more invested in using the findings as a basis for their action projects.

Our work also involved many visual and interactive tools. These tools were key to our research success. Pictures and interactive activities allow people to understand a topic and contribute to a conversation in many ways, tapping into multiple areas of knowledge. They are inherently more experiential than words and give a wide range of people entry points into a topic in ways that can transcend boundaries of age, race, class, language, and literacy. Similarly, photographs are always open to interpretation and often elicit responses that are less formally directed than surveys or traditional interviews.

One of our most effective tools was a series of visual collages depicting photos of a variety of climate action strategies (see Figure 31.2). Each collage depicts a different type of climate action strategy, such as improved transportation options, energy efficiency, etc. Many of the photos relate specifically to the Chicago

Figure 31.1 Collecting stories: as part of our research study in Pilsen, a Latino community on Chicago's Near West Side, a staff member from the Mexican hometown association Casa Michoacán collected stories from residents passing by on the street

Source: Image credit: © The Field Museum

region and the diverse cultures represented here. Some actions are traditional, such as carpooling or turning off the faucet. Others are creative or rooted in cultural heritage, such as a Polish greeting card that depicts a young girl hanging her doll's clothes out to dry on a clothesline or a youth recycling program run by a community development corporation on the South Side of Chicago. Interviewers showed these collages to interviewees and asked, 'Do you do any of these things? Do you know others who do these things?'

In response to these collages, residents shared stories about practices that they likely had never thought of as climate action but which served to personalize a seemingly impersonal issue. For example, during a focus group with African Americans living on Chicago's West Side, a lively discussion about gardening broke out in response to our visual collage on climate-friendly practices related to the land (representing the 'Adaptation' strategy of the Chicago Climate Action Plan). One elderly woman who had grown up in Arkansas was reminded of all the crops her family grew and expressed pride in their farming knowledge and self-sufficiency. She commented, 'You're dealing with a country lady. I know everything and can do everything.'

Visual practices were also a key strategy for creating accessible materials based on the research, all of which aim to help community leaders take ownership of climate action. For example, our reports include visual story packages that pair research photos and stories highlighting specific climate actions as central to residents' lives—literally changing the face of climate change from the polar bear to your next-door neighbor. Relatedly, asset maps highlight the assets identified in the research by our ethnographers and research participants. They visually depict the large quantity of climate-related work already happening in a community, providing an expanded frame for thinking about community identity and work that likely has not been thought of to date in terms of climate change. Our intention in creating and sharing these types of visualizations was twofold: to present the assets in an easily accessible way, and to prompt new thinking about potential actions and partnerships.

Finally, our multimedia engagement materials rely heavily on visuals, as demonstrated by our Chicago Community Climate Action Toolkit. Launched online in May 2012, the Toolkit comprises over sixty multimedia tools that communities can use to develop and implement local climate action projects in ways that also advance their ongoing work for improving quality of life, for example, around issues such as education, food access, and economic development. The Toolkit documents four projects that The Field Museum created and implemented with community partners in the research communities of Pilsen, South Chicago, Forest Glen, and Bronzeville. It also includes education and engagement tools created by Field Museum scientists, with input from community and environmental partners around the region. The projects and the tools built on and addressed our community research findings.

Toolkit materials use visuals towards multiple ends. They help people understand complex scientific concepts—such as the difference between climate change and the problem of the 'hole' in the ozone layer, two issues that research participants often conflated. They also help people see climate change and climate action as related to Chicago—including our extreme weather events such as a major snowstorm in 2011 that stranded people in their cars for over 24 hours. Finally, they help local groups develop their own ideas for action tailored to their communities that will simultaneously advance the region's climate action strategies at the local level—such as the Council of Islamic Organizations' 'Green Ramadan' campaign, which promotes local action among Muslim residents as part of a long-term solution to drought and famine in Somalia.

Community projects: using visual and interactive practices for action

Visual and interactive practices have also emerged as key components of our community climate action projects themselves. Specifically, they have helped with: 1) integrating climate action into a community's identity and vision; 2) communicating the project to residents and broader audiences; 3) making climate action fun and engaging; and 4) connecting climate action to other community issues and co-benefits. In this final section, we use examples from the four community projects that we worked on as part of the Toolkit to demonstrate each of these points.

Together, these communities represent a significant portion of Chicago's diversity. Pilsen, just west of Chicago's downtown, is largely Mexican and working-class and is known as the hub of Chicago's Mexican community and an artistic enclave. South Chicago, on the far South Side, is known for its industrial history, proximity to natural areas, and immigrant populations. It is racially and ethnically diverse (African American, Latino, and white) and largely working class. Bronzeville, just south of Chicago's downtown, is an area rich with black history that is often compared to Harlem in New York. It is almost wholly African American and is class stratified, with many subsidized public housing residents living next to affluent homeowners, most of them African Americans who have 'returned' to help revitalize the area. Forest Glen, located on the far North Side, is largely affluent and white. It is known for its natural areas and suburban feel.

Point 1: Integrating climate action into a community's identity and vision—South Chicago project

The South Chicago community project was highly visual and aimed to create a community-wide exhibit that celebrates local green practices and promotes the community's vision for a green future. Titled 'Retrofit Your Neighborhood,' the exhibit includes a mural, two large outdoor planters made from recycled sewer pipes, ten displays made from old windows and shutters, and a website and video. The various pieces all highlight stories collected by a group of local youth using The Field Museum's visual collages. The stories describe how people are already caring for the environment, in small and large ways, individually and together. Taken together, they show the many different paths that can lead people to see themselves as stewards of their physical environment and to view their community as 'green.' The fact that the exhibit represented community members' stories also makes it interactive, in the sense that it represents conversations held between exhibit creators and residents, thereby prompting more investment in the exhibit from the community.

The exhibit was part of a community-wide effort to increase residential retrofits, and each piece of the exhibit also advertised the retrofit project. According to Sarah Ward, Executive Director of the art center that created the planters, a visual strategy like this exhibit is important in a diverse community, including one like South Chicago where some speak limited English. She explained: 'Because the neighborhood has such a diverse mix of people…there's a lot of public art…I think the community responds to art as a vehicle for communication.'

Point 2: Communicating climate action to residents and beyond—Forest Glen Project

The Forest Glen community project was not as explicitly visual in nature. Initiated by a local Chamber of Commerce as a way to unite multiple stakeholders around a common vision, it included a variety of activities aimed at promoting climate-friendly practices. These included installing bat boxes in a local nature preserve as an alternative to pesticides, installing rain barrels in 40 homes for storm water management and water conservation, and planting native plant and food gardens using climate-friendly gardening methods. All of this work was done by local Scout troops.

It was exactly this broad scope of work that led the partners to decide to brand their project with a graphic identity that would allow them to communicate the project to the wider community and recruit other organizations to participate. The project team held a series of meetings to brainstorm a list of objects, places, and symbols that represent the project in relation to the community's broader identity. They then created icons representing the community and chose fonts and colors that captured the look and feel of the project. The icons represent important community assets. For example, a smiley face is the logo of a small grocery store called Happy Foods, which is one of two key community gathering places. Icons like this one situate the climate action project as part of Forest Glen's unique community culture. They are featured in local advertisements and large signs with QR codes that were installed at project sites and other popular places throughout the community, such as train stations.

Project partner Jennifer Herren, of the local chamber of commerce, led the development of this communications strategy. She explained, 'Any project, no matter how great it is for the community, or how great it is for the environment or anyone else, will stand alone as an island if you don't share what you're doing with other people. And that is what is so fascinating to me about this Community Toolkit, that it's designed to be shared with the rest of the city, as well as potentially the world.'

Point 3: Making climate action fun and engaging—Bronzeville project

Like the Forest Glen project, the Bronzeville community project brought together multiple activities, in this case aimed at building a local green economy centered on African cuisine. Project partners included a community developer, a local chef, and two community organizations. Two key activities took place in community gardens: vegan soul food cooking demonstrations and a youth horticulture program. Additionally, project partners led green tours for residents and visitors, which focused on the green economy and public art.

The project's focus on public art, and the visual and interactive elements in the gardens, made the climate action activities more accessible and inviting to the broader community by adding elements of beauty, fun, play, and physicality. These elements came together most successfully at the Bronzeville Community Garden, a space designed both for food cultivation and social interaction (see Figure 31.2). The garden's lead cultivator, Guadalupe Garcia, explained: "There aren't too many green spaces, and I think there are different things here that kind of make it an ideal meeting space: the cooking pavilion, the chess set. It's just connecting them with where their food comes from." The garden also includes

Figure 31.2 Bronzeville Garden: a father and daughter play chess in the Bronzeville Community Garden in Chicago's historic African-American community

Source: Image credit: © The Field Museum

colorful mosaics. Mecca Brooks, a project leader from the Bronzeville Alliance community organization, shared: "A lot of people that are into public art are also into urban gardening. So the two almost come hand in hand. So I think the use of public art allows the people that are installing the gardens to provide a more appealing backdrop."

Indeed, the garden was included as a stop on both of the community tours run by the Toolkit partners. At the conclusion of both, it served as a meeting place where tour attendees celebrated Bronzeville's African-American heritage with food, music, dancing, and play. Said otherwise, it served as a central venue for the multi-sensory engagement of community members young and old in Bronzeville's project. Public art, music, games, and food help create vibrant public spaces and foster appreciation for outdoor environments–-key ways to bring awareness to issues like climate change that can often feel distant otherwise.

Point 4: Connecting climate action to other issues—Pilsen project

Like the Bronzeville project, the Pilsen community project took place at the intersection of gardens and public art. Project partners included a Mexican hometown association, an environmental justice group, and a local daycare center. The partners worked together to turn a vacant lot into a climate-friendly native plant garden that would serve as a play space for the daycare and an outdoor classroom where the two other partners could engage constituents and hold workshops. As with the Forest Glen project, this project included important icons, and one in particular: the monarch butterfly. This garden is part of a larger effort in Pilsen to plant milkweed throughout the community. Milkweed provides habitat for monarch butterflies, and the goal is to turn all of Pilsen into a monarch sanctuary. Like many of the Michoacanos who live in Pilsen, monarchs migrate between Michoacán and Chicago. They serve as a cultural symbol and as a powerful symbol of the ability to freely cross borders. Meanwhile, climate change is expected to spur massive migrations in the years to come, including movement between Mexico and the United States as Mexico continues to see the effects of climate change, such as landslides.

The Pilsen garden includes many milkweed plants, and artwork in the garden, created by a renowned local muralist with local youth, prominently features the Monarch butterfly. Signage in the garden explains the connection in both English and Spanish, establishing the Monarch as 'a symbol of peace and freedom across borders' (see Figure 31.2). As it is employed in the garden's visual content, the Monarch connects the issues of play, education, health, industrial pollution, climate change, and immigrant rights.

This connection invites multiple constituents to the table, and points towards the need for holistic solutions.

Conclusion: visual and interactive practices are key in creating community climate narratives

In our work at The Field Museum, we have seen the power of visual and interactive practices to engage diverse sectors of society in larger climate action efforts, such as the Chicago Climate Action Plan and the Climate Action Plan for Nature. These practices help communities understand and relate to climate change and integrate climate action work into their ongoing efforts to improve local quality of life. Perhaps most significantly, we have found that they help communities create their own narratives about climate change as it relates to their assets, concerns, and local cultures. These narratives situate the scientific discourse of climate change, and the technological solutions that dominate the climate action movement, within the fabric of local life. This visual, interactive—and narrative—approach to community engagement is resulting in local efforts that advance broader climate action strategies, such as energy-efficient buildings or climate-friendly gardens, but in unique ways that resonate with local communities. In turn, these community approaches suggest new models for asset-based approaches to climate action.

Unlike surveys, traditional scientific presentations, or policy discussions, visual practices adapted from fields such as applied visual anthropology can create bridges for involving new allies and leaders in the pressing work of climate change. As described above, the tools that can be created within this emerging sub-discipline are well suited to support widespread engagement projects like collaborative research and community-based interventions led by micro-coalitions. These approaches take time, but they will build the leadership and relational infrastructure that will be needed for successful, long-term adaptation efforts. In contrast, initiatives that seek to transform communities without involving their local leadership can happen very quickly. Yet such externally-driven approaches often do not leave the areas where they work with stronger ideas, local leaders, or partnerships.

To successfully respond to the challenges of climate change, we will need to develop transformation projects that can both manifest quickly and inspire new local leaders and bonds. Doing so will require the creation of new tools and methods, and the creative work of ongoing experimentation. In this chapter, we have attempted to show how the work of applied visual anthropology can help with the important task of climate-conscious community building that can serve as the basis for broadening climate action to be a movement led by and benefitting us all.

References

Ageyman, J., 2007. The climate-justice link: Communicating risk with low-income and minority audiences. In: S.C. Moser and L. Dilling, eds, *Creating a Climate for Change: Communicating Climate Change and Facilitating Social Change*. Cambridge: Cambridge University Press, pp. 119–138.

Barnes, J., Dove, M., Lahsen, M., Mathews, A., McElwee, P., McIntosh, R., Moore, F., O'Reilly, J., Orlove, B.,Puri, R., Weiss, H. and Yager, K., 2013. Contribution of anthropology to the study of climate change. *Nature and Climate Change*, *3*, pp. 541–544.

Corner, A. and Randall, A., 2011. Selling climate change? The limitations of social marketing as a strategy for climate change public engagement. *Global Environment Change*, *21*, pp. 1005–1014.

Crate, S.A., 2011. Climate and culture: anthropology in the era of contemporary climate change. *Annual Review of Anthropology*, 40, pp. 175–94.

Crate, S.A. and Nuttall, M., 2009. *Anthropology and Climate Change: From Encounters to Actions*. Walnut Creek, CA: Left Coast Press, Inc.

Marshall, G., 2012. *Hearth and hiraeth – Building values-based climate change narratives in the Celtic heartland.* [video] Available at: www.garrisoninstitute.org/climate-mind-behavior-project/cmb-video-presentations/cmb-video-2012/1495-george-marshall [Accessed 31 August 2012].

Michaelis, L., 2007. Consumption behavior and narratives about the good life. In: S.C. Moser and L. Dilling, eds, *Creating a Climate for Change: Communicating Climate Change and Facilitating Social Change*. Cambridge: Cambridge University Press. pp. 251–265.

Pike, C., 2011. Seven reasons why the public is not engaged on climate. *Climate Access*. [online] Available at: www.climateaccess.org/blog/seven-reasons-why-public-not-engaged-climate [Accessed 31 August 2012].

Pink, S., 2009. Applied visual anthropology: Social intervention and visual methodologies. In: S. Pinked, *Visual Interventions: Applied Visual Anthropology*. Oxford: Berghahn Books. pp. 3–28.

Pink, S., 2011. Images, senses and applications: Engaging visual anthropology. *Visual Anthropology*, 24, pp. 437–454.

Roncoli, C., Crane, T., and Orlove, B., 2009. Fielding climate change in cultural anthropology. In: S.A. Crate and M. Nuttal, eds, *Anthropology and Climate Change: From Encounters to Actions*. Walnut Creek, CA: Left Coast Press, Inc. pp. 87–115.

The Field Museum, 2007. *Collaborative research: A practical introduction to participatory action research (PAR) for communities and scholars*. Available at: http://archive.fieldmuseum.org/par/ [Accessed 31 August 2012].

The Field Museum, 2009–2012. *Engaging Chicago's diverse communities in the Chicago Climate Action Plan*. [online] Available at: http://fieldmuseum.org/climateaction [Accessed 31 August 2012].

The Field Museum, 2011–2012. *Chicago Community Climate Action Toolkit*. [online]. Available at: http://climatechicago.fieldmuseum.org/ [Accessed 31 August 2012].

Westphal, L.M., and Hirsch, J., 2010. Engaging Chicago residents in climate change action: results from rapid ethnographic inquiry. *Cities and the Environment*, 3(1), Article 13. [pdf]. Available at: www.nrs.fs.fed.us/pubs/jrnl/2010/nrs_2010_westphal_002.pdf [Accessed 31 August 2012].

32 Ohio University State Museum of Ice

Emily Hinshelwood

It's not just a freezer
with floor to ceiling shelves of ice –
like a ship of Frederic Tudor's
packed with pieces of the Hudson
and freighted to Cuba.

It's not a state of the art Zanussi
with stores of San Miguel,
boxes of ready-made lasagne
and anything from mascara
to cod liver oil tablets.

It's not a 'load up the trolley darling'
walk in it like it's a cathedral
'we've got the biggest fridge freezer
on the planet – it's bigger than the arctic!'

No – This museum is the Arctic,
Mount Kilimanjaro,
Rwenzori,
Nevado Huascaran,
Dasuopu.

Cut cores of ice
like small gravestones in a white cemetery

"In loving memory of our Glaciers".

33 Global Water Dances

Embodying water solutions

Marylee Hardenbergh, Laura Levinson and Karen Bradley

Global Water Dances (GWD) is a multi-national, community oriented dance event that takes place at water-related sites around the world on the same day, biennially. GWD inspires action, leadership and international collaboration for environmental water solutions, using the universal language of dance to empower global communities and leaders. Its collaborative process has brought together 60-80 cities on six continents, using art (dance) to illuminate water issues. GWD has taken place on alternating years since its inception, with global events in 2011, 2013, and 2015. The project grew out of Artistic Director Marylee Hardenbergh's 2006 multi-site performance, One River Mississippi. This performance connected seven sites along the Mississippi River, from the headwaters to the Gulf, where each site performed the same movements to the same music at precisely the same moment.

Collectivity and environmental education

> Not everyone can talk at the same time; that just creates bedlam. But everyone can dance at the same time, and they feel closer.
>
> (Marylee Hardenbergh)

The GWD initiative begins with communication across continents. Without the internet, such a project would not be possible. But international communication is not the sole purpose of the project. The primary goals are to affect community-based change regarding attitudes toward and stewardship of local water resources.

Beginning with a gathering of movement experts in 2008, plans were laid for a global experience. The template would be based on Hardenbergh's project, "One River Mississippi," which centered on a dance with four sections: An opening sequence honoring the water, a locally choreographed work about a local water issue, a shared dance that every participating community would do, and an inclusive and empowering section that the audience could join.

The project is overseen by a committee consisting of artistic leaders in Germany, Colombia, Canada, and the USA. Six women meet online every few weeks to discuss next steps, to divide tasks and to facilitate the project. All planning takes place via Skype, with the far-flung steering committee researching options and making decisions under the direction of Artistic Director Marylee Hardenbergh.

The GWD planning committee has developed a framework for a model of how to use participatory art that can be expanded to include other environmental or social issues. The methods of site-specific, place-based environmental education are a powerful technique for environmental activism; the strategies used engage the imagination of a community and help them to envision active solutions.

Because artists, environmentalists, and community leaders come together to realize the event, an important exchange of skills, ideas, and approaches takes place. Artists deepen their understanding of how water issues affect the community they live in on multiple levels. Environmental activists and scientists learn how powerful artistic expression can be as a tool for change, especially when it uses the body in a conscious way (an embodied practice). Community leaders build their capacity for reaching out to different groups within a community in order to affect change.

One River Mississippi: the source

GWD grew out of a 2006 project, One River Mississippi, which connected seven sites along the Mississippi River through movement. As in GWD, this river project included sections that centered local music and issues and sections that were performed simultaneously across sites. The project succeeded in getting all ten governors of states bordering the Mississippi River to create a proclamation for "One River Mississippi Day," thus creating the intention of togetherness and goodwill toward the river on the part of these political leaders.

One River Mississippi emphasized accessible participation of dancers, choreographers, and audience members at each site, ensuring that the event was reflective and inclusive of the local community and not only those community members privileged with greater access to resources and traditional arts venues. In St. Louis, Memphis, and New Orleans over one third of the dancers and the majority of the choreographers were African-American, an important valorization of the Black community in these areas. One important story from the One River Mississippi project took place in St. Louis, where a distinct effort was made to include East St. Louis. East St. Louis, directly across the bridge from St. Louis, is a predominantly Black city that has traditionally been underserved and under-resourced in comparison to its larger counterpart. Referring to the river that flows between the two cities, one East St. Louis resident said, "To you, it's a river. To us, it's an ocean." In preparing for the event, Hardenbergh visited a dance studio in East St. Louis, and strongly encouraged the dancers to participate, and many did. On the day of the performance, the Eads Bridge (between St. Louis and East St. Louis) was closed to traffic and the audience walked out on to the bridge to watch the performers on the bridge, barges, and riverbanks. The mayor of East St. Louis, who identifies as Black, and the mayor of St. Louis, who identifies as White, strolled along the bridge together, a welcome occurrence that brought East St. Louis and St. Louis together in a harmonious way.

The timing of the One River Mississippi event was also significant, as it took place less than a year after Hurricane Katrina. The New Orleans dance community was still reeling from the events. Hardenbergh was able to secure some funding to pay the dancers and choreographers, which encouraged some of the most accomplished dancers in the city to participate where they might otherwise have been unable to do so. The New Orleans site ended up with a diverse group of dancers who profoundly enjoyed working together, despite any prior separations based on class or race. The event provided a space for healing from the trauma that had recently devastated not only the dance community, but every single community along the Gulf Coast.

Dance as a feminist approach

Dance can be seen as a gendered modality. Professional dance is, and has historically been, a predominantly female field. The aesthetics and practice of dance have been labeled by mainstream society as feminine or feminizing: self-expressing through movement, displaying a wide range of emotions, and moving in curvilinear ways are stereotypically limited to women. Dance as an embodied practice also relies on types of kinesthetic knowledge and ways of connecting to and engaging with the body that are often devalued in capitalistic and patriarchal systems. Dance provides a unique opportunity to connect individuals to a more holistic sense of the body-mind connection, enabling a powerful approach to social change that is driven and perceived by the whole self.

GWD reflects dance as both a female-driven and a feminist form. Most of the choreographers are women. The few communities where men take charge may be reflective of patterns of male dominance that still play out in many societies; however, male participation in what is often considered "women's work" is also an important aspect of the feminist power of dance. Dance utilizes the body as the primary entrée into the world, and critiques the dichotomous separation of the body-mind. Dance also challenges singular forms of knowledge, embracing the full range of the human potential for expression and relationships, for pattern recognition and invention. Whereas patriarchal ways of knowing emphasize logic and academic thought, embodied practices such as dance uplift kinesthetic and emotional knowledge as valid and essential.

Because dance as a feminist form is inclusive and challenges the limits of patriarchal approaches, it is particularly amenable to creating social change. Global Water Dances offers multi-modal forms of learning in order to execute place-based environmental education. GWD events include site-specific viewing experiences, participatory kinesthetic experiences, as well as verbal information on environmental issues. This structure allows different approaches to reach different kinds of learners. The water issues highlighted at each event not only impact a wider audience, but affect each viewer on multiple levels, allowing them to engage with rivers, lakes, and reservoirs in a more holistic way. The body-based, place-based level is particularly noteworthy because it allows people to develop a profound connection to and personal knowledge of their environment.

GWD couples the art of dance – a nontraditional means of promoting learning and taking action – with other tactics for environmental education. Partnering with environmentalists and water education consultants, several dance sites did invite environmental organizations to set up tables or displays to enhance educational aspects concerning the local citizens. The organizers of GWD believe that the primary benefit is not only to encourage responsible stewardship of the environment but also to inspire each participant to relate to their community in a more engaged and informed way. The dances enhance the quality of the spaces themselves, creating beauty that gives the event intrinsic as well as instrumental value. A post-event survey was conducted after the 2013 event. Three quarters of the audience respondents stated that the performance gave them a definitive sense of community, and helped them to see the location in a new and positive way.

Gender dynamics

The use of dance brings to light a particular set of gender dynamics. Not only is dance a field with a disproportionate number of female practitioners, dance as a discipline is coded as feminine – ephemeral, intuitive, emotional, relational. Throughout dance history, dance artists have understood and approached this in a variety of ways, from capitalizing on these feminine qualities, to excising feminine elements to present a gender-neutral body, to portraying wide ranges and combinations of genders in their work. GWD's lineage has a feminist perspective, and uses the corporal self to embody concerns about water. Rudolf Laban, a key influential figure in Hardenbergh's work, felt strongly that everyone is a dancer. In his world, dance stopped being a consumer object like ballet, which objectifies the dancer in relation to the audience and sets up a gender-based power dynamic onstage. In Laban movement, everyone was invited to join in equally; no one was a spectator, and there were no prescribed gender roles. In GWD, we remember that everyone consumes water, and everyone can participate in the dance. The core value of inclusivity is demonstrated at every site, where the community is invited to participate; no one who wants to perform is turned away, and active efforts are made to reach out to as diverse an array of community members as possible. The choreographers are encouraged to find a valorizing role for everyone.

Hardenbergh's methods utilize horizontal power structures, rather than hierarchical relationships. This model reflects feminist ways of creating, communicating, and ultimately realizing a project. By nature, such an approach makes the work more flexible – a characteristic that is also reflected in the use of outdoor sites that are subject to weather and other natural changes. There is an aspect of fluidity to the work that creates room for different needs and perspectives; each choreographer is given a dance score and invited to create their own interpretation.

The impact of a feminist modality that is community-centered and geographically aware is quite different from forms of environmental activism premised on masculinist logics of ownership and protection. Ecofeminist environmental actions such as those undertaken by GWD aim to subvert the capitalistic and patriarchal setup of a subject-object relationship, where humans are the intended beneficiaries of a commodified Earth (Diamond and Orenstein 1990). The event draws attention to the plight and the voices of different communities, not just those who hold power through control of environmental resources. It also centralizes the worth of the environment for its own sake; we are not keeping it pristine merely for further use by humankind.

Empowerment through art

Site-specific dance as it is used by Global Water Dances is an especially potent tool in that it connects the hearts of the audience to the site in a new way. Each dance in each community creates deep associations between water, place, and personal experience, motivating participants to become stewards of their local water resources. These dances engender in the participants and audience a bodily-felt sense of place, remembered kinesthetically for decades. According to Powers (2004), when individuals become attached to the place in which they live, they become more actively engaged in their community. Audiences and dancers alike learn to see the site differently, taking a familiar landscape and reframing it, so that it can be seen as a resource to value and protect. This process of recontextualization fundamentally changes people's view of familiar places: many audience members have remarked that they will never see the site in the same way again. These performances are designed to alert, arouse and inspire both participants and their audiences to take action on behalf of irreplaceable water sources. These effects were shown in the 2013 audience survey, which yielded data from four sites around the world: Bangladesh, Peru, Germany, and Minnesota. Nearly two thirds of the respondents said that the performance increased their interest in water issues a great deal, and that it very much inspired them to take action regarding water issues. Three quarters said that, as a result of the dance performance they had just seen, they were likely to make more efforts to conserve water in their personal use. In this context, pro-environmental action is not just about distributing information or encouraging people to act, it is also about creating new patterns of thought.

A model of participatory art

The GWD events invite a wide range of participation from the community. Because of the geographic scope, constituents are diverse in race, ethnicity and nationality. There has also been a vast range in socioeconomic status amongst the performers and audience members; for example, in Savar, Bangladesh, the participants ranged from individuals with PhDs to those who were illiterate. This is encouraged by the project's strong emphasis on community inclusivity; when choreographers are sent to the GWD webpage to learn about the overall goals, they read that it is a strong intention of Steering Committee to facilitate as wide a range of participation as possible.

Figure 33.1 Joan van der Mast, choreographer, Netherlands. Audience participation next to the sea for Global Water Dances in Netherlands, 2013

Source: Photographer: Ray Hartman

The performance is the lens through which the process of learning about and coming together to address water issues takes place. It is the seed of awareness and change, a gift given to the community. These performances take place in the public domain. This means that people are putting their bodies in visible places to make a statement. The audience also experiences an art form that would normally be viewed in a theater, a setting that is economically inaccessible to many. GWD provides a free service and invites inclusion. Presenting a work of art is a gift, and we are giving in the spirit of consciousness raising and to create positive change. The method of presentation utilized by Hardenbergh ensures that every individual involved in the work – from the participants, to the audience members, to other collaborators – is receiving something valuable from the experience. The performances bring inspiration through their use of creativity, and they allow people to experience a different way of looking at the world. Participants are lending their physical self to a demonstration about the need for safe water for all, a message that benefits the community as a whole.

Creating experiential connections with our natural global environment deepens our sense of being human, especially when we are simultaneously made aware of our own geographical, community, and cultural contexts. As Gruenewald (2003) states, this can lead to considered social and ecological awareness of a sense of belonging and the importance of connecting to nature in place-based education. Dance is a powerful medium for the very reason that we are using our bodies, we are creating a message that transcends words and inherently incorporates local and global contexts. The body can be used to communicate amongst people who speak different languages. But because different bodies carry different historical and cultural contexts, embodied performance is also a powerful way to bring forth that which is locally relevant. Dance builds a collective community outside the box where art gives activists new ways of working and thinking about change. It has a particular ability to rejuvenate and empower communities.

In the work of Artistic Director Hardenbergh, a key value is inclusivity: everyone who has a body can dance, so everybody is invited to participate, regardless of ability or expertise. The movement in the audience participation section is simple, repetitive, and easily modified by individuals to suit their own needs. The invitation to dance is open to all at each site, and audience members who would not have considered themselves performers are invited to dance as equally important parts that make up the whole experience.

Water: the universal solvent, accessible to all?

Water serves as a unifying force because it affects every human being. While pollution and environmental crises are not evenly distributed across the population, everyone deals with access to water in some way. The simultaneous local and global focus of Global Water Dances allows the project to address both the specific and the universal: to show our global interconnectedness, while allowing each site to highlight a particular water issue that affects community members. Because GWD organizes around specific places, all the constituents of a site share similar connections to the water there. This creates a powerful connection among and across sites. However, GWD focuses its intention on geographic-specific issues that are concurrently social issues: not only has human behavior wreaked havoc with many of our water resources, but communities are often disproportionately affected by water pollution based on dynamics of privilege and power. It is important to provide a platform for communities experiencing particularly acute water shortages or contamination to speak out, when these voices are not often uplifted or even heard in dominant communication mediums. The merging of global connection with local contexts allows for this to happen.

It is important to note that, while this project would not be feasible without the Internet, the digital divide does present challenges regarding access. People from participating sites register via the project website, where they also access the music and video of the shared choreography sequences, templates for publicizing the performance in their local communities, and environmental education materials addressing water issues. The performances are broadcast throughout the day on an online live-streaming network, increasing the global interconnectedness. In this case, technology works as a means of increasing access and as a barrier. Choreographers without access to the Internet—or without a working knowledge of English—would be extremely unlikely to learn about the project and prevented from connecting with the worldwide network of participants. GWD choreographers in China face different challenges in accessing information and making a performance with a political message than do the professional choreographers in Europe.

Geographical considerations

As an organization, GWD has found a challenge in clarifying and deepening its connections to existing environmental groups. This project incorporates forms of environmental activism such as public assembly and educational handouts, while simultaneously offering an embodied alternative to these other methods. Because we empower all choreographers to choose their own geographic site and water issue, themes have ranged from site to site, and relate to their community's own experience and social origins. In Guinea, for example, the choreographer's home village, Tshalbonto, is inundated by the rising ocean. Villagers are moving reluctantly, and slowly; they have no government help as they lament the loss of their ancient village. The performance was titled "Au Revoir to Tshalbonoto et nos ancestres" (Goodbye to Tshalbonoto and our ancestors) and drew upon the traditional movements of the Baga and Susu cultural groups. In Savar, Bangladesh, the textile factories collapsed a mere two months before the performance, killing nearly a thousand people. The performance here drew attention to the Bangshi River, where industrial and chemical waste of hundreds of factories in the city drains directly into the river.

On one hand, GWD has been working to integrate existing environmental groups and their education techniques. Some sites create panel discussions with researchers and environmental leaders, hand out environmental literature, or ask representatives of environmental organizations to speak to their audience at the performance. These activities are meant to focus on a local water issue of current concern to communities at the site, including drought, floods, 'red tides,' pipelines threatening the groundwater, desalinization discharge, hydraulic fracturing, and destructive runoff. The organization also conducted an audience/participant survey on four continents, in an effort to speak to these groups using their own standards and rhetoric, leveraging the existing knowledge and activism around water issues, and providing a measure of legitimacy in established environmental circles.

At the same time, GWD offers a different method of outreach, the effectiveness of which may not be measurable within conventional rubrics. Some observers cannot see a connection to tangible forms of activism and change—results that would be validated within the logic of hard science. The methods of place-based education, participatory art, and kinesthetic learning can be framed as a critique of established forms of environmental activism. The structure of this project makes space for alternative modes of knowledge production and multiple ways of making meaning. The local/global structure of the project allows it to adapt to a variety of political contexts and to encompass many forms of political action. Body-based, community-focused, and locally grounded ideas are brought to the forefront, as opposed to top-down, empirical knowledge. GWD exemplifies a collaborative process. Each site has its own issues around which the dance helps to raise community awareness. The dance offers an embodied experience, taking issues from unconscious to conscious. Each site is empowered to address river/water issues in a way that is not hierarchical from the top down, but rather uses grassroots initiative and inclusivity in the choreographic process. This means that there is room for the expression of differing experiences that are based on geographical placement; gender, race, and/or class; and cultural contexts for relating to water and to the chosen site. We are engendering "environmental empathy" which is different from stewardship. In a culture where gender is seen as a measure of power, and power is seen as hierarchical, these events change people's perspectives; they retrain our eyes, they change our view, and expand our empathic perspective.

Figure 33.2 Giselda Fernandes, choreographer and dancer, Rio De Janeiro, Brazil, 2011. Giselda Fernandes is in a dress made of plastic bags, helping another dancer climb out of a huge bag of plastic water bottles, all collected from the trash

Source: Photographer: Christina Almeida

This approach resonates with theories of intersectionality, as well as feminist critiques of science and mainstream environmental activism. The very structure of the event, weaving together over sixty geographic locations from around the globe, with each site encouraged to explore its own expression while being part of a whole, is the very opposite of an oppressive institution. Rather than being exploited for the focus of the dance, nature has become the star of the show, and is subtly counteractive to the systematic injustice by large private companies harming our water supplies and our communities.

Think global, dance local

The methods employed by GWD hold promise for weaving together environmental activism, education, and the arts. These events bring together local environmental experts and organizations, artists and members of the community together in a process that can build ongoing collaborations. The power of these performances lies in their ability to reframe issues in a way that encourages inspiration and activism, and allows the audience's perceptions to be altered. Place-based, community-oriented environmental education emphasizes an authentic experience of a place, which is a gateway to ecological understanding.

The main power of these events is the raising of community awareness, allowing each community to connect with its constituents at the actual place of a water issue. Kinesthetic learning and active participation provide an integrated foundation for informed political action. Global Water Dances enriches the field of environmental activism, and its methods have the potential to transform ongoing practice in the field.

References

Diamond, I. and Orenstien, G.F. 1990. *Reweaving the World: The Emergence of Ecofeminism*. San Francisco: Sierra Club Books.

Gruenewald, D. A. 2003. The Best of Both Worlds: A Critical Pedagogy of Place. *Educational Researcher*, *32*, pp. 3–12. Available at: http://faculty.washington.edu/joyann/EDLSP549Beadie_Williamson/gruenewald.pdf

Powers, A. L. 2004. An Evaluation of Four Place-based Education Programs. *The Journal of Environmental Education*, 35, pp. 17–32. Available at: http://meera.snre.umich.edu/sites/all/files/Powers_Detailed_Profile_Powers_EfSD.pdf

Sobel, D. 1996. *Beyond Ecophobia: Reclaiming the Heart in Nature Education*. Great Barrington, MA: The Orion Society and The Myrin Institute.

34 Mni

Candace Ducheneaux

Mni, the Lakota word for water, is an indigenous-led grassroots nonprofit, founded in response to the acute water crisis on the Cheyenne River Lakota Reservation in South Dakota. Mismanagement of tribal water sources has caused desertification of tribal homelands and a poverty among the people that has expanded and deepened with each new generation. Traditional fresh water sources are drying up and being depleted, leaving an impoverished tribal membership reliant on costly water piped through an obsolete, asbestos-lined system. The already overburdened water infrastructure will not support housing growth, which has led to a severe housing shortage and overcrowded, unhealthy living conditions for much of the population.

Compounding tribal health problems, is a water supply drawn from the stagnant waters of the Oahe Reservoir, the massive lake created with the damming of the Missouri River in the late 1950s. Reservation tap water is laced with heavy metals from mine tailings; industrial, pharmaceutical, agricultural and residential waste and chemicals from pesticides, herbicides and fertilizers that have washed and been dumped into the river system, as well as with EPA mandated fluoride. The toxic cocktail that winds up in the Lakota's drinking cup has been blamed for the staggering rise of crippling and deadly diseases on the reservation, where the average life span hovers at around fifty years.

The health of the entire Missouri River ecological community was degraded when its cleansing, life-giving flow was obstructed and the hydrologic cycle necessary to the replenishment of fresh water sources and the regeneration of all life in the prairie habitat was completely and wholly interrupted. Its unique biodiversity, which that had sustained the people and all manner of prairie animal and plant species for ages, was buried under the vast floodwaters. The loss of this vital watershed deprived the Lakota of a means to self-sustainability and independence and left their entire ecosystem vulnerable to the cataclysmic climate changes that come with a dysfunctional water cycle.

The current ongoing, decades-old drought, was spawned in and sustained by extreme weather events that are a direct consequence of a broken water cycle, climate change and global warming. This has been and exacerbated by inefficient agricultural, ranching and water use practices, and has heightened the water crisis for the Lakota people. The desiccated earth has hardened and is unable to absorb essential moisture. Ever-decreasing annual rains and snowmelts instantly leave the land, carrying away much needed topsoil as the runoff escapes into the river system and rushes away to add to the ocean rise.

Drawing from thousands of years of traditional ecological knowledge, the Lakota understand that their future water security and economic viability is inextricably intertwined and interdependent with all life within their specific ecosystem, and throughout the entire planet. All of humanity share a common responsibility to the Earth. A common destiny.

Coming from a culture that recognizes and respects the oneness of all things and requires reciprocity at all levels of existence, the Lakota understand that, to bring the planet back into balance, reverse climate change and return the earth to sustainability, we must give back. We must help restore the water to the land, regenerate the biodiversity, put carbon back into the soil and reestablish the small water cycle over tribal territories in order to repair the larger, worldwide water cycle.

Ancient and modern Indigenous prophecies have warned, and, the scientific community now confirms, that our window of opportunity time is short. Two hundred species are driven to extinction every day, up to some forty percent of all human deaths are due to pollution and global warming is accelerating beyond worst-case predictions. If we hope to avoid mass extinction, we must act globally and with utmost urgency or we will reach an ecological tipping point, beyond which the earth's capacity to, as it supports humanity, will be unable to survive.

Mni has sought and found solutions to achieving climate stability and returning the earth to its balance and sustainability. We can keep the water on and in the land through rainwater harvest techniques, we can sequester carbon emissions through holistic management of grazing lands and we can eliminate bare soils through reforestation and permaculture practices. The solution is simple, but it will take hard work and global participation.

Mni has mobilized a global effort to achieve spiritual unity in resuming natural life ways that can reestablish and maintain the balance which allows the existence of mankind (see mniwater.org). The Mni vision is solidarity among the Indigenous nations in water safety, security and sovereignty to achieve worldwide water justice for all of creation. Indigenous nations as sovereigns, many with vast territories and little or no water use regulations, have a unique opportunity to implement large-scale rehabilitation upon damaged tribal landscapes and, thereby, make a significant impact on the world water cycle and give the world an example to follow.

Since its inception in 2012, Mni has organized, collaborated on and participated in numerous projects to confront the ongoing water crisis, battling the degradation of water resources and tribal landscapes and working to restore the ecological and economic health on the Cheyenne River Reservation and throughout the entire planet. On Cheyenne River, Mni has taken the lead in promoting civic engagement to raise awareness of the magnitude of the water and climate crises and the urgent need to take positive action to avert global disaster and collapse of the ecosystem.

Mni has traveled to other reservations and states to speak at various venues with diverse audiences, including to hundreds of climate activists and experts who gathered at Tufts University in 2014 to discuss restoring ecosystems to reverse global warming and to a small band of Lakota and supporters gathered around the campfire in Oglala pine-covered hills near Wounded Knee to discuss ground water restoration on Fast Horse Creek, to aid the Fast Horse family in reclaiming and resettling their inherited land holdings.

At home, partnered with the Cheyenne River Youth Project, Mni has hosted several informational meetings, showing documentaries to raise awareness and feeding the people to foster a spirit of unity and common purpose. On two occasions, Mni has brought world-renowned Slovakian hydrologist, Michal Kravcik onto Lakota homelands to talk about his innovative method of water restoration, the Blue Alternative. In July, 2012, he came to address the Cheyenne River Tribal Council and in May, 2013 he presented at the Mni Indigenous Water Summit that was held at the tribal agency. The summit, organized and staged by Mni, was a three-day gathering of Indigenous leaders, water activists and water restoration practitioners from the US and Canada to discuss a safe, secure and sustainable water future for all nations.

Last summer, Mni, along with the Tatanka Wakpala Tiospaye, a reservation nonprofit building a model sustainable community on the family lands, hosted a two-week water sustainability training camp that brought Indigenous and non-indigenous volunteers to Cheyenne River for hands on training in rainwater harvesting and ecological restoration techniques to implement on their own lands. Currently, Mni is raising funds to offer the training experience to all volunteers again this summer and to support long-term volunteers for further restoration work.

In 2014, Mni began a collaboration with Biodiversity for a Livable Climate (BLC) and the Savory Institute to meld their method of ecological regeneration by returning atmospheric carbon to the soils through holistic management and planned grazing of the world's grasslands with the Blue Alternative method of rainwater harvest to accelerate the mending of the world wide water cycle. To this end, Mni has raised funds to bring representatives of BLC to Cheyenne River this spring to create an environmentally restorative grazing plan for a large cattle ranching operation on eight thousand acres of tribal land. Mni is also looking for funds to have water experts assess the same lands for the Blue Alternative method of water restoration.

Mni has received support from numerous organizations: two $5000 grants from the Seventh Generation Development Fund, Arcata, California to finance the 2013 summit and the work with BLC; a $8000 fellowship with a $1000 internship attached from The Center for Collaborative Conservation based at Colorado State University at Ft. Collins to fund the 2014 summer training camp; and a $15,000 matching grant from Tribal Ventures, a reservation-based poverty program, to establish a permanent base camp from which to do further restoration work on the reservation and beyond.

35 Whale prayer

Subhankar Banerjee

Figure 35.1 Prayer after an Iñupiat whale hunt. Barter Island, along the Beaufort Sea coast, Alaska, September 2002

In the last decade, the United States government made a strong effort to open up the Chukchi Sea and the adjacent Beaufort Sea in Arctic Alaska, to oil and gas development, initiating a second wave of such development, after a failed first wave that started in the late 1970s and ended by early 1990s. The Iñupiat communities all across the North Slope of Alaska depend on marine mammals of the Chukchi and the Beaufort seas, including bowhead and beluga whales, seals, and walrus, for nutritional as well as cultural and spiritual needs. Many members of these communities, who value their traditional culture, have opposed offshore oil and gas development, fearing that such industrialization would seriously impact the whales and their migration and, subsequently, the Iñupiat culture. Oil drilling in the Arctic Ocean is likely the most dangerous form of drilling anywhere on Earth, as no one knows how to effectively clean up a major spill from underneath the broken sea ice in the extremely harsh environment of the far North. Moreover, Arctic sea drilling would contribute significantly to global climate change. After nine years of trying to pry open the Arctic seas for oil, and spending more than $7 billion, and receiving all the necessary permits for exploratory drilling, Shell announced on September 28, 2015, after a brief season of exploration in the Chukchi Sea that, the company has abandoned its Arctic Alaska offshore program citing disappointing results from exploration and high operating costs. A few weeks later, the Obama administration announced that it is canceling future lease sales in those seas. The second wave of Arctic offshore development likely has come to an end, which is significant news for climate change mitigation, and a relief for the Iñupiat people who have opposed industrialization of the Arctic Ocean, working in close collaboration with the environmental organizations.

36 Forced in or left out

Experiencing green from community redevelopment to voluntary simplicity and the potential in-between

Janet A. Lorenzen and Daina Cheyenne Harvey

In this chapter we look at two very different approaches to mitigating climate change: mandatory community redevelopment in New Orleans and voluntary lifestyle change in the Northeastern United States. We compare social contexts, socioeconomic backgrounds and shared ideologies, and highlight distinct commitment styles and challenges in the way environmentalism is negotiated in everyday life. We offer these case studies to problematize current macro- and micro-level approaches to addressing climate change and argue in favor of an embedded meso-level approach that brings together local energies and ideas with the resources of proficient organizations.

The first case study is the Lower Ninth Ward in New Orleans. After Hurricane Katrina and the federal levee failures, the Environmental Protection Agency (EPA) declared 100 percent of the homes in the neighborhood uninhabitable. The Lower Ninth Ward is now committed to rebuilding sustainably and achieving carbon neutrality by 2020 and climate neutrality by 2030. Most of the drive, however, towards mitigating environmental problems has come from organizations outside of the community, comprised of middle-class volunteers, often college educated whites, modeling global best practices for resilience and sustainability. The result has been reluctance and even resentment among residents to adopting the proposed solutions.

The second case study investigates voluntary simplicity as a kind of green lifestyle. Voluntary simplifiers, primarily lower-middle class white women (Grigsby, 2004), are members of a loosely organized social movement centered on personal change at the micro-level that includes buying less and reducing waste. These are environmentalists who view their political activism as part of a larger project of personal change. We contrast these two studies in order to ground current climate change solutions, which typically focus on macro-level planning and policy or micro-level alternative consumption options, in the complex empirical realities of everyday life.

'Experiencing green' as a form of environmental sustainability, in these case studies, amounts to interaction with top-down programs or implementation of bottom-up projects that attempt to address climate change and other environmental ills. Experiencing green differs based on the unique and multiple social positions in which actors are embedded. The extent to which these community redevelopment programs (Allen, 2011; Arena, 2012) and household projects (Dietz *et al.*, 2009; Maniates, 2002) actually address climate change and other environmental ills (such as over-consumption, ecological violence, and environmental inequality and injustice) is a topic of on-going debate.

Social location, social position, positionality, and standpoint are all ways to express the interconnections or intersectionality of race and ethnicity, social class, sex and gender, sexuality, age, religion, country of origin, able-bodiedness, etc. that shape our experiences of the world (Collins, 2000). In working with multiple categories of socially constructed difference it is important not to reify them as they are woven together. The definitions and boundaries of these categories, and the specific ways they interact, are constantly produced and reproduced by ongoing social processes. Thus, categories continually change over time and from region to region. While environmental disparities between whites with class privilege and people of color lacking class privilege are real and systematically related to a legacy of bias (Bullard and Wright, 2012), the ways in which people experience environmental privilege (Park and Pellow, 2011) or eco-apartheid (Checker, 2008) differ enough to warrant interrogation. We seek to highlight the experiential aspects of greening rather than studying reified categories. An understanding of intersectionality assists in this project by breaking down and integrating categories of race, class, and gender – stressing the way they are experienced simultaneously within particular historical and social contexts. We argue that meso-level projects that cross race, class, and gender boundaries, and address issues of social justice (Rees and Westra, 2003), are more likely to be embraced and create meaningful environmental change than either macro-level interventions or micro-level incentives given the subjective experiences of resentment, exclusion, and uncertainty we document here.

Case study 1: community redevelopment in New Orleans

For urban planners, architects, and community and environmental groups, Hurricane Katrina seemingly presented a tabula rasa for the urban-environmental ills of New Orleans. While 80 percent of the city flooded and the number of blighted structures (which were present throughout the city before the storm) rose to over 50,000, the aftermath presented different challenges and opportunities for different neighborhoods. The first city-wide solution, initiated by the Urban Land Institute on behalf of the Bring New Orleans Back Commission (BNOB), called for a drastic reordering of space. The plan recommended that those neighborhoods that suffered minimal damage and were above (or near) sea level would be allowed to rebuild, while those neighborhoods which suffered the most damage and were comparably further below sea level would be turned into green space or become part of an experimental environmental flood control system. The plan was met with great resistance across racial and class lines as it was clear that middle-class, white communities would largely be spared and black communities, such as the Lower Ninth Ward, would be 'greened' (Allen, 2011).

While the Lower Ninth had a history of poverty (36.4 percent of households), drugs, gangs, and crime (a homicide rate ten times the national average), 20,000 people nonetheless called it home when the levees were breached on August 29, 2005. Residents remember their neighborhood as one of the few majority black communities (91 percent) that featured both a high rate of homeownership (67 percent) and no public housing. In an effort to preserve the distinct qualities of their neighborhood, residents began to rebuild as soon as possible, but were stymied along the way.

In the aftermath of Katrina, when residents realized, in the words of one community member, that 'the cavalry was not coming,' they welcomed numerous non-profits into their neighborhood. Many of the non-profits came with greening strategies, which ranged from simply reusing building materials (Denhart, 2010), to weatherizing homes, to building Leadership in Energy and Environmental Design (LEED) certified dwellings. As Allen (2011) notes, what has followed can only be explained as the 'laboratorization' of the neighborhood. This laboratorization has resulted in Holy Cross, for instance, one of the two neighborhoods in the Lower Ninth Ward, committing itself to achieving carbon neutrality by 2020 and climate neutrality by 2030. When current plans are finished the Lower Ninth Ward will have more LEED certified homes on a per capita basis than any other neighborhood in the country. The reception and the result of the greening, however, have been quite mixed.

Much of the 'green' aid from non-profits has come at a price, and some residents believe that they and others have acquiesced to a green lifestyle when they had little choice. One resident explained the years since Katrina as 'being taken hostage by environmentalists.' Others, however, welcome it as a new narrative for the community. In what follows we describe contrasting efforts at rebuilding the community in terms of environmental/climate sustainability, the response of residents, and what we might learn from marginalized communities about commitments to sustainability efforts. These descriptions come from one of the authors living in the Ninth Ward of New Orleans for fourteen months, where he did participant observation in the Lower Ninth Ward with a consortium of nonprofits and community groups, had informal conversations with hundreds of residents, formally interviewed 37 residents, and worked with a nonprofit to help rebuild homes.

Brad Pitt's Making It Right (or wrong)

Make It Right (MIR) was the first rebuild non-profit established in the Lower Ninth Ward after Hurricane Katrina. It was set up to right the wrongs of grant programs that were not providing adequate funds for rebuilding purposes and as a project for social and ecological justice. The goals of MIR were two-fold: (1) to demonstrate to New Orleans and the rest of the world that the Lower Ninth Ward could be rebuilt [in the words of Tom Darden, executive director, 'If you can rebuild here, you can rebuild anywhere' (personal communication, November 13, 2010)]; and, (2) to 'search for answers to the problems of global warming within architecture's most repetitive module – the family residence' (Feireiss and Pitt, 2009, p. 9). In part, the later goal has been accomplished through building strategies modeled from McDonough and Braungart's (2002) cradle to cradle concept – which calls for ecologically intelligent design that seeks to mitigate the conflict between the building industry and the environment. At MIR this has included using 'friction piles' – movable posts sunk forty feet into the ground, using disaster resistant and recyclable materials – such as putting Kevlar on all the windows and using recyclable carpet in the homes, raising homes above hundred year flood levels, and using both passive energy strategies and photovoltaic panels, as well as designing structures like floating houses. MIR has also focused on the landscape by installing rain gardens, xeric plantings, edible gardens, and experimenting with pervious concrete. While the project has been mildly successful – rebuilding 75 of the 150 scheduled homes – public support has been mixed at best.

First and foremost, very few residents knew what MIR was doing despite a professional public relations team and communication from MIR. There were numerous community meetings and symposiums where non-profits and community groups came together for brief presentations that focused on their recent work and their future goals. Likewise, because the project is attached to Brad Pitt it was by far the most publicized rebuild project in New Orleans. The most significant aspect of the misunderstanding between MIR and the community involved money. The average cost of a home was $150,000, with the average contribution of the resident being $75,000 – made up mostly of rebuilding grants. MIR made up the difference with forgivable loans (if the owner stays in the home for a set period of time) and MIR financing. The public perception, however, was that MIR was saddling the elderly and poor with burdensome mortgages. More troubling was the widespread belief that Pitt was somehow making money off of the community, a common trope aimed at all non-profits and outsiders.

Second, the project has taken up a significant amount of space and is the most visible rebuild effort in the Lower Ninth. The MIR homes are the first thing you see when you cross into the Lower Ninth Ward from the Judge Seeber Bridge. Additionally, the homes are all elevated much higher than other homes in the neighborhood and are painted in bright colors (as are many homes in the city, but not to such a degree). Likewise, they all have solar panels on their roofs, which offers a tragic contrast to homes on the other side of the street which still need to be gutted and are often missing portions of their roofs. A common complaint was that the MIR homes seem out of place.

These two complaints, the economic and the aesthetic, often came together to form a succinct argument against MIR.

> You see I just don't like it. I think it's good he [Pitt] did something. I mean he didn't have to do that you know. He ain't from here. But those houses don't belong here. They ugly. And even more is they cost, for what you paying or, they could of built five houses, you know... Now that one is real

> nice, I been in 'em. A few. But we could of had four more of our neighbors back, and that's what's wrong with those houses.
>
> (Charles, 50s, black male)

Others note, like Charles, that the MIR project has done some favorable things for the community, but that the bad far outweighed the good intentions.

> Nah, I don't support them. On the one hand it's good that he's gotten fifty, or whatever it is, families back. There is a real good chance those people wouldn't be back if it were not for Make It Right. But on the other hand, we just don't need that here. That's not us. … You got, how much money they got? Millions and millions. And that's all they can do?'
>
> (Paul, 60s, black male)

Many of those who are suspicious about MIR have had little to do with the project and struggled to rebuild their own homes. Furthermore, what MIR sees as ecological urbanism, residents interpreted as urban experimentation. This soured many in the community on the MIR project. They simply want their old neighborhood back and do not enjoy being part of an urban laboratory (Allen, 2011).

While MIR has acknowledged missteps and miscommunications with the community, they continue to irritate community stakeholders by not hiring local contractors and craft people. While there is disagreement in cited figures, several stakeholders believe that Pitt and others promised them that at least sixty percent of all workers would be hired locally. This has not been the case. Finally, many residents are opposed to MIR because they have increased the presence of tourists in the neighborhood. Before Katrina very few tourists to New Orleans made their way to the Lower Ninth Ward. Today, hundreds of thousands track through the neighborhood to see the urban decay caused by Katrina, but also the urban spectacle of the MIR homes. Residents have long complained that tourists represent interlopers who come to witness the community's suffering from the safe confines of their $45 bus seats (residents of the Lower Ninth Ward have never received any remuneration from the daily tour companies). Additionally, they complain that the buses are unsafe for children playing in the streets, they cause pollution, and ruin the streets. While both the physical and psychological damage the tours cause is arguable, the result, that they've increased the dissatisfaction of some residents with MIR, is not.

Other groups working locally toward achieving climate sustainability goals are Global Green, Preservation Resource Center (PRC), and the Center for Sustainable Engagement. Similar to MIR, the PRC and Global Green are seen as developing properties. However, as of 2016, the signature projects of PRC and Global Green are still unfinished. The lesson taken from these projects for some residents has unfortunately been that commitments to sustainable building and environmentally responsible lifestyles are too difficult.

In contrast, The Lower Ninth Ward Center for Sustainable Engagement and Development (CSED) was founded in 2006, in part to restore the ecology of the neighborhood and to help it become one of the first carbon neutral communities in the US. Despite their stated emphasis on community resiliency, consisting of food security, coastal sustainability, and the built environment, their outreach in the community has been quite modest. Their most ambitious project to date has been the restoration of Bayou Bienvenue—a local bayou that many residents remember fondly. Therefore, they are not seen as trying to develop the neighborhood or create a green branded space in the neighborhood. And while they support other green rebuilding projects (not mentioned here), they do so in an understated way. As such, they remain 'from here' as one resident noted, whereas MIR and Global Green, in particular, seem to be imposing 'outsider' ideas on the neighborhood.

Case study 2: voluntary simplicity and lifestyle change

Environmentally concerned 'voluntary simplifiers' are members of a loosely organized social movement that uses frugality to address climate change and other environmental ills such as the growing size of landfills (Grigsby, 2004; Lorenzen, 2012a; Schor, 1998). They focus on buying less and reducing waste to conserve resources. Jane (white, 70s), for example, views simplicity and sustainability as "overlapping," where living a simpler life is also more sustainable. However, voluntary simplicity has been critiqued as a middle class approach to addressing climate change that directs attention away from broader structural changes (Chitewere and Taylor, 2010; Maniates, 2002).

This section draws on a set of fifteen interviews with voluntary simplifiers conducted in the Northeastern United States. This set of interviews was among 45 interviews conducted with different kinds of environmentally responsible groups, including religious environmentalists and green home owners (Lorenzen, 2012a). The data uncovers a clear distinction between green lifestyles that save money by reducing overall consumption (i.e. voluntary simplicity) and those with high adoption costs that focus on investing in innovative technologies, which increase consumption in the present (of solar panels for example) only to reduce the consumption of energy in the future. This stark distinction between affordable and unaffordable green lifestyles highlights the shortcomings of the micro-level approach in addressing consumption and climate change.

Voluntary simplifiers

In practice, voluntary simplicity is the least costly way to adopt a green lifestyle and can serve to revive working class practices of frugality that the people I interviewed grew up with. Ultimately, social location influences the extent to which environmental issues are defined as problems, the length of time it takes to transition to a greener lifestyle, and the type of green lifestyle one can afford.

The voluntary simplifiers interviewed were primarily white women who, on average, held a Master's degree and were working as, for example, elementary school teachers, social workers, nurses, paralegals, real-estate agents, or community advocates. Several were single or divorced, relying on their own incomes to care for themselves and, for some, their children; other women were retired and on a fixed income. Not only was voluntary simplicity a feasible choice for women with financial constraints, it was also an option for married women whose husbands did not support environmentally responsible practices and were unwilling to share the resources necessary for investing in innovative technologies. Studies find that single/divorced women without children, or married women with no children living at home, are the most likely to participate in voluntary simplicity (Grigsby, 2004; Huneke, 2005).

For example, Catherine is a full-time special-needs teacher (K-12) who is divorced with two children. Catherine describes adopting a green lifestyle as a long process of changing habits; for example: driving less, turning down the thermostat in winter, taking shorter showers, using home-made cleaning products, eating less red meat, and composting food waste.

> I see some of the people at the [community group] that I'm involved with, they have – You know, they've made major changes like getting hybrid cars and putting solar panels on their homes and made a lot of changes more quickly because they had the resources. And I'm you know, a single parent, and I really haven't been able to do that.
>
> (Catherine, white, 50s)

Voluntary simplifiers could often not afford innovative, efficient technologies like solar panels or hybrid vehicles regardless of state and national financial incentives. Several of the simplifiers wanted to purchase a Prius hybrid ($5,000–$10,000 more than a compact car) or a green home (at least 5–15 percent more expensive than a standard home) but were priced out of the market on these efficient technologies.

At one point, the state in which these simplifiers lived was paying as much as 70 percent of household solar panel installations, yet barriers to adoption remained. They tended to live in older neighborhoods where the lot size was small and homes were situated close together. Mike, an adjunct professor and green business consultant, explained:

> We [my wife and I] have, I thought, a really, really nice, southern exposure here. But then it turns out, the company that I called came, and they did a quick audit, and they said we've got two monster trees on both neighbor's front yards. So they wouldn't even put it in because I wouldn't qualify for the prime, the minimal rebate – unless I can get my neighbors to chop down half their trees, and I couldn't ask my neighbors to do that.
>
> (Mike, white, 50s)

Thus, for this constituency, the process of making solar technology more accessible is not simply a matter of financing, but also of taking into account the geographical and social context in which these people live. Those most able to take advantage of the state solar program owned more than an acre of land (enough space for a solar ground installation), were building a new home (and could orient it to face south), or had the resources to build a new garage that faced south to accommodate a solar roof installation. Voluntary simplifiers were not a part of this group; in fact, limited access to efficient technology was part of what shaped an individual's interest in the ethic of simplicity (Lorenzen, 2012b).

As a group, simplifiers focus more on changing their practices than adopting efficient technology. Some simplifiers return to practices that they grew up with in conservative, white, working class households (Lorenzen, 2012a). Hence, green lifestyles do not arise in a vacuum; rather, they rely on past materials, practices, and relationships. Angela, who recently retired from her job as a program coordinator for a local community center, explains that she was born at the end of the Depression and 'I learned from my poorer relatives what simple living was like. They made do with a lot less, and yet kept their dignity' (Angela, white, 70s). Similarly, Martha acknowledges that she was able to return to lessons her mother, who grew up in Hungary, had taught her when she became interested in environmentalism (e.g. 'waste not, want not'). Previous knowledge of frugality, learned through personal experiences or social connections, is not necessary to living a simple life, but it does offer a toolkit of practices to draw on for lifestyle change. This awareness of inequality influenced Martha's ideas about who should benefit from environmental policies—namely people who need economic assistance, rather than those who want to turn a profit. Martha continues,

> The rich are getting richer, and the poor are getting poorer. But I'm for anything [incentives] that's legitimate, will save the planet, provide something, clean air, clean water. Windmills, solar, I'm all for that. But you know, it has to – it has to benefit where the benefit needs to be, not just somebody's pocketbook, the corporations or whatever.
>
> (Martha, white, 70s)

Martha is suspicious that businesses influence the creation of environmental policy so as to benefit from it financially (Cap and Trade), while wealthier individuals seek to take advantage of subsidies that are not directed at them (Cash for Clunkers), both of which she deems 'illegitimate.' Whereas, by her definition, 'legitimate' environmental policies would not only help the environment but also the poor and working class as well.

Voluntary simplifiers think that in order to avoid the political controversy over climate change they must work on smaller and smaller scales (regions, communities, and households). This trend means that voluntary simplifiers will spend many years greening their lives and homes, but make only a limited contribution to addressing climate change. Projections show that changing household practices and increasing efficiency can reduce overall US carbon emissions by 7.4 percent over ten years (Dietz *et al.*, 2009). This hardly comes close to the 80 percent greenhouse gas emissions reduction that is recommended for industrialized countries by 2050 (or about 2 percent a year) (Hassol, 2011). Government policies that focus on changing household practices will not only fail to reach this goal, but they also concede a whole arena of policy options (regulating business) that go far beyond what individual agency can accomplish.

Conclusion

What does it mean for academics to ask: 'how can we make people be more environmentally responsible?' Policy outcomes are experienced differently depending on intersecting positions of race, class, and gender, which produce advantage (more individualized policies that respect consumer sovereignty) and/or disadvantage (top-down community redevelopment). Our studies demonstrate that attempts to mitigate climate change must recognize that social positionality and social context influence the success or failure of these and any given strategies.

While post-Katrina New Orleans represents a unique case, the lessons learned from the various ongoing projects to rebuild the city are instructive for other urban spaces in crises—both those suffering from chronic stressors like poverty and marginalization and those suffering the acute stressor of a disaster or catastrophe. Because of decades of abandonment by planners and politicians, residents are suspicious of altruistic efforts on the part of outsiders. Residents who lived in urban decay most of their lives see the dramatic greening of their community as something to be wary of, as a gift with strings attached. And because the neighborhood has been inundated with volunteers (upwards of one million), residents often feel that they are told what to do rather than being asked what needs to be done. Additionally, they feel that lay knowledge is less valued than the expert, 'green knowledge' of ecologists or environmentalists (who tend to be white and college educated). Residents have thus come to resent many sustainability efforts because they originate with 'green outsiders.' While most of MIR's sustainable building practices are employed elsewhere, for the residents of the Lower Ninth they are strange and unfamiliar, encroaching on the autonomy of the neighborhood. Thus, environmental practices are seen as part of a middle-class or the 'other's' sensibility. Those institutions, like the CSED, which have taken smaller, less visible steps, have been more successful in attaining community support for their environmental projects. Such smaller projects, however, go unnoticed or seem to originate 'from here.' CSED's work on Bayou Bienvenue therefore represents something sustainable for the whole community.

In contrast, the most popular policy interventions to come out of the micro-level 'personal consumerist paradigm' (Chitewere and Taylor, 2010, p.148) include information campaigns and policies that combine (1) taxes on goods that release greenhouse gases, with (2) incentives to support energy efficiency (i.e. weatherizing homes), and (3) the adoption of more efficient technologies (i.e. hybrid vehicles or solar panels). These policies center on improving access to alternative 'choices' in the marketplace that are more environmentally responsible; yet these policies alone have had little success in changing behavior (Jackson, 2005) and the diffusion of efficient technology, aimed at the shrinking middle to upper-middle class, excludes many. Without access to these incentives, voluntary simplifiers work on changing their practices by embracing working class resourcefulness and old world frugality. Aware of the hierarchy between green lifestyles that save money and those that cost money, voluntary simplifiers advocate for broader, more economically inclusive policies to bridge this gap. Voluntary simplifiers have begun working together as part of meso-oriented projects to support changes in schools and communities, like edible gardens at the local high school paired with a pay-what-you-can cafe. They set aside the question of what is the greenest path and instead respond to the financial constraints of the community with plans they deem feasible to reduce waste and use resources more efficiently.

These two case studies show a lack of trust in expert knowledge and top-down policies that benefit (or are perceived as benefiting) the wealthy rather than the economically disadvantaged. Top-down programs like redevelopment or subsidizing innovative technologies are out of touch with the current economic landscape and are inadequate in reaching people. These case studies reveal the need for a meso-level approach to climate change that speaks to everyday lived realities. More contextually embedded policies would avoid blaming the victim—whether that person wants their old neighborhood back or wants to avoid the divisive politics of climate change—and instead help us to understand their standpoint and work with them, not against them. What works in one community, or perhaps more importantly, what is embraced in one community, may not work in another. People interpret environmental problems and solutions in radically different ways, some in accordance with deep racial and class divisions. Sustainability projects at the community or regional level should take into account the diversity of actors, their socio-economic and cultural needs, and the historical context that shapes policy outcomes. Only then can we understand the complexity of how environmental sustainability is experienced in everyday life and why some projects are embraced and others are rejected.

Acknowledgments

Both authors would like to thank the editors of this book for their comments on an earlier draft of this chapter and the Rutgers University Initiative on Climate and Society for funding portions of this research. Janet Lorenzen would like to acknowledge support from the AAUW American Fellowship program. Daina Harvey would also like to acknowledge support from The Horowitz Foundation for Social Policy and The Natural Hazards Center at the University of Colorado at Boulder and the Public Entity Risk Institute with support from the National Science Foundation and Swiss Re.

References

Allen, B., 2011. Laboratorization and the 'Green' Rebuilding of New Orleans's Lower Ninth Ward. In C. Johnson, ed, *The Neoliberal Deluge: Hurricane Katrina, Late*

Capitalism, and the Remaking of New Orleans. Minneapolis, MN: University of Minnesota Press, pp. 225–244.

Arena, J., 2012. *Driven From New Orleans: How Nonprofits Betray Public Housing and Promote Privatization*. Minneapolis, MN: University of Minnesota Press.

Bourdieu, P., 1984. *Distinction: A Social Critique of the Judgment of Taste*. Translated from the French by Richard Nice. Cambridge, MA: Harvard University Press.

Braungart, M. and McDonough, W., 2002. *Cradle to Cradle: Remaking the Way We Make Things*. New York: North Point Press.

Bullard, R. and Wright, B., 2012. *The Wrong Complexion for Protection*. New York: NY University Press.

Checker, M., 2008. Eco-apartheid and Global Greenwaves: African Diasporic Environmental Justice Movements. *Souls*, 10(4), pp. 390–408.

Chitewere, T. and Taylor, D.E., 2010. Sustainable Living and Community Building in Ecovillage at Ithaca: The Challenge of Incorporating Social Justice Concerns into the Practices of an Ecological Cohousing Community. *Research in Social Problems and Public Policy*, 18, pp. 141–176.

Collins, P.H., 2000. *Black Feminist Thought: Knowledge, Consciousness, and the Politics of Empowermen*, 2nd edn, New York: Routledge.

Denhart, H., 2010. Deconstructing Disaster: Economic and Environmental Impacts of Deconstruction in post-Katrina New Orleans. *Resources, Conservation and Recycling*, 54(3), pp. 194–204.

Dietz, T., Gardner, G.T., Gilligan, J., Stern, P.C., Vandenbergh, M.P., 2009. Household Actions can Provide a Behavioral Wedge to Rapidly Reduce U.S. Carbon Emissions. *Proceedings of the National Academy of Sciences*, 106(44), pp. 18452–18456.

Feireiss, K. and Pitt, B., 2009. *Architecture in Times of Need: Make it Right-Rebuilding New Orleans' Lower Ninth Ward*. New York: Prestel Publishing.

Grigsby, M., 2004. *Buying Time and Getting By: The Voluntary Simplicity Movement*. New York: State University of New York Press.

Hassol, S.J., 2011. Questions and Answers: Emissions Reductions Needed to Stabilize Climate. *Presidential Climate Action Project* [pdf]. Available at: www.climateactionproject.com/docs/Hassol_PPM_rev.pdf [Accessed 1 December 2013].

Huneke, M.E., 2005. The Face of the Un-consumer: An Empirical Examination of the Practice of Voluntary Simplicity in the United States. *Psychology & Marketing*, 22(7), pp. 527–550.

Jackson, T., 2005. Motivating Sustainable Consumption: A Review of Evidence on Consumer Behaviour and Behavioural Change. *Sustainable Development Research Network*. [online] Available at: www.sd-research.org.uk/post.php?p=126 [Accessed 1 February 2010].

Lorenzen, J.A., 2012a. Going Green: The Process of Lifestyle Change. *Sociological Forum*, 27(1), pp. 94–116.

Lorenzen, J.A., 2012b. Green and Smart: The Co-construction of Users and Technology. *Human Ecology Review*, 19(1), pp. 25–36.

Maniates, M., 2002. Individualization: Plant a Tree, Buy a Bike, Save the World? In T. Princen, M. Maniates and K. Conca, eds, *Confronting Consumption*. Cambridge, MA: MIT Press, pp. 43–66.

Park, L.S.H. and Naguib, D.P., 2011. *The Slums of Aspen: Immigrants vs The Environment in America's Eden*. New York: NY University Press.

Rees, W.E. and Westra, L., 2003. When Consumption Does Violence: Can there be Sustainability and Environmental Justice in a Resource-limited World? In J. Agyeman,

R.D. Bullard, and B. Evans, eds. *Just Sustainabilities: Development in an Unequal World.* Cambridge, MA: MIT Press, pp. 99–124.

Schor, J.B., 1998. *The Overspent American: Why We Want What We Don't Need.* New York: Harper Perennial.

Part V
Æther

Figure V.1 Æther

37 Softly Walking

Sufia Giza Amenwahsu

I walk softly on this Earth because Energy carries vibrations
that ripple across the Universe, like waves on approaching Low-country Tides.

I walk softly on this Earth because I hear the sweet whisper of
My Ancestor's wise voices reasoning with us in blowing winds …

I walk softly on this Earth because SHE is my Mutha.
Her Love & Sacrifice birthed the Land we must honor to survive and
The Sea we need to breathe in vital negative Ions.

I walk softly on this Earth because my Ancestors did and were ordained
To build Grand Civilizations in exchange for their committed care of
Mutha Nacha …

I walk softly on this Earth because my people swung like strange fruit from her
Poplar Trees, so that we could be FREE to change Realities, impacting societies …

I walk softly on this Earth because Energy carries vibrations
that ripple across the Universe like waves on cascading Solar Rays.

I walk softly on this Earth because I hear the sweet whispers of
The Most High's wise voice in blowing leaves.

I walk softly on this Earth because SHE is my Mutha
Her Wisdom & Inspiration birthed the air we must honor to survive and
Our RELATIONS we need to breathe in perfect TRANQUILITY …

I walk softly on this Earth because my Ancestors left a blueprint etched in stone
For us to follow home, in exchange for our committed care of
Mutha Nacha …

So, I must Love & Honor ALL my RELATIONS and
Walk softly on this Earth very softly
So that my KARMIC & CARBON Footprints are
Virtually - Invisible.
WALK
SOFTLY
ON
THIS
EARTH …

38 Reach

Vanessa Lamb

Figure 38.1 Reach. Acrylic paint on canvas, 2014

39 A pilgrimage for hope revisited

Grieving together (*caminamos preguntando*)

Ryan Pleune

> And so long as you haven't experienced
> this: to die and so to grow
> you are only a troubled guest on the dark earth.
>
> (Johann Wolfgang von Goethe, in Bly, R. n.d.)

I used to hate activists. In high school and college I scoffed at people with petitions and ridiculed protestors in the streets. "Caminamos preguntando" literally means "we walk asking." It is in reference to the Zapatistas, activists who have resisted military, paramilitary, and corporate incursions into Chiapas Mexico since 1994. I use it here acknowledging I do not know the way forward; I hope anyone reading can take a long walk with their soul and ask: What do I love most in this world? How can I act with the highest integrity for that love and embrace both outcomes of success and failure? I invite you to walk and talk with your soul, questioning how the world-wide economic system in power harms our life and puts future generations at risk.

My soul's work consists of co-creating life-affirming connections by acting with integrity, demonstrating my love towards others, and honoring my spirituality. I am a walking contradiction: both status quo, white, middle-class, male that ridicules activists; and fringe activist channeling anger and critique at the status quo. Typing my master's thesis during the BP oil spill in the Gulf of Mexico in 2010, I gained what Brazilian educator and revolutionary, Paulo Freire, called *conscientização* or conscientization: It's unjust to educate about systemic crises without my own attempts to intervene.

Home – privilege and pain

My psyche is deeply rooted in the intermountain west of the United States; though, like all white people in the Americas, I am from an immigrant family. My parents are from Wisconsin and New York and my ancestors immigrated from Holland, Ireland, and France. I grew up in Denver, Colorado and graduated from a public school desegregation program where whites were bussed into black neighborhood schools.

The pain of historic oppression was beat into me at an early age. Whites were a minority but had access to a majority of the academic and material success because we shared mainstream cultural, political, and economic power. Despite the intent of desegregation and being in the racial minority, the classes I was placed in on the 'accelerated' track were segregated; whites held the majority. Even as young people, unaware of the institutional racism controlling us, there was powerful racial suffering that was 'acted out' through playful teenage games like 'Jump the White Boy' each day at lunch. Years later at a predominantly white college, I reflected on my own privilege in the 'accelerated' classes, and the privilege of generations before me who accumulated social and economic capital while the ancestors of my peers were enslaved or exploited.

With the same education, I got accepted into (and could afford) a private liberal arts college while many of my black and Latino peers did not. Today I have access to higher paying jobs while many of my peers, who are people of color, still struggle to make ends meet. My privilege of being a white, college-educated male with the ability to willfully quit a job to go on a pilgrimage contrasts the painful inequities that I have witnessed. I can only attempt to understand this through empathy and being in relation with communities that are oppressed.

The rift of inequality I bared witness to is also visible on a global scale. The majority of the world is composed of poor people of color and they suffer the most from environmental degradation and the effects of climate change. The Climate Vulnerability Monitor, commissioned by the governments of twenty countries, estimates that if the world stays on its current path, 90 percent of the 100 million climate-related deaths between now and 2030 will occur in developing countries – a.k.a. the "Global South" (Climate Vulnerability Monitor 2012). In contrast, the primarily white Western European people of privilege (a.k.a. the "Global North") control and benefit from acquiring most of the world's resources while suffering few of the consequences. Fifty of the ninety entities producing the world's highest carbon emissions since the industrial revolution are investor-owned, multinational corporations. Four of the top six are from the Global North: Chevron Corporation (United States multinational), ExxonMobil Corporation (United States multinational), BP plc (British multinational), and Royal Dutch Shell plc (Dutch multinational) (Carbon Majors, 2013).

My belief and hope is that as the climate justice movement responds to escalating conflicts from climate change we have an opportunity for global anti-oppression work to transcend barriers that have been institutionalized through a competitive global economic system. As this opportunity arises I will push for a renewed sense of empathy and cooperation. Focusing on this aspect of human nature can feed a paradigm shift towards a new global economic system that does not accept inequality as an inherent part of society. Working to transcend barriers of my own dichotomous thinking, I am learning how to do this systemically with others. Addressing the root causes of climate change will require answering questions that many social justice leaders, including scholar Patricia Hill Collins, have posed for over two decades. "How can we re-conceptualize race, class and gender as categories of analysis?" and "How can we transcend barriers created by our experiences of race, class and gender oppression in order to build the types of coalitions essential for social exchange?" (Collins, 1993, p.27).

Departure – "this changes everything"

From 2000 to 2010, I worked as a wilderness therapy instructor with adjudicated youth, taught high school biology and earth systems, and worked in the non-profit sector as an environmental educator and green schools coordinator. During this decade of working as an educator, I felt increasingly depressed about the dire forecast for human sustainability. This mirrored the social injustices I experienced in my life. My understanding of privilege and pain was translated into empowering youth to share their stories and highlight inequities. Then I prayed that someone else would act to create solutions.

I taught science at East High School in Salt Lake City, Utah from 2004–2007. We had 2200 students: 880 identified as other than white; 700 spoke a language other than English at home; and 90 were living in the United States as political refugees. Many of the remaining 610 English language learners were economic refugees: persons fleeing a life of economic genocide, as compared to political refugees who enter the United States with legal status after cultural genocide. For example Abdi, a Somalian student, was given amnesty by the US government to live here after tribal warfare wreaked havoc on his family. In contrast Eduardo's family risked their lives crossing the border from Mexico. Free trade laws like NAFTA flooded Mexican markets with subsidized corn and grain, undercutting Mexican agriculture workers, left them unemployed and seeking migrant work in the US even if it was illegal. Abdi is a political refugee and Eduardo is an economic refugee.

Working to empower students, the trauma of inequality continually smacked me in the face. The American Civil Rights Movement passed substantive laws, yet equality was far from reality over 50 years later. One of my Latina students, Brenda, screamed at me in mixed Spanish/English: "Pleune, if all us brown people aren't dumb, porque no hay brown teachers in this school?" I explained that while discrimination was illegal, institutional racism still permeates our economic and political systems. That wasn't enough and we were unable to come to a conclusion together or co-create knowledge together; the rift between us seeming too big to transcend the barriers between our different experiences of race class and gender oppression.

Shared power and equal outcomes that many movement leaders died for was not visible at this school or the one I graduated from a decade earlier. I hoped my students would lead the way into action. I was part of school integration and was born after the Civil Rights Act and Voting Rights Act, so why weren't my Black and Latino peers in teaching or administrative positions at a school like East 50 years later? Why was Wells Fargo still allowed to do business in 2012 after settling a lawsuit in Baltimore Maryland where they were accused of unfairly steering racial minorities seeking home ownership into subprime mortgages and giving them rates less favorable than whites? These inequities are not unique to Baltimore or the Salt Lake City School District. They seem to be inextricably supported by capitalism (Badger 2015; Rosthstein 2015). Discrimination from realtors and bankers is illegal; segregated schools are illegal. However, profit-seeking that segregates neighborhoods and, ultimately, schools is simply a matter of business.

Similarly, while walking my dog in neighborhoods less than one mile from three oil refineries, I was reminded of ecological and climate collapse despite progress from the Environmental Movement that passed the Clean Air and Clean Water Act almost 50 years ago. The crude oil pipelines of Chevron and Tesoro flow underneath our city and across beautiful, open-space land along the Bonneville Shoreline trail. In the summer of 2010, a crack in the pipeline leaked nearly 800 barrels of oil into our watershed and through urban parks. Just six months later, another spill leaked between 400–600 barrels in the same drainage. In the spring of 2013, a third leak in less than three years left over 600 barrels in a migratory bird refuge. Our watershed is not an aberration. An Energy Wire Analysis released in early 2014 shows about twenty spills per day in the United States (Sorgam, 2014). Polluting the air and water are illegal, but also a matter of business.

All I knew how to do was teach people about the problem and hope they would find the solution. I grew up recycling, and had options to buy environmentally friendly products. I was born after the Clean Air and Clean Water Acts were passed, so why did ecological destruction increase at twenty oil spills per day and climate change still occur at an unprecedented rate almost fifty years later? Why was BP allowed to profit and maintain its business model after blatant negligence in the Gulf oil spill of 2010?

My theory of change—'planting seeds' in young people and then expecting they will find solutions to create more equity and stop ecological destruction—was too late for Brenda and would not halt the Chevron and Tesoro spills. I realized that waiting for my students to come into social and political power would not create change at the rate scientists show we needed to halt carbon dioxide emissions and stabilize the climate.

Nearly fifty years have passed since significant social and environmental legislation was enacted, and we are still fighting for equity and a healthy world. I asked myself: If laws aren't enough to catalyze change, what is at the root cause of these inequities? What systems are strengthening and accepting inequality at the same time that we passed laws to 'level the playing field'?

From 2008–2010, I was personally introduced to the geo-political and economic systems that run our country while living in Washington D.C. with my best friend and former wife Jamie Pleune. I became more aware of the global complicity in the climate change crisis and the immense collective action needed to create a paradigm shift. In response to our lobbying efforts to act on climate change, an energy staffer for one Utah senator insisted we elect a new representative because this congressman was elected on a climate-denial platform. Citizen lobbying or letter writing were fruitless as he will always vote against regulating carbon emissions, even when they are harming our food and water systems.

In 2009, 167 countries, including the United States, signed an agreement to maintain a global temperature rise of two degrees Celsius above 1990 levels. The 2015 Paris Agreement maintained that target and acknowledges that 1.5 degrees is desirable (Conference of the Parties 2015). Even with a two degree rise in global temperatures we will lose many Pacific Islands due to sea level rise; the Northeast United States infrastructure, agriculture, fisheries and ecosystems will be compromised; the Northwest will be threatened by sea level rise, erosion, inundation, and increasing ocean acidity; the Southeast will suffer extreme heat affecting health, agriculture and water availability; in the Midwest, heat, heavy downpours and flooding will affect health, agriculture, air and water quality; and in the Southwest there will be increased wildfires, insect outbreaks and drought (The current and future, 2014). Science shows us that to limit the negative effects we can burn 565 gigatons of carbon between 2011 and 2049. However, the fossil fuel industry has business plans to burn 2795 gigatons in their proven reserves. Despite the international agreement and climate threshold numbers, the multi-national fossil fuel industry plans to burn five times the global limit. In simple terms, roughly 80 percent of the proven reserves must stay in the ground to maintain the two degree limit (Leaton, 2013). The multi-national corporations would be harmed if world governments passed laws to effectively keep that 80 percent in the ground.

The US senator I met with received over $600,000 from fossil fuel corporations between the time our country signed on to the two degree limit and when I met him; He is not alone (*Dirty Energy Money*, 2015). Until our constitution is amended, corporations are allowed to spend as much money as they want on elections and have legal personhood rights. This also means citizens are not technically allowed to pass laws that could harm profits. For example, a federal judge recently upheld a lawsuit by Royal Dutch Shell against Mora county citizens who had passed a law to ban drilling (Matlock, 2015). Why were historic social and environmental laws being eroded and recent citizen ballot initiatives creating new laws being overturned?

In her book *This Changes Everything: Capitalism vs The Climate*, Naomi Klein challenges climate activists to critique capitalism rather than act as self-flagellating martyrs. Many environmentalists, including myself, are focused on a low carbon diet via individual life changes like growing local food and using bikes for transportation. Jamie and I even decided not to have kids to reduce our impact. At the same time Klein gave birth to her first child, I denied myself that desire knowing the average lifestyle in the United States needs five planets to sustain it (Global Footprint Network 2015). I longed to be a father but this selfish desire seemed unfair to the potential child and to the planet. What I didn't calculate is that despite my individual low carbon diet, capitalism relied on profiting and growing from this five planet lifestyle.

Unaware of the ensuing metaphorical death, I did recognize the suppression of reproductive desires felt like an obsession with darkness that Goethe refers to in The Holy Longing.

Tell a wise person, or else keep silent
because the mass man will mock it right away.
I praise what is truly alive,
what longs to be burned to death.
In the calm waters of the love nights
where you were begotten, where you have begotten,
A strange feeling comes over you,
when you see the silent candle burning.
Now you are no longer caught in the obsession with darkness,
and a desire for higher love making sweeps you upward.
Distance does not make you falter.
Now arriving in magic, flying,
and finally, insane for the light,
You are the butterfly and you are gone.
And so long as you haven't experienced
this: to die and so to grow
you are only a troubled guest on the dark earth.

(in Bly, R. n.d.)

Initiation – burned to death

In the fall of 2010, Jamie invited me to walk 350 miles with her through Utah, raising awareness for the most important number in the world. We called it our Pilgrimage for Hope. Science shows that, for human sustainability on this planet, 350 parts per million is the safe upper limit of carbon dioxide in the atmosphere. We have exceeded that at an unprecedented rate and hit a record high of 400 ppm in May 2015.

Jamie and I decided to do what humans have been doing for thousands of years. We walked to contemplate our existence. Stopping every mile we took a picture to practice humility and ask our souls for guidance. Sometimes we made vows of silence for several days and walked without food, the hunger digging deeper into our raw emotions. We encountered evidence of collapsed civilizations such as the Ancestral Puebloans, who went extinct from genocide or resource scarcity. We contemplated our relationship to death and how we could prepare for it, literally and metaphorically.

One morning in the darkness, turning to walk back into civilization, we heard a kit fox sing. The beguiled sunrise song sent my mind reeling about dark moments in human history. Musical notes through eerie cries in the predawn shadows of moonlight also sounded beautifully wild and full of mystical possibilities. Remembering a mindfulness mantra "breathing in – everything is getting worse; breathing out – everything is getting better" I wondered, "how could I put this feeling into action?"

Some indigenous traditions teach about paying attention to the wild world. In the book *Medicine Cards*, authors Jamie Sams (Choctaw decent) and David Carson (Cherokee, Seneca and French decent) describe medicine as "anything that improves one's connection to the great Mystery and to all life" (Sams, 1999, p. 13). Approaching animals with humility and intuitiveness, people encountering them can learn certain wisdom. The fox archetype explains that, as if invisible, a fox can observe *with* its surroundings as opposed to looking *at* its surroundings (Sams, 1999). For me, the kit fox song is a reminder to pause and observe the tension of seemingly opposite sides of phenomena. Are they antithetical? What are the overlaps and intersections?

Similar to the fox observing *with* its environment, Brazilian educator and activist, Paulo Freire encourages us to think *with* instead of thinking *for* or thinking *about* people in a process called dialogical thinking. Freire describes it as a means for action and liberation (Freire [1972] 2002). Walter D. Mignolo, an Argentine theorist, expands on dialogical thinking with the concept of border thinking. He describes it as the process of thinking *from* dichotomies instead of thinking *about* them. (Mignolo [2000] 2012). Mindfully like a fox, I notice my own ridicule towards activists, and I am judged similarly by others. I am frightened by racist narratives that percolate my thoughts, and I work diligently to infuse anti-racist curriculum in my teaching.

Most conflicts in the world reveal an intersecting and contradicting polarity in our individual psyche and collective unconscious: oppressor and oppressed. Author and activist Audrey Lourde reminds me that, "The true focus of revolutionary change is never merely the oppressive situations which we seek to escape, but that piece of the oppressor which is planted deep within each of us" (Lourde, 1984, p. 123). It is a common error to create either/or dichotomies of good versus evil when the reality is that both exist together in a dynamic tension. In a Cherokee legend, an elder explains to his grandson about a constant fight inside him. It is between two wolves of good and evil. When the grandson asks which one wins the reply is "the one I choose to feed" (The tale of two wolves, 2015). When working for climate justice I can see opposing sides within myself: Both the oppressive/competitive and the empathic/cooperative.

Instead of working with both, our global economic system creates an either/or choice between the two perspectives and feeds, with money and material gain, only the competitive aspect of human nature. As a result we have institutionalized oppression/competition as a better choice through neo-liberal 'free market' trade rules. The overlap between oppressive/competitive with tendencies of empathy and cooperation is framed in a dichotomy that many people view as antithetical.

In addition, most governments are dominated by the profit interests of multinational corporations encompassing only the oppressive/competitive aspects of human nature. Those justifying neoliberal global trade policies and institutions such as NAFTA (North American Free Trade Agreement), the WTO (World Trade Organization), or the TPP (Trans Pacific Partnership) create rules that prohibit our cooperative and empathetic nature. This part of our nature would prioritize trade rules with equitable distribution of resources and reparations for historical trauma before prioritizing 'free market' competition. According to the neoliberal free market rules, cooperation and sharing are seen as the antithesis of competition and there is an acceptance that inequality will always exist.

The global trade rules' seemingly auspicious foundation is creating a marketplace where there is an assumed equal access. This attempt at equality results in clear winners and losers. Instead of working towards equal outcomes and repairing historical imbalances of capital and exploitation, the only measure is equal opportunity. Winners look like CEOs of many western clothing companies; losers look like garment workers in Bangladesh. Winners are CEOs in the fossil fuel industry; losers are one quarter of the world's population in South Asia, relying on water from glaciers that are disappearing due to climate change (Lagahri, 2013).

On this Pilgrimage for Hope Jamie and I made sacred contracts with our souls and I vowed to make public my commitment to work for climate justice.

Returning – truly alive

Returning from the pilgrimage, Jamie and I co-founded the iMatter Utah campaign with teens, parents, and grandparents in our community. iMatter, a national campaign filing lawsuits to address climate change, was unsuccessful in the judicial realm, but it helped us find multi-generational communities of change makers.

In the winter of 2011, after grueling hours typing an academic paper, I was awarded a Master's of Science and planned to start a youth development program called Wasatch Wilderness Rites of Passage. I became deeply involved with Peaceful Uprising, a climate justice group formed to promote and escalate civil disobedience in the climate movement. We educated the public and demonstrated support for an economics student from the University of Utah, Tim DeChristopher, who was arrested for disrupting an oil and gas auction. I realized that if I want people to act, I could not teach them about it – *I* had to act.

Working as a bus driver later that winter, I drove a school bus full of 5th graders and six of their teachers to see *A Tale of Two Cities*. The play depicts revolutionary chans. The play depicts revolutionary change that influenced the American Revolution, and when I asked why they were seeing it, the teachers explained that they had been discussing dissent in democracy and sharing current events such as the Wisconsin Occupation of the Capital and the Arab Spring. After leaving the play, I drove the bus full of students and teachers by the US courthouse, where fellow activists rallied to support Tim during his trial. I explained to my passengers that he was on trial for registering as a bidder during an oil and gas auction and won several leases without the intent to pay, thus nullifying all other transactions. My impromptu detour involved me texting a press release stating, "5th Graders studying democracy and revolution visit Bidder 70s rally". When the story exploded in the news, the school district received calls asking if they endorsed this type of teaching.

It is unfortunate that we teach about civil disobedience as if it ended with Rosa Parks and the American Civil Rights Movement. I decided to show the students and teachers that people today, even in our own city, are using similar tactics and creating history that is not yet in textbooks. Peaceful Uprising gained national attention while campaigning around this trial to encourage more civil disobedience in the climate movement. One teacher smiled and said "This will go down as the most memorable day in the fifth grade year. That we studied the rights in our democracy and then got to see them in action." The next day I was fired from the transportation department and was forever banned from teaching in the Salt Lake City School district. After several emails and meetings with the human resources director, his response was, "When someone is terminated, they are ineligible for re-hire for any position in the district. And in addition I see you are involved with these anti-tar sands campaigns, and I think what you are doing is inappropriate."

Without any remorse I continue organizing and acting in as many creative and powerful ways as I can. I am still proud of risking my career in the same way I imagine my grandpa felt as he left a family and 'career' to serve his country in World War II. Many still don't think the climate crisis is as urgent as WWII, but I think it is worse because it is a slow violence that goes unnoticed. I am not insinuating that I take the risk of being shot at or suffer the emotional trauma of witnessing mass killings like people in the military do. However, I think people working for climate justice should be honored.

Along with my activism, I am confronting an important failure of mainstream environmental education: the question of children. Instead of being a martyr of my own longing to be a father, I am shifting my focus. Like Klein, "I'd rather fight like hell than to give these evil motherfuckers [fossil fuel CEOs] the power to extinguish the desire to create life. We don't all have to do it. But if we want to do it, if we want to be part of this amazing process that we share in common with all living things, I'm certainly not going to give these guys the power to take that away from me" (Klein, 2012).

Five years after my pilgrimage, I am reclaiming my desire to become a father. Looking back on the decision Jamie and I made to not have kids I recognize the slow violence of climate change led to our worst selves being exposed, even as we tried to act with integrity for what we loved by reducing our carbon footprint. An obsession with darkness burned us to death. Instead of telling Jamie of my desire to have kids, I created an either/or choice, "stay married or have kids", and asked for a divorce. Dissolving our marriage and ending a fifteen-year partnership was painful for both of us and the psychological damage is permanent. I wish I had remembered the kit fox singing and employed border thinking to co-create a new paradigm. Our divorce may still have occurred, but we might both understand each other better if I had proceeded more mindfully. Now we are both left grieving loss in the same way we would for a death. I don't know if my decision was right. I know it was catalyzed by confronting the root causes of the slow violence of climate change yet I remain uncertain if the loss and grief will have a positive outcome.

Similarly, I don't know if there will be positive outcomes from the risks taken by campaigns using social media hashtags like #idlenomore, #BlackLivesMatter,

#occupy, #notarsands, #NoKXL, #blockadia, #sHellNo, #FloodTheSystem, Climate #Redlines, #Breakfree2016 and others using direct action tactics. There is a lot of urgency and logic in agitating for change, yet grief exists knowing those changes will bring the loss of what we are used to. I believe that employing strategy that disrupts dichotomous thinking will be important. Using border thinking Klein points out "As communities move from simply resisting extractivism to constructing the world that must rise in its rubble, protecting the fertility cycle is at the heart of the most rapidly multiplying models, from permaculture to living buildings to rainwater harvesting." (Klein 2014 p. 446)

A well-known strategist, Gene Sharp, has written many books on the power of non-violent resistance over the last few decades, ranging from the *Politics of Non-Violent Struggle* in 1973 to *Sharp's Dictionary of Power and Struggle: Language of Civil Resistance in Conflicts* in 2011. I value non-violence and I also value armed strategies used by the Zapatistas or Nelson Mandela with the armed branch of the ANC. Using border thinking I can recognize non-violence as good strategy and acknowledge that violence-shaming is one way existing power structures subvert social movements.

The more I learn about achieving climate justice and confronting systems of oppression, the more committed I become to the core principles of Peaceful Uprising:

1. We refuse to be obedient to injustice
2. Our human stories are extremely powerful, and genuine sacrifice has the ability to awaken and inspire others
3. We are connected to something much greater than ourselves, which has an incredible power to change the world
4. We are steadfast in our commitment to the truth
5. Our allies and strategies align with and create the healthy and just world we want to see
6. A powerful movement originates with personal transformation and a commitment to being an agent of change
7. Creating a better world is not only necessary, but makes us authentically happy people
8. We are committed to building a supportive community that empowers our members to realize their potential
9. The best response to intimidation is joy and resolve
10. We recognize a nonviolent movement as the most effective means of creating a just and healthy world
11. We respect the inherent worth and dignity of every person[1]
12. Protecting Peaceful Uprising as an institution will never take precedent over our commitment to the fight for a healthy and just world

(Peaceful Uprising, 2015)

Every time I sing with Peaceful Uprising, I remember Tim DeChristopher's message in 2010 as he was convicted and later sentenced to two years in prison.

He did not ask for sympathy; he simply stated that, "We will begin to act like a movement when we can sing like a movement." In the 5 years since he was sent to prison we followed his lead to escalate actions, and now we need to practice singing like a movement. This must include rich harmonies with diverse voices. At the very least it must include ally-ship with indigenous communities around the world and using what Tim calls 'the accountability of privilege'. Accountability to follow through with fossil fuel divestment at religious, academic, and social institutions worldwide and to force our government to halt all new fossil fuel leases on federal land, doing our part at leaving 80 percent of the fossil fuel reserves in the ground.

Appreciating escalation and growth in the climate movement since 2010, we still must "transcend barriers created by our experiences of race, class and gender oppression in order to build the types of coalitions essential for social exchange" (Collins, 1993, p. 27). Activists in the climate justice movement are not the same faces as those in the Black Lives Matter movement. Laying down to blockade a road and close the federal building during a "Die In" protest in Salt Lake City I was gently dragged off the street by police. During escalated actions to protest the Keystone XL tar sands pipeline, I and thousands of activists nationwide were cited with resisting peace officers and often never faced trial.

Conversely a black teenager named, Freddie Gray never took another breath outside the police van he was shoved into. Michael Brown's body shot by police lay in the street for 6 hours. Protests against police brutality towards people of color are erupting around these and other police killings. At the heart of the Black Lives Matter movement is also a critique of our economic system that has segregated neighborhoods and schools despite laws prohibiting segregation.

I feel the rage of racially segregated poverty beating me down during "Jump the White Boy Day" and screaming in my head during my gentle treatment from police 25 years later. Is one of these my struggle and the other not? Do I choose to lay my body down for one and not the other? I see clearly that our economic system has exacerbated both.

My *conscientização* continues even writing this story. Spanish Literature scholar, feminist, Mexican and friend, Keri Gonzalez, shared in peer review of this essay how border thinking and one of its skill sets called bi-languaging unites and erases "the borders between knowing *about* and knowing *from*" (Mignolo 2012, p. 310). Our dialogue in mixed Spanish/English combines our diverse experiences as activists and co-creates knowledge on raising kids in an age of climate change and institutional racism. Bi-languaging between social movements and between borders from our different experiences of race, class, and gender oppression is part of the theoretical answer to a paradigm shift. A concrete and short term political step forward that would help address injustices from climate change and beyond is amending our Constitution to state that corporations are not people and money is not speech, thus reducing corruption of our current system while we co-create new ones. With this established, a law to curb the damage from capitalism and keep 80 percent of the fossil fuels in the ground might actually pass.

Using border thinking to move beyond the binary of exchanging socialism for capitalism, let's create a new system that ensures equity, cherishes interdependence and feeds the cooperative and empathetic side of human nature. Parting from what we know and grieving together, let's grow into something we've never imagined.

Note

1 Corporations are not people.

References

Alder, J., 2014. The reality of a hotter world is already here: As global warming makes sizzling temperatures more common, will human beings be able to keep their cool? New research suggests not. [online] Available at: www.smithsonianmag.com/science-nature/reality-hotter-world-already-here-180951172/#Ob7YvuOA3gLMrEVr.99 [Accessed 3 August 2015].

Badger, E., 2015. The long painful and repetitive history of how Baltimore became Baltimore. [online] Available at: www.washingtonpost.com/blogs/wonkblog/wp/2015/04/29/the-long-painful-and-repetitive-history-of-how-baltimore-became-baltimore/ [Accessed 3 August 2015].

Bly, R., n.d. *The Holy Longing – Johann Wolfgang von Goethe*. [online] Available at: www.yorku.ca/lfoster/documents/The_Holy_Longing_Goethe.htm [Accessed 3 August 2015].

Carbon Majors, 2013. *Climate Accountability Institute*. [online] Available at: www.climateaccountability.org/carbon_majors.html [Accessed 3 August 2015].

Climate Vulnerability Monitor, 2nd Edition: A Guide to the Cold Calculus of a Hot Planet, 2012. *DARA*. [online] Available at: http://daraint.org/climate-vulnerability-monitor/climate-vulnerability-monitor-2012/report/ [Accessed 3 August 2015].

Collins, P.H., 1993. Toward a new vision: race, class and gender as categories of analysis and connection. *Race, Class and Gender Journal* vol. 1, no. 1, pp. 25–45. [online] Available at: www.jstor.org/stable/41680038 [Accessed 3 August 2015].

Conference of the Parties 21st Session, 2015. *Adoption of the Paris Agreement*. [online] Available at http://unfccc.int/resource/docs/2015/cop21/eng/l09r01.pdf [Accessed 1 January 2016].

Dirty Energy Money 2015 *Oil Change International*. [online] Available at http://dirtyenergymoney.com/view.php?searchvalue=84115&com=&can=&zip=&search=1&type=search#view=connections [Accessed 3 August 2015].

Freire, P., (orig. 1970) 2002. *Pedagogy of the Oppressed*. New York: Continuum International Publishing Group.

Global Footprint Network, 2015. *Footprint Basics*. [online] Available at: http://footprintnetwork.org/en/index.php/GFN/page/basics_introduction/ [Accessed 22 September 2015].

Klein, N., 2012. *Naomi Klein on The Failures of the Environmental Movement* [online]. Available at: www.nationinstitute.org/featuredwork/fellows/3081 [Accessed 3 August 2015].

Klein, N., 2014. *This Changes Everything: Capitalism vs The Climate*. New York: Simon and Schuster.

Laghari, J., 2013. Climate change: melting glaciers bring energy uncertainty. *Nature* [online] vol. 502, no. 7473 pp. 617–618. Available at: www.nature.com/news/climate-change-melting-glaciers-bring-energy-uncertainty-1.14031 [Accessed 3 August 2015].

Leaton, J., Ranger, N., Ward, B., Sussams, L., and Brown, M., 2013. *Unburnable Carbon 2013: Wasted Capital and Stranded Assets* [pdf]. Available at: www.carbontracker.org/wp-content/uploads/2014/09/Unburnable-Carbon-2-Web-Version.pdf [Accessed 3 August 2015].

Lord, A., 1984. *Sister Outsider*. Berkeley, CA: Crossing Press.

Matlock, S., 2015. Federal judge overturns Mora County's drilling ordinance. *Santa Fe New Mexican* 20 January. Available at: www.santafenewmexican.com/news/local_news/federal-judge-overturns-mora-county-s-drilling-ordinance/article_dddd444a-6ae8-56ea-b8a7-999c562a77b8.html [Accessed 3 August 2015].

Mignolo, W. (orig. 2000) 2012. *Local Histories/Global Designs: Coloniality, Subaltern Knowledges, and Border Thinking*. Princeton, NJ: Princeton University Press.

NASA, 2014. The current and future consequences of global change. [online] Available at: http://climate.nasa.gov/effects/ [Accessed 3 August 2015].

Peaceful Uprising, 2015. *Peaceful Uprising Core Principles*. [online] Available at: www.peacefuluprising.org/about/peaceful-uprisings-core-principles [Accessed 3 August 2015].

Rothstein, R., 2015. *From Ferguson to Baltimore: The Results of Government Sponsored Segregation*. [online] Available at: www.epi.org/blog/from-ferguson-to-baltimore-the-fruits-of-government-sponsored-segregation/ [Accessed 3 August 2015].

Sams, J. and Carson, D., 1999. *Medicine Cards*. New York: St Martin's Press.

Sorgam, M., 2014. *Oil and gas spills up 17 percent in U.S. in 2013*. [online] Available at: www.eenews.net/stories/1059999364 [Accessed 3 August 2015].

The tale of two wolves, 2015. *The Nanticoke Indian Tribe*. [online] Available at: www.nanticokeindians.org/tale_of_two_wolves.cfm [Accessed 3 August 2015].

40 Of the necessity and difficulty in working across borders

Race, class, gender, and transnational environmental organizing

Rachel Hallum-Montes

The Alliance for International Reforestation (AIR) was founded in Guatemala in 1991 with the aim of working with indigenous communities to establish community-based reforestation programs. AIR advertises its services – including free training on agroforestry techniques, as well as free provision of tools and resources – in local community centers. AIR has worked with nearly 3000 community members and families throughout Guatemala to plant over 4.2 million trees. Approximately 70 percent of the community members who have partnered with AIR have been indigenous women.

While the organization was founded by a North American woman (my mother), its staff is based entirely in Guatemala and includes both indigenous and *ladin@* women and men. The Guatemalan-based Director of Operations is a Ladina[1] woman, while all of the five agroforestry technicians (*tecnicos)* are indigenous men. My mother and I, both gringas, are based in the United States (US) and have primary responsibility for securing the resources to keep the organization afloat, primarily through grant-writing and individual donors (most of whom are also based in the US). As a transnational environmental organization with diversity in race, class, and gender across its leadership and staff, AIR provides a valuable opportunity for examining both the necessity and difficulty of working across borders to combat global climate change (GCC). This chapter offers the case of study of AIR to examine some of the challenges inherent in mobilizing across both geographic and socially constructed borders – as well as strategies for addressing some of the challenges the organization has experienced in its more than two decades of operations.

Background

Social movement scholars have long argued that transnational mobilizing is critical for addressing the most pressing social, economic, and environmental challenges of our time (Alvarez, Dagnino, and Escobar, 1998; Keck and Sikkink, 1998). In the case of GCC, communities that live within a context of poverty or who have otherwise historically been marginalized and oppressed are often the ones to be most strongly impacted by the consequences of environmental degradation. However, these same communities often lack access to the economic and/or political resources needed for widespread and effective mobilization to combat the local effects of GCC. For these groups, transnational organizations and advocacy networks represent bridges to the funding and political connections necessary to support their mobilization and realize their goals. As Keck and Sikkink (1998) contend, these organizations are key to "mak[ing] international resources available to new actors in domestic political and social struggles" (p. 1).

In this chapter I draw from interviews with leadership and staff of AIR, as well as members of communities with whom AIR has worked, to examine how the organization's experiences and 'lessons learned' in addressing the difficulties associated with transnational mobilization to combat local deforestation, and, by extension, GCC. I argue that in spite of difficulties, certain shared commitments, values, and ideals have helped to bridge social and cultural divides and have solidified the alliances between AIR and local actors. While the story of AIR's partnership with indigenous women and men may be unique in some ways, it is also an instructive case that offers important lessons about the challenges, strategies, and rewards, involved in transnational mobilization.

Building bridges and forming alliances

AIR was founded by political science professor Anne Hallum after volunteering to lead a group of college students on a study abroad trip to Guatemala. Anne has acknowledged that taking that trip was a "reckless" move on her part; she had never travelled outside the US, did not speak any Spanish, and knew nothing about the political history of Guatemala—a country that was in the midst of a civil war at the time of her first trip. Nevertheless, upon returning to the US, Anne decided that she had to "do something." She met with a former student who had NGO experience, and he suggested she start an organization. Over multiple meetings, they drafted a set of bylaws and articulated the aims of the organization as working with local farmers in "designing, implementing, and promoting community-based reforestation projects" in Central America (AIR 1991)

The student also knew a few activists in Guatemala and put Anne in contact with Father Andres Girón, a Ladino priest who was heavily involved in a national peasants' movement in the latter years of Guatemala's civil war. According to Anne, Father Girón had a profound impact on influencing the organization's community-based approach and ensured that the founders of AIR had an understanding of the salience of race and class in their agroforestry work.

In late 1993 Anne met and hired her first two staff members: Eladio, an indigenous Guatemalan with a degree in agroforestry; and Chris, a North American with a degree in agroforestry and prior experience working to develop water resources in Guatemala. Together, began working in five communities in the highlands of Chimaltenango and within only 10 weeks had developed close relationships with local farmers and worked with them to cultivate and plant

20,000 seedlings. Word of AIR's work quickly spread to other communities, and the organization had to hire new staff to meet the growing community demand for its services.

In 1995, AIR hired two ladin@s, Guillermo, an agroforestry technician, and Cecilia Ramirez, as secretary. Two years later, AIR hired Luis, Eladio's son, as an agroforestry technician. In 1998, Chris left AIR and Cecilia was promoted to Director. By 2005, AIR was working in a total of twenty-six communities with staff consisting of seven full-time employees, including five tecnicos, an administrative assistant, and the Director. Each tecnico works with between thirty and fifty communities every year; in some communities they may work with only one or two families, while in others they may work with larger community groups—resulting in 200,000–275,000 trees planted each year.

AIR is truly a transnational organization. While the tecnicos and director live and work in Guatemala year-round, President Hallum and the board of directors are based in the US. AIR also has networks with other transnational organizations, including the Peace Corps, and has hosted volunteers from the United Kingdom and Japan, as well as the US.

Power, privilege, and conflict

Transnational mobilization of individuals and resources is challenging. To achieve both short and long-term goals, the multiracial, multinational *equipo AIRE (AIR team)* must mobilize across highly contested borders of race, class, gender, and nationality—and associated negotiations of power, privilege, and conflict in its everyday work. Given my position within AIR (as a board member and the daughter of AIR's founder), I was aware that throughout this research, there may be difficulty in trying to elicit stories of conflict or discontent from AIR team members. However, I shared with interviewees my knowledge of internal and external challenges that AIR has had to overcome. Participants acknowledged this, and many expressed a desire to share their own experiences to 'help others learn.' Broadly, these stories revolve around themes of negotiating power; the "emotional labor" critical for building and maintaining cross-border alliances; and the difficulties in securing funding.

Negotiating race, class, and gender in a transnational organization

In countries like Guatemala with histories of ethnic or racially-based violence, and with clear stratification along racial lines, mobilizing *across* these lines to achieve common goals can be particularly difficult. Centuries of exploitation and the appropriation of indigenous land and resources have resulted in a distinct "racial hierarchy" in Guatemala, a "sharp differentiation among distinct strata along the lines of power and privilege, with Ladinos generally occupying a higher stratum and Indians a lower one" (Hale 2006, p. 209). The persistence of this hierarchy has led to "pervasive Maya distrust" of Ladino efforts to build solidarity with indigenous groups (Hale, 2006, p. 174).

To successfully build coalitions or alliances in a highly stratified society, it is imperative that activists and organizations have an understanding of the historical, political, and social context in which they operate. While AIR's founder Anne Hallum was not well-versed in Guatemalan history or politics, she made sure to locate individuals who were, and deferred to their suggestions on organizational structure and strategy, especially in regards to hiring both indigenous and Ladina/o staff. As Anne recalled,

> [Father Girón] said that if our goal was to reach out to all sectors of Guatemalan society, then we needed to reflect that in our organization. And that … we needed to have indigenous staff, to build trust with indigenous communities. So I said, "OK".

Having indigenous staff has helped AIR in two key ways: First, it has enabled the organization to build relationships with communities, as staff have numerous contacts and networks that they can leverage in expanding AIR's work. Secondly, indigenous staff both reflect and help to solidify AIR's commitment to working in solidarity with indigenous communities. Through its staff-community connections and word-of-mouth, the organization has established a reputation as an ally of indigenous communities. This reputation, in turn, is what has led additional communities to seek out AIR's help. As indigenous activist Mona, 42, from Puebla explained,

> We work with AIR because we trust you all. In contrast we laugh if the government says it is going to help, because the officials say many things, they have beautiful words, but they do not do anything! But AIR works with people … we have friends in other communities and they told us that we can trust AIR.

The communities and AIR staff also form alliances across borders of gender. All of the organization's tecnicos are men—a fact which reflects the lack of educational and employment opportunities available for women in a patriarchal society. For this research, I spoke with several members of indigenous women's groups, including the group Mujeres Unidas, one of the largest women's groups to work with AIR. Overall, the indigenous women interviewed spoke in very positive terms about their experiences working with male agroforestry tecnicos; however, some noted that the tecnicos' presence in the women's group may compromise the "safe spaces" that are important for indigenous women to cultivate a sense of self-empowerment and critical consciousness. As community leader Elena explained,

> They [the tecnicos] are very good people, very good. But they are not gender experts. So I was a little worried at first. In our group we encourage women to share their experiences, to make connections between their experiences and the experiences of others. Because that is how we learn, right? So at the

> beginning I was thinking, "Ay, I don't know what will happen, what will happen if a woman's husband hits her, and she tells the tecnico about it, what will he tell her? To be patient or, or to not worry or to not talk about it?" No, that is not what the woman needs to hear. We are trying to teach the women to talk more about these issues, and to find help if they need it.

The male tecnicos also admitted that as men it is difficult for them to talk with women community members about "*cosas familares*," or intimate subjects, which oftentimes relate to domestic violence. Guillermo described the initial difficulty he experienced upon beginning work with the women's group in Itzapa:

> And for me it was difficult at first. Many women came to work in the nursery [in Itzapa] because they had problems at home. Sadly their husbands were drinking a lot, they beat [the women], [there were] many problems. They [the women] came to the nursery to leave their problems, to forget their problems. And to me, they told me, "Ah, Don Guillermo, I have problems". And it is difficult because I do not want to involve myself in family problems.

Elena and Guillermo have developed a strategy for how to address gender issues in their agroforestry work. Guillermo has decided that he will offer a sympathetic ear to women if they want to talk with him about any topic; however, if he feels that he is not qualified to address a problem, then he will refer the women to Elena or another leader of Mujeres Unidas. As he said, "What I do is listen. I listen to them and I refer them to Elena [the leader of Mujeres Unidas] if I cannot help."

While it is important that organizations guard against 'mission drift'—the tendency of organizations to become involved in activities and programs that distract them from their original aims and objectives—because gender, race, and class are ubiquitous in shaping the life experiences and daily realities of all involved parties, it is inevitable that at some point transnational organizations will have to address these issues in their work. It is therefore important that they have a strategy in place for doing so. As the case above illustrates, it is helpful for an organization to have connections with individuals and/or other organizations that are equipped to address such issues. In this way, an organization does not ignore issues when they arise, but also does not overstep its boundaries and area of expertise—and perhaps do more harm than good.

Interestingly, AIR has experienced the most conflict within the organization itself. I argue this stems in large part from the gendered makeup of AIR's organizational structure and how this structure defies traditional norms of gendered power and privilege. Kanter (1993) observes that organizational power or "the ability to get things done" is largely structured along lines of gender, with men occupying the upper rungs of the organizational structure. When this traditional gendered arrangement is challenged, it can lead to conflict or outright backlash against women in leadership positions (Acker, 2006; Martin, 2003).

Both the President and Director of AIR reported occasional conflicts with staff, which they attributed in part to being women leaders of an all-male staff. Anne recalled how her decision to promote Cecilia to the position of Director of AIR was met with resistance from both the tecnicos and the former Director of the organization. As Director, Cecilia has acknowledged that she has to be "very strong" to successfully manage AIR, and this often clashes with the roles and behaviors she is expected to assume as a woman in Guatemala:

> I have to be strong to manage AIR, to do everything that we as an organization need to do. But it is not common to have very strong women leaders in Guatemala. To have a woman as a boss, it is almost … it is very rare. So sometimes to be a woman leader also brings conflicts. Some here in AIR have told me that it makes them nervous to have a woman boss, or that I am too strong. So I have to be strong to manage the work that we do, but not too strong because I do not want people to fear me!

Both Anne and Cecilia noted that their success in managing AIR has depended in large part on their abilities to mediate conflict and appease staff, donors, and community members throughout the years.

Emotional labor in transnational mobilization

The highly gendered (and often overlooked) work of emotional labor (Hochschild, 1983, p. 7), or "the management of feeling" to produce a particular emotional state in another person, has been key to resolving conflicts within AIR and ensuring the long-term success of the organization. Although industries may exploit emotional labor, I argue it is a significant part of transnational mobilizing, essential for negotiating power relations and differences of race, gender, and class on a day-to-day basis. It is also key to building relationships with donors in order to secure funding. Yet, it is reserved for those who I refer to as 'mediators'— individuals who must work to manage relationships within the organization and between the organization and its donors and constituents. This type of work is 'gendered' in that it "involves creating in others feelings of well-being or affirmation, responsibilities typically assigned to women" (Wharton, 2009, p. 149).

Anne acknowledged that emotional labor is a regular part of her work as President. For instance, when donors visit Guatemala to tour AIR's projects, some may experience a form of 'culture shock' and may be hesitant to eat local foods, visit certain areas, or may express an urgent desire to return to the US. When this occurs, the donors must be put at ease while also ensuring AIR staff and community members are not offended. All of this requires a delicate balancing act in which everyone's emotional needs are met and relationships are not damaged. Thus, in addition to being important for maintaining cohesion and alliances within an organization, emotional labor is also an important part of being able to mobilize and make connections across organizational and national boundaries.

The never-ending search for funding

Social movement scholars have documented the many difficulties transnational NGOs face in funding their work (Peña, 2007; Staudt and Coronado, 2002). Many organizations opt to pursue grants from large foundations; however, this strategy often leaves them vulnerable to donor discipline, as "granting agencies or foundations ... set the agenda for what will be funded" (Peña, 2007, p.136). Peña notes that in order to avoid compromising their goals and objectives for donors, organizations will often develop alternative strategies to find funding.

One strategy that AIR has developed is to cultivate relationships with individual, private donors. Approximately half of AIR's donations are from individuals. The large portion of AIR's additional funds are received from churches or foundations that respect AIR's mission and take a very 'hands-off' approach to shaping the organization's agenda vis-à-vis its funding. This strategy has both positive and negative aspects. On the positive side, it means that AIR is not wholly dependent upon the desires and demands of heavy-handed donors and funding agencies, and that it can retain the integrity of its original mission and community-based strategy. On the negative side, this approach entails a great deal of time and energy to solicit donations.

Building solidarity: the importance of unifiers in cross-border activist work

As the experience of AIR's leadership and staff has demonstrated, transnational organizing is a process that involves the daily negotiation of power, privilege, and difference across race, class, gender, culture, nationality, and myriad other social boundaries. For AIR, identifying and leveraging key 'unifiers' – shared values, beliefs, and practices – has been instrumental in facilitating cross-border collaboration, fostering solidarity, and maintaining long-term alliances across borders. Specifically, religion, dialogue, and active listening were cited by AIR leadership, staff, and community members as the most important unifiers.

Religion

Scholars have noted that religion "has always played a big role ... in mobilizing collective groups to pursue social change" (Hondagneu-Sotelo, 2007, p.18). As Hondagneu-Sotelo argues, religion can unite activists by providing them with a shared belief system that gives moral weight to their causes, while at the same time serving as a common "social movement language" of ritual and shared cultural practices (2007, pp.18–23). In discussing their environmental work, indigenous women community members, as well as the staff and leaders of AIR, acknowledged that their religious beliefs and spirituality both motivate and unite them. For instance, most of the members of the women's group in Itzapa identify as Catholic. They explained that religion has helped them to develop close ties with AIR staff, as well as its US-based leaders and volunteers. One member pointed out that she appreciated the fact that the President of AIR is "very religious." As she explained, "this [religion] unites us, because we can pray together. It the same God, right? And it is the same Creation. So ... yes I think this helps us in our work because it is like a link."

Religion has also provided AIR with a set of core values that unite its members and guide their work. Anne elaborated on how religion informs these values:

> Humility, would be a core value ... That we can't keep doing this work by ourselves...that is that we need each other ... So humility in the sense of community. ... And then the other [value] is compassion, which that of course also has a religious basis. And then for the nature. We are true believers in Creation as something to be treasured, protected ... Gratitude, I have to put that. Gratitude is a core value. It's a motivating value.

By providing activists with a shared set of beliefs, values, and practices, religion can help to bridge social and cultural divides and build the solidarity necessary to motivate and sustain transnational activism.

Dialogue and listening

Community members and AIR staff emphasized that dialogue informed by humility and mutual respect are key to building relationships both within AIR and between AIR and the communities—with humility on the part of those in positions of social and/or economic power being important. As Tripp (2005) notes, effective transnational mobilization requires that international actors "consider that local actors have the most intimate knowledge of issues, other players, conditions, laws, and cultural sensitivities" (p. 306). When recalling her experiences starting AIR, emphasized the importance of her adopting a 'humble' approach. She recounted how her approach to letting local experts lead the organization has led to its growth and success:

> [there was] kind of a learning curve for me [at the beginning] ... that my gift was, ironically, that I didn't know what I was doing. So I did the right thing: I stood back, and let it grow ... so it was out of my own inexperience that AIR became what it is today.

AIR has developed a reputation for privileging the voices of local community members—a reputation which has in turn helped AIR to network with other communities, and extend its reforestation work through highland Guatemala.

Hemos aprendido muchas cosas tambien: developing an organizational gender consciousness

As described by AIR leadership and staff, one of the most important components of the organization's growth has been its development of an organizational 'gender consciousness'—a general, organization-wide recognition of the importance of gender and gender relations in its environmental community work. This awareness is not restricted to one or a few individuals, but is shared by AIR's leaders and staff and has become embedded within the organization and its programs. Here, it is important to reiterate that AIR did not begin as a 'women's organization;' however, through the organization's willingness to listen, recognize, and address the needs and concerns of both women and men in its work, AIR has developed an awareness of the importance of incorporating gender-specific programs. Luis explained:

> I think … that AIR has a gender consciousness. And now, we can say that the women have even taught us. We have learned a lot of things, too. Many things, right? And that is part of gender, that not only men can work. No, women also have space [to work] … And they [the women] have done great work. And we, as an institution, we know that effort, that work.

Guillermo explicitly connected the development of AIR's gender consciousness with the organization's willingness to recognize women's work and to listen to their concerns.

> We do practical work on the issue of gender. Even though many organizations apart from AIR say that they work on gender, but for me, we have done it. Without having a gender program … But we listen. Yes, we listen to the women, right, they talk and we listen. They tell us about their problems, or, or their suggestions and we try to help them.

Thus, for both Luis and Guillermo, a big part of AIR's gender consciousness involves recognizing the value and potential of *both* women and men as allies in struggles to protect the environment.

AIR's gender consciousness has in turn fundamentally shaped the organization. First, in recognizing the value and potential of women as important allies in its environmental projects, AIR formed strong and lasting alliances with numerous groups of women farmers throughout Guatemala. These groups constitute the majority of the farmers that choose to work with AIR, and their collective efforts have led to the planting of millions of trees in communities throughout highland Guatemala.

Secondly, the organization developed additional environmental programs that it likely would not have pursued had it failed to incorporate the concerns and suggestions of indigenous women. For instance, one of the most successful projects—the construction of fuel efficient stoves—was developed when women community members pointed out the inefficiency and environmental costs of traditional open-fire methods of cooking, as well as the health problems. AIR, with community members, built fuel-efficient stoves with chimneys that use between one-half and one-third of the firewood of open-fire cooking methods. Thus, beyond preserving firewood, they reduce the number of trips women must take to find firewood. Since 1995, 1000 fuel-efficient stoves have been constructed throughout highland Guatemala, conserving an estimated 1,000 tons of firewood yearly. As Anne acknowledged, "the stove project—that was all [from] the women. If [AIR] hadn't paid attention to them, that would have been a missed opportunity."

Discussion

By forming cross-border alliances with AIR, indigenous women and men throughout Guatemala have been able to access the knowledge and resources necessary to implement and maintain environmental programs in their communities. This cross-border collaboration has also been fraught with challenges and obstacles. As enumerated by both indigenous women and AIR staff, some of the main challenges include the negotiation of power and privilege across race, gender, class, and nationality; the exhausting nature of emotional labor in transnational organizing; and the difficulties in securing funding. AIR's experiences reveals that it is indeed possible to negotiate the daunting obstacles inherent in transnational activist work. Specifically, it is important that local and international actors identify and make use of key 'unifiers' that can bridge social and cultural divides. While such unifiers by no means dismantle the power structures in which local and transnational activists are embedded, they *do* help these activists to bridge seemingly insurmountable divides and work across borders of race, class, gender, and nationality to achieve a shared goal and vision.

AIR's ongoing partnership and dialogue with numerous indigenous women's groups has created a gender consciousness that has helped to inform and shape its work – leading to not only more partnerships with indigenous women's group, but also the development of its stove-building programs. True indigenous inclusion has resulted in recognition by the UN Permanent Forum on Indigenous Issues (UNPFII) and the UN Framework Convention on Climate Change, where in 2013 AIR was awarded the Lighthouse Prize for its work in empowering indigenous women in local efforts to combat GCC. Most notable perhaps, is that AIR's work has been sustained over the course of 20+ years: relationships with community members have endured, reforestation programs have continued and grown to other areas, and seedlings planted in the early 1990s are now full-grown forests. Thus, the case study of AIR presents a valuable learning opportunity for both researchers and practitioners alike in understanding both the importance and challenge in working across borders in efforts to combat GCC. Through ongoing dialogue, careful listening, and the identification and cultivation of unifying values and beliefs, it is possible to overcome divides and develop the solidarity needed to realize our shared vision for a more sustainable future.

Note

1 The term 'ladino/a' is distinct from 'latino/a' and is widely used throughout Guatemala to distinguish non-indigenous groups from indigenous groups. The term

is laden with class privilege, however, as members of the indigenous community who may have higher socio-economic status and dress in non-indigenous clothing may also be referred to – and claim – the ladino/a identity.

References

Alliance for International Reforestation [AIR]. 1991. Alliance for International Reforestation: Organization Bylaws. Primary source document.

Alvarez, S. E., Dagnino, E. and Escobar, A. 1998. *Cultures of Politics/Politics of Cultures: Re-visioning Latin American Social Movements*. Boulder, CO: Westview Press.

Hondagneu-Sotelo, P. 2008. *God's Heart Has No Borders: How Religious Activists are Working for Immigrant Rights*. Los Angeles, CA: University of California Press.

Keck, M. E. and Sikkink, K. 1998. *Activists Beyond Borders: Advocacy Networks in International Politics*. Ithaca, NY: Cornell University Press.

Peña, M. 2007. *Latina Activists Across Borders: Women's Grassroots Organizing in Mexico and Texas*. Durham, NC: Duke University Press.

Hochschild, A. R. (1983). *The Managed Heart: Commercialization of Human Feeling*. Berkeley, CA: University of California Press.

Tripp, A. M. 2005. Challenges in transnational feminist mobilization. In *Global Feminism: Transnational Women's Activism, Organizing, and Human Rights*, edited by Myra Marx Ferree and Aili Mari Tripp. New York: New York University Press, pp. 296–312.

Wharton, A. S. 2009. The sociology of emotional labor. *Annual Review of Sociology* 35: 147–165.

41 Examining the environmental injustices of clean development mechanism and reducing emissions from deforestation and forest degradation schemes in South Asia

Vincci Cheng

In the wake of increased awareness about the pressing dangers of global climate change (GCC), international cooperation on climate mitigation has become extremely important. Two schemes that have arisen in response to these international efforts are the Reducing Emissions from Deforestation Degradation Plus (REDD+) and Clean Development Mechanism (CDM), which claim to reduce greenhouse gases in the atmosphere by having developed countries subsidize climate mitigation projects in developing countries. The South Asia region (SAS), which encompasses the developing countries of Afghanistan, Bhutan, Nepal, India, Pakistan, Bangladesh, and Sri Lanka, has great potential to contribute to these international climate mitigation schemes, because of abundant forests and foreseen industrial development. Forty percent of Nepal (a total of 5.8 million hectares) is covered by forest (Dhital, 2009). India's rapid industrialization is causing a depletion of energy and natural resources that is exacerbated by a rapidly rising population and a growing middle class. However, as the world's second poorest region that contains some of the world's least developed countries, including Bangladesh, Burma, Afghanistan, and Nepal, SAS might need the foreign aid offered in REDD+ and CDM to execute climate mitigation initiatives (The World Bank, 2011). In this chapter I review some of the environmental injustices caused by REDD+ and CDM, which affect marginalized populations including the poor, ethnic minorities, women, children, and elderly. These injustices may be a barrier to the longevity and successful execution of these international environmental schemes. I explore how an intersectional blend of marginalized traits makes individuals experience heightened levels of environmental injustices. I pay special attention to the countries of Bangladesh, India, Nepal, and Sri Lanka.

Background on climate mitigation and environmental justice

Traditionally, developed countries of the Northern Hemisphere such as Australia, Canada, the United States (US), and those in Western Europe could afford to devote attention to environmental issues as they have already reached a certain level of social, human, and economic standards of living. These countries generally have better social welfare, more extensive education, higher GDP per capita, higher life expectancy, greater gender equality, and generally rank higher on the Human Development Index (HDI) compared to developing countries (UNDP, 2013). Developing countries such as China, India, and Brazil, which rank lower on the HDI (UNDP, 2013), have devoted resources towards other pressing social problems through means such as achieving a certain level of economic prosperity first, in order to proceed towards similar standards of living. They reached this goal by repeating the same environmental-degrading methods that developed countries have used in the past—especially during the Industrial Revolution between 1730 and 1850—such as logging to pave away land for urban development, and using fossil fuels to power factories and transportation. However, by the 1900s, Western nations established many national parks to protect the wilderness (Reynolds, 2002). Rachel Carson's 1962 book, *Silent Spring*, jumpstarted public awareness on environmental issues by informing the public on how chemicals such as insecticides and pesticides harmed not only wildlife, but also human health.

The UN's first international Earth Summit, held in 1972, kick-started international cooperation on environmental issues such as protecting wildlife, mitigating climate change, and preventing the transnational spread of toxic pollutants. The 1992 Rio Earth Summit addressed climate change, and gave birth to the Kyoto

Box 41.1 Ethnic minorities in South Asia: who are they?

Dalits, also named the "untouchables", are the lowest caste in South Asian society. Their low status has restricted them from engaging in any kind of work that involves contact with food, water, temples, public offices, health services, cremation grounds and the villages of non-Dalits (IDSN 2009). Dalits may need to resort to arduous unsanitary jobs such as "manual scavenging" and removal of dead animals (IDSN 2009). Dalits' access to market facilities such as loans are limited, and REDD+ payments and revenue earned in CDM projects that involve land ownership are not applicable to the Dalits since they are prohibited from owning land (IDSN 2009).

Dalits and indigenous people have few political rights, and those who protested for their rights or tried to vote at election polls and run for political office have been threatened, imprisoned, and faced violence (IDSN 2009; Karunakar 2011). These South Asian ethnic minorities have never attained important positions in committees, commissions, and taskforces in the parliament, and instead were in charge of less significant ministerial portfolios like domestic affairs (Karunakar 2011). Moreover, many indigenous people of India find it hard to be involved in the government since they live in the forests, separate from mainstream society (Karunakar 2011).

Protocol, under which countries agreed to reduce carbon emissions by 5 percent from 2008 to 2012. As of 2013, 193 countries worldwide have ratified the Kyoto Protocol, although some of the developed countries put their short-term interests first, such as the US, who signed but didn't ratify the Protocol, possibly because their economy depends on oil trade. Protocol deadlines were less stringent for China and India, two of the fastest industrializing developing nations with the highest populations in the world.

In Kyoto Protocol discussions, developing countries have argued it is a violation of global environmental justice for developed countries to expect them to curb their emissions at the expense of local needs. Environmental justice is "an ideal of accountability and fairness in the protection and vindication of rights and the prevention and punishment of wrongs related to the impacts of ecological change on the poor and vulnerable in society," whether this be of certain nations, class, race, gender, age, or any other marginalizing factors (Khoday and Perch, 2012, p. 1). Faced with other pressing development priorities such as combating poverty, sustaining health, and improving education, these developing countries may not have financial capacity to prioritize environmental considerations. Because many polluting industries in developing countries are a result of outsourcing from developed countries, developing countries shouldn't be entirely blamed for building these polluting industries. Recent international efforts have addressed this issue by having developed countries subsidize environmental projects in developing countries, but these schemes work in the interest of developed nations, as they help market developed nations' green technologies and make developing nations further dependent on foreign assistance.

Previous literature on global environmental justice documented developed countries exporting their polluting industrial factories and waste to developing countries, while importing and depleting natural resources such as the textile, electronics, and raw materials manufactured from China, India, or South America (Adeola, 2008). In SAS, GCC generates a range of problems including sea level rise, increased frequency of natural disasters, precipitation variability, and increased pests and diseases. South Asian countries have low financial capacity for buying technologies that would allow them to adapt to effects of GCC (Sivakumar and Stefanski, 2011). GCC will bring about lower water quality, frequent floods and droughts, lower agriculture productivity, and infestation of waterborne diseases such as cholera in SAS (Kelkar and Bhadwal, 2007).

Understanding the claims and criticisms of REDD+ and CDM

Reducing emissions from deforestation and forest degradation

REDD+ aims to reduce greenhouse emissions through conserving, sustainably managing forests, and boosting global carbon stocks by providing monetary incentives (UN-REDD, 2013). Besides reducing carbon emissions, REDD+ efforts to protect forestland could also prevent habitat loss, strengthen biodiversity, and maintain ecosystem services such as water quality and regulation (Acharya *et al.*,

2008). Many SAS countries have high deforestation rates, but could ideally adopt reforestation, afforestation, and conservation projects in order to counteract the impacts from land-use change, urban development, agriculture, timber harvesting, and gem-stone mining that have led to forest degradation. The UN-REDD programme currently has partnered with 47 developing countries worldwide, including Bangladesh, Nepal, Pakistan, and Sri Lanka, in SAS (UN-REDD, 2013). In SAS, the REDD+ program could potentially govern community forestry, which involves collaboration between government, non-governmental organizations, and local communities on forestry management, and has been the most effective proven method so far that has lowered Nepal's deforestation rate since its adoption in the 1990s (MSFP, 2012).

However, I think those who see REDD+ as a grand solution for climate mitigation are being overly optimistic. For every ton of carbon saved in the REDD+ program's intervention area, anywhere from 0.42 to 0.95 tons of carbon arises elsewhere, as the original deforestation and urban development activity has relocated to other parts of the world (Gan and McCarl, 2007). REDD+ may not be able to prevent further environmental degradation in the long-run because of global policies relating to biomass and biofuel production, increased demands for forestry products, agricultural products and living space in the future, and poor governance due to the absence of clear and secure tenure rights (Eliasch, 2008). Nevertheless, I think reforestation, afforestation, and conservation projects are preferable in SAS's least developed countries, such as Nepal and Bangladesh, where people may not be able to afford to pay for the electricity from renewable energy.

While REDD+ claims to generate income from agricultural production for host countries that could be spent on approaches and technologies to reduce emissions and alleviate poverty, I am critical of its ability to help needy South Asian populations. REDD+ monetary payments are only given to people with secure land tenure, so poor farmers who don't own land, indigenous people, and women may not receive REDD+ payments because they often lack secure land tenure (Chapagain, 2007; Schroeder, 2010; Upreti *et al.*, 2012).

Other than public land owned by the government, secure land tenure would more often be possessed by wealthy farmers who own livestock, and private corporations such as timber industries and property developers (JLL, 2014). Many rural and indigenous people living in the forests, who are usually poor and illiterate, have been forced to relocate to urban areas because of urbanization and corporate domination. While REDD+ saves forests in intervention areas by means of commodifying the forest, I think it is important to not let this monetary value replace the forest's cultural, spiritual, and religious values for traditional peoples, and its ecosystem values for everyone.

Clean development mechanism

The goal of CDM, developed under the Kyoto Protocol, is to reduce global greenhouse gas emissions while at the same time promoting sustainable development (SD), which is "development that meets the needs of the present without compromising the ability of future generations to meet their own needs" (UN-WCED, 1987, n.p.). Developed countries fund environmentally related projects in developing countries to earn certified emission reduction (CER) credits, which can be traded and sold in the carbon market so countries struggling to meet their carbon reduction requirements could buy CERs from countries with a surplus. By providing energy to rural areas, transferring technology, creating jobs, and generating income to agricultural sectors, the CDM is supposed to touch on the UN's Millennium Developmental Goals (MDG) to alleviate the burdens of the world's most impoverished. Most CDM projects around the world are renewable energy projects using wind power (30 percent), hydropower (26 percent), and biomass (9 percent) technologies (UNEP, 2014). Michaelowa and Michaelowa (2007) found that renewable energy projects, which make energy cheap and readily available in rural parts of SAS, are actually more easily introduced in rural areas than conventional energy generation from burning of fossil fuels, coal, and natural gas.

However, there are several problems with the CDM's regulation, governance, inclusivity, and distribution of benefits that not only prove against its SD claim, but, if left unsolved, may make it merely a moneymaking mechanism that transfers money from the powerful and wealthy in developed countries to wealthy polluting industries in developing countries. Although CDM projects should operate under an 'additionality' requirement, which means that projects receiving CDM subsidies wouldn't have otherwise taken place without the aid of CDM carbon payments, researchers found that 75 percent of CDM projects were actually non-additional since they were already constructed by the time of approval (Haya and Oranstein, 2008). Besides this, validators can be biased to approve CDM projects because they are hired by the directors themselves, and want to be hired again for these favors. However, this is a hard problem to address because assessing project development, lending, and the investment criterion for project additionality includes a degree of subjectivity (Haya and Oranstein, 2008). A third of the CERs granted until 2010 were suspected to be fraudulent, as some firms in developing countries generated more carbon emissions than necessary for the sole purpose of grabbing CDM funds (Gronewold, 2010).

One of CDM's major problems for the least developed countries of SAS (Bangladesh, Burma, Afghanistan, and Nepal) is that they are too poor to attract project funding from CDM investors. It is more desirable for developed countries to instead invest in CDM projects in developing countries with higher energy consumption and a middle class capable of paying for electricity services, as this would reap greater profits. Brazil, China, India, and Mexico alone have hosted over 70 percent of the number of CDM projects as of 2015 (UNEP). The scant business and investment climate, inadequate institutions and infrastructure to govern CDM projects, uncertainties and inequalities present in international investments, and lack of carbon experts, which have all contributed to shortfalls of CDM in Sub-Saharan Africa, may likewise hinder CDM development in SAS's least developed countries (Esambe, 2011). Many of the poorest countries lack domestic capability to create their Designated National Authority (DNA) proposal required for implementing CDM activities (Michaelowa and

Michaelowa, 2007). Claims for SD could just be an empty promise, as there is a tradeoff between reaching carbon reduction and SD goals, so countries may aim to gain CERs at the expense of SD. A study using the Multi-Attributive Assessment (MATA) of CDM approach found that only one percent of registered CDM projects contributed significantly to SD, whereas 72 percent were effective in reducing emissions (Michaelowa and Michaelowa, 2007;). Marginalized groups in SAS (ie. the poor, indigenous peoples, Dalits and other ethnic minorities, women, children, and elderly) are less able to benefit from CDM renewable energy projects, and this undermines the SD claim of many CDM projects worldwide. For example, the poor may not be able to afford the grid-connected power that renewable energy CDM projects supply, nor be able to afford secure land rights to renewable energy projects, pay for opportunities that allow them to avoid environmental injustices, or pay for healthcare expenses in cases where CDM projects threaten health.

Role of marginalized communities in REDD+ and CDM

The role of indigenous populations, Dalits, other ethnic minorities (see p. 305, Box 41.1), poor people, women, children, and elderly in REDD+ and CDM is currently limited, and I speculate that the lack of power marginalized groups have in climate mitigation schemes may derive from various forms of cultural, social, and political discrimination, and social problems such as poverty and illiteracy. These forms of discrimination and social problems lead to a lack of knowledge and respect, thus making it hard for marginalized groups to have their voices heard in REDD+ and CDM processes, and for them to become informed about and communicate to others about environmental problems they face. For example, Dalit women in Nepal are at a severe disadvantage compared to both Nepalese men and non-Dalit women in terms of lacking education, so they would be more helpless to speak out and become informed about environmental problems compared to both other groups. While Nepal's literacy rate in 2009 was 70.7 percent for males, it was more than a third lower for females at 43.4 percent, and even lower for the female Dalit population, at 24.5 percent (IDSN, 2009; Government of Nepal, 2009). It would be difficult for marginalized groups to win power in REDD+ and CDM without support from the public, government, and higher administration to implement various policies and programs promoting equality. SAS's rising social inequities have fostered a climate of social and political dissatisfaction that results in political instability and intra-state conflict, which make investments in this region particularly risky and thus could potentially scare developed countries away from investing in REDD+ and CDM projects in SAS (Hashmi, 2013).

Environmental injustices for REDD+ in South Asia

The discrimination and social problems that women, indigenous people, and the poor face have led these marginalized groups to suffer a lack of representation in the REDD+ decision-making processes, lack of secure land-tenure to receive REDD+ payments, and negative alterations to their lifestyle. Because the government ignores informal or customary rights to forestland possessed by poor, rural communities and indigenous people, these marginalized groups don't receive any of the REDD+ payments, although some of the most pristine South Asian forests today can be accredited to indigenous peoples' use of traditional forestry practices for sustainable management (Schroeder, 2010). Voices of poor, rural and indigenous communities have been excluded from participatory processes. For example, indigenous peoples in Nepal complain that they had not been sufficiently consulted with about the design of the Forest Carbon Partnership Facility. While national leaders blame this partly on the difficulties in communicating with remotely located communities, it is also reflective of the powerlessness that rural and indigenous groups have over government decisions (Schroeder, 2010). Indigenous groups have used the special status they possess under international law to reclaim forestland to benefit from REDD+ payments at intergovernmental climate negotiations (Schroeder, 2010). However, the other rural poor and ethnic minorities are excluded from this special status, so cannot use this method to gain secure land tenure. One such ethnic minority is the Dalits, the 'untouchable' caste of lowest status in SAS who are often limited to jobs of low-paying manual labor, and don't have a chance to gain revenue earned in REDD+ projects because they aren't allowed to own land (IDSN, 2009).

Despite women's vital role in sustainable forest management, many haven't had opportunities to influence forestry management decisions or participate in REDD+ design and implementation. Discrimination and social problems arising from gender inequality remain significant barriers for South Asian women to hold higher-level management positions. Like indigenous and poor rural communities, women have played an important part in sustainably growing, managing, and preserving forests to the current date. In Nepal, women are in charge of collecting forest products for household use—such as fuel wood, leaf compost, non-timber forest products, natural medicines, and livestock bedding—while men reap the moneymaking commercial rewards of forestland, usually by degrading the environment as a by-product through extracting commercial products such as timber (US AID, 2011). Yet men hold a majority of the REDD+ governance and executive positions, represent 68.5 percent of participants in Nepal's Community Forestry Executive Committee groups, and civil society organizations invited to REDD+ discussion forums are dominated by males (Upreti *et al.*, 2012). Nepal's REDD+ Readiness Proposal Plan (R-PP) doesn't guarantee women's participation in decision-making processes. One of the problems for South Asian women who wish to participate in REDD+ is that if they aren't notified about REDD+ meetings sufficiently beforehand, they may not be able to adequately prepare for these meetings, as their schedule may already be packed with household chores. This is because most women in SAS need to complete an abundance of daily household chores, such as taking care of children and cooking. Women's inability to influence REDD+ means that they can't receive the compensatory benefits from loss of livelihood resulting from forest restrictions, and may have insecure access and rights to resources (Upreti *et al.*, 2012; US AID, 2011).

REDD+ schemes maintaining these systems, which unjustly treat ethnic minorities, indigenous people, women, and other rural populations living in the forest, will put these often-poorer populations at a further economic disadvantage. These marginalized populations wouldn't benefit from REDD+ monetary benefits, and also may lose their original jobs in the industrial, food-processing, and food services sector if land was put under REDD+. Marginalized groups may decide to not cooperate with REDD+ initiatives if their rights to forests aren't respected. These marginalized communities' input on how REDD+ should sustain and preserve South Asian forests is valuable, since forest communities, indigenous people, and women possess traditional knowledge and expertise in forest management (Schroeder, 2010; Upreti *et al.*, 2012). The current forest management system would benefit from the different values that women place on forests compared to men, by prioritizing fodder and grass benefits, and subsistence consumption, over commercial products such as timber and non-timber forest products (US AID, 2011).

Environmental injustices in CDM in South Asia

Many environmental justices in the CDM also evolve from poverty, public neglect, illiteracy, and cultural and political discrimination faced by South Asian marginalized individuals. Wind power and hydropower are the most numerous CDM projects worldwide. Hydropower dams generate significant amounts of constant electricity perennially, and wind power generates electricity at a low cost that can compete with the cost of electricity produced by natural gas. SAS could also greatly benefit from CDM projects like biomass that target the agriculture sector because of its wealth of forest resources and productive agriculture. However, an analysis of these CDM projects shows that not only do marginalized communities lack secure tenure or representation in CDM decision-making processes, but they also don't receive payments for leasing land to electricity generation sites or gain long-term job opportunities for various CDM renewable energy projects. This undermines the CDM's SD claim.

a. Wind power and hydropower

Marginalized communities in SAS may be displaced from their homes as a result of activity in the project area, are especially vulnerable to health problems that wind power and hydropower projects pose, and lack secure tenure rights to receive CDM payments. Also, not only are the rural poor unable to afford electricity generated from hydropower plants, but some of the long-term job opportunities such as maintaining power plants may require skills that the rural poorest don't possess (Michaelowa and Michaelowa, 2007; Sirohi, 2007). These marginalized populations have minimal power to influence the placement of wind and hydropower projects in their neighborhoods because, compared to interests of people with status and the wealthy (who don't want these renewable energy projects placed in their neighborhoods either), their interests are a lower priority to the administration and the government. The poor may not be able to afford to relocate to other neighborhoods where there is a cleaner environment, yet even if they were able to move elsewhere, sanctity and non-monetary values of the natural environment, such as the cultural, spiritual, and religious meaning that indigenous people place on the forests, can't be replaced (Acharya *et al.*, 2008; Schroeder, 2010). Renewable energy projects that displaced rural, indigenous, and other local populations from their homes have led to resettlement, social unrest, and conflict between governments and societal factions over land and clean water, such as tensions that have arisen between Pakistan and India over the placement of hydropower dams that violate the 1960 Indus Water Treaty (Nosheen and Toheeda, 2011). Social unrest may bring about further hunger for SAS's impoverished populations.

Wind power and hydropower projects disturb the water and forest ecosystems where rural communities live, threatening local ecosystem health and stripping away the sanctity from these natural environments. They do this by altering local night-time temperatures and weather, causing soil erosion, degrading land, and leading to habitat loss for the many species residing in waters and dense forests where hydropower dams are usually built (Grumbine and Pandit, 2013). Activity from CDM renewable energy projects may bring about loss of fish and forest products, which indigenous communities depend on for food, trading, and medicinal purposes. Dammed reservoirs are a potential breeding ground for waterborne diseases such as malaria, and these poor, rural, and marginalized communities may not be able to afford or have access to better healthcare (Keiser *et al.*, 2015). Cholera and typhoid are also common waterborne diseases in SAS.

b. Biomass

Biomass makes use of heat from burning crops, wood, waste, alcohol fuels, and landfill gases such as manure to use as cooking fuel and to convert heat into electricity in biomass plants. Biomass is common in many SAS countries, and is the primary source of cooking fuel for more than 78 percent of Sri Lankans (Elledge *et al.*, 2012). Biomass use has been controversial because the process is energy-intensive and water-intensive. Likewise, unsustainable fuel wood practices may lead to deforestation, and burned crops could instead be used to eradicate the widespread hunger faced by many of SAS's impoverished. SAS's rural poor are landless or marginal and small farmers, so they may not generate significant income from supplying biomass to and leasing out land to energy production plants or engaging in other self-employed agricultural activities related to CDM (Olsen, 2007).

Marginalized populations are especially vulnerable to health problems of biomass. Burning dung-briquette and fuelwood releases toxic fumes that could lead to acute lower respiratory infection, chronic obstructive pulmonary disease, asthma, lung cancer, cataract, adverse pregnancy outcomes, and eye diseases.

Accompanying healthcare expenses, which might exceed the amount of money saved by using biomass as an energy alternative, could become too costly, especially for poor populations (Pant, 2012). There also may be few healthcare facilities in such remote rural areas. A safe alternative, that relieves users from breathing toxic fumes of biomass, is biogas, but in Bangladesh biogas was most often adopted by educated elites owning many cattle and not by poor communities (Kabir, Yegbemey, and Bauer, 2013).

Bringing South Asian marginalized communities into REDD+ and CDM governance and implementation

An unwillingness of society, national, and international governance to address these environmental injustices could threaten some people's livelihoods, disrupt parts of the natural environment, cause conflict between communities, and would ultimately limit levels of national and international collaboration in REDD+ and CDM. Marginalized groups who are excluded from REDD+ may decide not to participate in the REDD+ and instead turn to illegal methods of harvesting fuel wood or other products, and conflicts over REDD+ could lead to intentional forest destruction. Conflicts arising over discontent with the REDD+ and CDM schemes that turn violent could scare investors away from funding green projects in SAS. Inequalities in SAS between all age groups, genders, and ethnicities aren't specific to REDD+ and CDM, and may be difficult to tackle because addressing these inequalities requires not only government action but also a change in societal norms that involves perspective shift from all South Asian people. While norms of social inequality remain deeply embedded in some SAS countries, I think it is hopeful that others are starting to move towards a society more equitable for marginalized groups, especially with the aid of organizations and non-profits. Cooperation between nations and all sectors in society is critical to success of the REDD+ and CDM. Given the costs of excluding marginalized groups from REDD+ and CDM, and potential benefits of their involvement in these schemes, I think it would be strategic for these climate mitigation schemes to raise gender, class, age, and ethnic equality, and listen to what marginalized groups have to contribute to these schemes. As these global climate mitigation schemes become more widespread around the world and expand the scope of initiatives and types of projects that they undertake, it is important to reconsider how inclusivity and social equity in governance and implementation of the REDD+ and CDM will improve their environmental and social effectiveness.

References

Acharya, K., Dangi, R., Tripathi, D., Bushley, B., Bhandary, R., and Bhattari, B., 2009. *Ready for REDD: Taking Stock of Experience, Opportunities and Challenges in Nepal*. Kathmandu: Nepal Foresters' Association.

Adeola, F.O., 2000. Cross-national environmental injustice and human rights issues. *American Behavioral Scientist*. 43(4), p. 686–706.

Carson, R., 1962. *Silent Spring*. Boston, MA: Houghton Mifflin.

Central Intelligence Agency, 2013. *The World Factbook*. [online] Available at: www.cia.gov/library/publications/the-world-factbook/ [Accessed 5 October 2014].

Chapagain, D., 2007. Land Tenure and Poverty: Status and Trends Land Systems in the Hills and Mountains of Nepal. In M. Banskota, T.S. Papola, and J. Richter (eds) *Growth, Poverty Alleviation, and Sustainable Resource Management in the Mountain Areas of South Asia*. Nepal: ICIMOD. pp. 407–432.

Dhital, N., 2009. Reducing emissions from deforestation and forest degradation (REDD) in Nepal: exploring the possibilities. *Journal of Forests and Livelihood*, 8(1), pp. 57–62

Eliasch, J., 2008. *Climate Change: Financing Global Forests*. Oxford UK: Routledge.

Elledge, M., Phillips, M., Thornburg, V., Everett, K., and Nanadasena, S., 2012. A profile of biomass stove use in Sri Lanka. *International Journal of Environmental Research and Public Health*, 9(4), pp. 1097–1110.

Esambe, N.G., 2011. *Why are there few Clean Development Mechanism investments in Africa? A study of private actors' involvement in global climate governance*. [online] Available at: www.diva-portal.org/smash/get/diva2:463332/ATTACHMENT01 [Accessed 5 October 2014].

Gan, J., and McCarl, B., 2007. Measuring transnational leakage of forest conservation. *Ecological Economics*, 64(2), pp. 423–432.

Government of Nepal, 2009. *Report on the Nepal Labour Force Survey* [pdf]. Available at: www.ilo.org/wcmsp5/groups/public/@asia/@ro-bangkok/@ilo-kathmandu/documents/publication/wcms_118294.pdf [Accessed 4 June 2015].

Gronewold, N., 2010. CDM critics demand investigation of suspect offsets. *The New York Times* [online]. Available at: www.nytimes.com/cwire/2010/06/14/14climatewire-cdm-critics-demand-investigation-of-suspect-63522.html?pagewanted=all [Accessed 5 October 2014].

Grumbine, R., and Pandit, M.K., 2013. Threats from India's Himalaya Dams. *Science*, 339 (6115). pp. 36–37.

Haq, K., 2000. Human Development Challenges in South Asia. *Journal Of Human Development*, 1(1), pp. 71–82.

Hashmi, A.S., 2013. Internal conflicts and regional security in South Asia. In: H. Singh (ed) *Pentagon's South Asia Defence and Strategic Yearbook*. New Delhi: Pentagon Press, pp. 37–44.

Haya, B., and Oranstein, K., 2008. *Problems with the Clean Development Mechanism (CDM)*. Friends of Earth & International Rivers [pdf]. Available at: www.internationalrivers.org/files/attached-files/foe_ir_cdm_fact_sheet_final3_10-08.pdf [Accessed 5 October 2014].

IDSN, 2009. *Caste-based discrimination in South Asia* [pdf]. Available at: http://idsn.org/wp-content/uploads/user_folder/pdf/New_files/EU/EU_StudyWithAnnexes_Caste_Discrimination_June2009.pdf [Accessed 5 October 2014].

JLL, 2014. *Nepal Property Investment Guide* [pdf]. Available at: www.joneslanglasalle sites.com/investmentguide/uploads/attachments/2014AP_PropertyInvestmentGuide-Nepal_6g04rbna.pdf [Accessed 4 June 2015].

Kabir, H., Yegbemey, R. and Bauer, S., 2013. Factors determinant of biogas adoption in Bangladesh. *Renewable and Sustainable Energy Reviews*, 28, pp. 881–889.

Keiser, J.U, Tanner, M., Caldes de Castro, M., Maltese, M., Bos, R., Bartram, J., and Haller, L., 2015. *The effect of Malaria on Large Dams on a Global and Regional Scale*. World Health Organization [online]. Available at: www.who.int/water_sanitation_health/publications/STImalaria.pdf?ua=1 [Accessed 4 June 2015].

Kelkar, U. and Bhadwal, S., 2007. *South Asian Regional Study on Climate Change Impacts and Adaptations: Implications for Human Development* [pdf]. Available at: http://hdr.undp.org/sites/default/files/kelkar_ulka_and_bhadwal_suruchi.pdf [Accessed 4 June 2015].

Khoday, K. and Perch, L., 2012. *Green Equity: Environmental Justice for More Inclusive Growth*. [online] Available from: www.ipc-undp.org/pub/IPCPolicyResearchBrief19.pdf [Accessed 4 June 2015].

Michaelowa, A. and Michaelowa, K., 2007. Climate or development: Is ODA diverted from its original purpose? *Climatic Change*, 84(1), pp. 5–21.

MSFP, 2012. Some of the Forestry Facts of Nepal [pdf]. Available at: www.msfp.org.np/uploads/publications/file/Some%20of%20the%20Forestry%20Facts%20of%20Nepal_20120727044419.pdf [Accessed 4 June 2015].

Nosheen and Begum, T., 2011. *Indus Water Treaty & Emerging Water Issues* [pdf]. Available at: http://aupc.info/wp-content/uploads/2012/12/V4I2-2.pdf [Accessed 4 June 2015].

Olsen, K., 2007. The clean development mechanism's contribution to sustainable development. *Climatic Change*, 84(1), pp. 59–73.

Pant, K., 2012. Cheaper fuel and higher health costs among the poor in rural Nepal. *AMBIO – A Journal Of The Human Environment*, 41(3), pp. 271–283.

Reynolds, A., 2002. *A Brief History of Environmentalism* [online]. Available at: www.public.iastate.edu/~sws/enviro%20and%20society%20Spring%202006/Historyof Environmentalism.doc [Accessed 5 October 2014].

Schroeder, H., 2010. Agency in international climate negotiations: the case of indigenous peoples and avoided deforestation. *International Environmental Agreements: Politics, Law and Economics*, 10(4), pp. 317–332.

Shakya, I., 2002. Development of biogas in Nepal. *International Energy Journal: Special Issue*, 3(2), pp. 75–88.

Sirohi, S., 2007. CDM: Is it a 'win-win' strategy for rural poverty alleviation in India? *Climatic Change*, 84(1), pp. 91–110.

Sivakumar, M., and Stefanski, R., 2011. Climate Change in South Asia. In Lal, R., Sivakumar, M.V.K., Faiz, M.A., Mustafizur Rahman, A.H.M., and Islam, K.R. (eds) *Climate Change and Food Security in South Asia*. London: Springer, pp. 13–30.

UN-DESA, 2012. *World Population Prospects The 2012 Revision* [pdf]. Available at: http://esa.un.org/wpp/documentation/pdf/WPP2012_%20KEY%20FINDINGS.pdf [Accessed 5 October 2014].

UNDP, 2013. *Human Development index and its components* [online]. Available at: http://hdr.undp.org/en/content/table-1-human-development-index-and-its-components [Accessed 4 June 2015].

UNEP, 2014. *CDM Projects by Type* [online]. Available at: www.cdmpipeline.org/cdm-projects-type.htm#1 [Accesssed 5 October 2014].

UNEP 2015. *CDM Projects by Host Region* [online] Available at: www.cdmpipeline.org/cdm-projects-region.htm#7 [Accessed 4 June 2015].

UNFCCC, 2009 *Further Guidance to the Clean Development Mechanism* [pdf]. Available at: https://unfccc.int/files/meetings/cop_12/application/pdf/cmp_8.pdf [Accessed 5 October 2014].

UN-REDD, 2013. *About REDD+* [online] Available at: www.un-redd.org/aboutredd [Accessed 5 October 2014].

UN-WCED, 1987. *Our common future*. Oxford: Oxford University Press, p. 43.

Upreti, H., Gurung, D., Szomor, S., Magar, R.A., Bohara, G., KC, R. and Onta, N., 2012. *An Assessment of Gender and Women's Exclusion in REDD+ in Nepal* [pdf] Available at:

www.forestrynepal.org/images/publications/Gender_REDDplus_Nepal_WOCAN.pdf [Accessed 5 October 2014].

US AID, 2011. *Getting REDD+ Right for Women*, Gender Climate [pdf]. Available at: http://gender-climate.org/wp-content/uploads/docs/publications/Gender_REDD_Asia_Regional_Analysis.pdf [Accessed 5 October 2014].

World Bank, 2010. *World Development Indicators: Energy Dependency, Efficiency, and Carbon Dioxide Emissions* [online]. Available at: http://wdi.worldbank.org/table/3.8 [Accessed 5 October 2014].

World Bank, 2011. *Poverty and Equity Data* [online]. Available at: http://povertydata.worldbank.org/ppoverty/region/SAS [Accessed 4 June 2015].

42 Regenerate

Beehive Collective

Figure 42.1 Regenerate

43 Call to action for system change and the solar commons

Terran Giacomini

Young people are inheriting a world in need of massive changes within a tight timeframe. The capitalist market cannot do the job (Anderson, 2012, p. 25). Three actions highlighted by social movements at the 2015 climate talks in Paris can help us to 'change the system, not the climate'.

Support convergences among movements

Movements seek 'convergences' among struggles, to unite in One Big Struggle. This recognizes that the struggle against climate change is intersectional; meaning that social and ecological crises are rooted in the same system that puts profits before life; a system of domination and exploitation of nature, including people.

Action: Because climate justice is not a single-issue struggle, it is crucial to connect to what other people are doing, and then to act collectively based on that interconnection. Ask: Who are my allies? How do the things I care about—whether it be justice for migrant workers or queer rights—relate to what others are doing, and how can our work be supported by alliances and joint actions?

Strengthen the struggles of women

In Paris, movement activists overwhelmingly acknowledged that indigenous women are at the forefront of struggles for system change because they are the most exploited or threatened by capitalism, yet they have immense power. This power is rooted in the for-life, 'commoning' political economy within which specific women have significant control. Commoning social relations can be quickly scaled up and generalized to preserve life itself. According to Kenyan Sophie Huot:

> Women know very well that if you destroy Mother Nature then we are all destroyed. And so women are taking charge. We have the solutions because it is women who deal with land, … [and] the water. … Indigenous women, grassroots women … are not relenting.

Action: We seek not 'green' capitalist reform but rather the break-up of the racialized, gendered class hierarchy that produces ecocide. We prioritize the struggles of frontline women. Ask: How can my community's campaigns strengthen the struggles of the women and their allies amongst men who are already defending the commons? How can we prioritize the intrinsic value of life support?

Build new institutions, expand the solar commons

Movements for system change are simultaneously saying 'no' to capitalism and 'yes' to the commons by defending territory and building new institutions that promote the solar commons—collective control over the means of life, relying on solar not fossil energy. In December 2015 in Paris, the International Rights of Nature Tribunal did both: it simultaneously indicted ecociders while documenting pathways to well-being and Earth restitution (Global Alliance, 2015).

Action: As individuals and groups, we can both deny the capitalist system its power and expand the commons. Ask: what kinds of actions can we take to remove our support from life-destroying production and consumption activities? To what life-affirming activities can we contribute creativity and power?

These three guidelines help us to choose system changing activism that swiftly and peacefully moves us away from fossil capitalism toward the solar commons.

References

Anderson, K. 2012. Climate change going beyond dangerous: brutal numbers and tenuous hope. *Development Dialogue* 61(1), p. 16.

Huot, S. 8 December 2015. 'Women and Gender Justice Assembly'. World March of Women in Kenya at the World March of Women panel, 21 Rue Voltaire, 72011 Paris [personal recording].

Global Alliance for the Rights of Nature. 2015. 'International Rights of Nature Tribunal-Paris'. Available at <http://therightsofnature.org/rights-of-nature-tribunal-paris/> [Accessed on 25 February 2016].

Figure 43.1 Solidarity, Beehive Collective

44 Food sovereignty or bust

Transforming the agrifood system is a must

Joshua Sbicca

The fight for food sovereignty is a response to the current corporate food regime, one predicated on industrial modes of production, corporate concentration in supply chains, the patenting of seeds and biotechnology, and neoliberal trade policies that displace agrifood systems in poor and developing countries (McMichael, 2009a). While peasants throughout the Global South fare poorly as their land and labor are integrated into the global agrifood system – often in the name of sustainable development – many are also at the forefront of new peasant-based agrarian movements (Borras *et al.*, 2008). Over the past few decades, the food sovereignty movement has fought to protect local food cultures, food economies, ecosystems, and political systems from neoliberal trade policies and capitalist modes of agricultural production. Therefore, after a brief review of the social and ecological relations perpetuated by the current global corporate food regime, I investigate this movement's potential to repair these effects, paying specific attention to the issue of anthropogenic greenhouse gases.

Social and ecological consequences of the corporate food regime

An agrifood system can be thought of as "an interconnected web of activities, resources and people that extends across all domains involved in providing human nourishment and sustaining health, including production, processing, packaging, distribution, marketing, consumption and disposal of food … and can be identified at multiple scales" (Grubinger *et al.*, 2010, p. 2). The globally pervasive corporate agrifood system contributes to ecological degradation, food related human health problems, economic destitution, and cultural and political exploitation (Magdoff *et al.*, 2000; Shiva, 2008; Holt-Giménez, Patel, and Shattuck, 2009). It also contributes around 33 percent of global greenhouse gas (GHG) emissions – 14 percent from industrialized agriculture and 19 percent from change in land use for agricultural production (De Schutter, 2010a). However, estimates vary depending on how much animal agriculture is attributed to GHG emissions, from 18 percent (Stehfest *et al.*, 2009) to as high as 51 percent (Goodland and Anhang, 2009) of global totals. The United Nations (UN) Millennium Ecosystem Assessment (2005) has noted that globally, agriculture threatens biodiversity and ecosystem function more so than any single human activity. Simultaneously, the global agrifood system contributes to climatic changes that threaten its own survival, such as drought, heat waves, heavy rainfall, and flooding. For example, with an increasingly hotter and drier climate, large parts of Africa may face reduced crop yields of over 50 percent by the year 2020 (Intergovernmental Panel on Climate Change [IPCC], 2007).

Additionally, this arrangement perpetuates hunger and malnutrition as the Global South simultaneously exports large portions of the world's proteins, vitamins, and minerals to the Global North (Roberts and Parks, 2009). North and South refer to the political and economic power differentials between 'core' and 'peripheral' nations (Wallerstein, 2004). With already high levels of food insecurity (Food and Agricultural Organization [FAO], 2010), global hunger is expected to grow by 10 percent to 20 percent by 2050 if conditions remain the same (World Food Program 2011). An ecological debt has emerged whereby lifestyles of rich countries necessitate high levels of consumption and GHG emissions that disproportionately contribute to global climate change (GCC), ozone depletion, and fishery impacts for people living in poor countries (Srinivasan *et al.*, 2007). In short, GCC exacerbates and is caused by problems associated with the current corporate food regime, leading to 'double exposure' for the world's peasants (O'Brien and Leichenko, 2000).

To mitigate these ecological and social impacts requires transforming the global agrifood system (De Schutter, 2010b). Instead, many professionals, activists, and scholars have sought to link the language of sustainability to the practice of development through 'sustainable development,' a now common solution advocated by many powerful countries and multi-lateral institutions. The notion of sustainability as development converges in the World Bank's framework of green neoliberalism: "[F]ew development practices, beliefs and truths can be expressed today outside the parameters of environmentally sustainable development, on the one hand, and neoliberalism, on the other" (Goldman, 2005, p. 7). The imperative of neoliberalism is economic growth through increased privatization of every public sector, supported by a belief in the individualization of social change through consumer choice. Neoliberalism, the supporting ideology and global policy regime of capitalism, is inherently unsustainable (Harvey, 2005; McMichael, 2008b). Moreover, the history of colonialism – largely responsible for the uneven development of capitalism – benefits the Global North, while undermining the ability of much of the Global South to direct, let alone resist, the development process. Hence, the corporate food regime supports eaters in the Global North, as agrarian peasants, many of whom labor in commodity export fields, struggle "for cultural survival through the food sovereignty movement" (McMichael, 2009a, p. 163).

Food sovereignty: Principles and practice

In 1996 at the World Food Summit, La Vía Campesina, a transnational peasant-run organization working for agrarian reform, first coined the term 'food sovereignty.' They define this as:

> the right of peoples, countries, and state unions to define their agricultural and food policy without the 'dumping' of agricultural commodities into foreign countries. Food sovereignty organizes food production and consumption according to the needs of local communities, giving priority to production for local consumption. Food sovereignty includes the right to protect and regulate the national agricultural and livestock production and to shield the domestic market from the dumping of agricultural surpluses and low-price imports from other countries. Landless people, peasants, and small farmers must get access to land, water, and seed as well as productive resources and adequate public services. Food sovereignty and sustainability are a higher priority than trade policies.
>
> (La Vía Campesina, 2010, n.p.)

Although neoliberal policies and industrialized agriculture proliferate globally, their impact on agrarian communities in the North and South varies by region and time, so responses by the food sovereignty movement also differ.

The heterogeneity of the food sovereignty movement reveals the intersectional complexity of geography, race, ethnicity, class, and gender, and the value placed on developing strategies that offer widespread social and ecological benefits. In short, the food sovereignty movement is imperative to developing strategies that ameliorate the global agrifood system's contribution to GHGs because it focuses on more ecologically sustainable alternatives, and it works to create new social relations that respect, attempt to understand, and build alliances across race, ethnicity, class, and gender. I now turn to clarifying the movement's principles and practices in order to evaluate the possibilities for what Shiva (2005) refers to as 'earth democracy,' that is, a world of peace, justice, and environmental sustainability.

Human rights, self-determination, social equity, healthy agro-ecosystems, counter-hegemonic narratives, and new subjectivities underlie the movement's discourses and organizing tactics. Table 44.1 is meant to be read as a guide to clarify how critiques of the corporate food regime provide the context within which food sovereignty principles, with their constituent discourses and strategies, evolved. The table does not imply a linear process whereby problems associated with the corporate food regime are the drivers of the evolution of the food sovereignty movement. Instead, the critiques and the principles are mutually constitutive both within and between themselves. The far left column helps to clarify the context within which the movement has developed. This then backlights the main principles driving food sovereignty discourse and strategy. In short, this table is meant to capture commonly articulated food sovereignty principles. My discussion focuses on how these principles can help contribute to building solidarities across race, ethnicity, class, and gender while also addressing the role that food sovereignty strategies must play in mitigating the impacts of GCC.

Table 44.1 Driving principles of the food sovereignty movement

Critique	*Principle*	*Discourse*	*Strategy*
Neoliberal policies open markets. Economic solution to political and social problems	Human rights	Right to shape food policy, not a privilege. Need for flexibility at local level	Develop systems of duty and obligation: local actors create policies
Food commodity exports and the patenting of nature displace diverse knowledge systems. Structural adjustment programs	Self-determination and/or self-governance	Sovereignty over local agrifood systems Biodiversity and cultural diversity. Delink from corporate industrialized agrifood system	Create local and regional agrifood systems. Establish locally controlled democratic governance arrangements. Ban biopiracy and dumping of transgenic food
Rugged individualism. Private property	Social equity/ egalitarianism	Respect for collective and individual rights. Solidarity between all groups	Create spaces and mechanisms for involvement of all members of a community
Monocultures. High input agricultural models. Biotechnological solutions (GMOs and pesticides)	Healthy agro-ecosystems	Proactive, responsible management of agro-ecosystems. Food production works within a functioning local ecosystem. Farming is culture	Agroecology: sees biodiversity and agriculture as intricately linked
Capitalist modernity views of peasants as disposable and agriculture as a problem to solve. Logic of development logic	Counter-hegemonic narratives	Center agriculture as foundation for social and ecological reproduction	Print media, television, radio and the internet. Develop transnational and local networks and coalitions. Educate to empower
Denial of diverse identities and culturally specific solutions in favor of market based and/or token political participation	New subjectivities	Respect non-capitalist and decolonial modes of thinking. Provide methods for the development of alternative modernity. 'Unity of diversity'	Work toward race, ethnic, gender, and class equality. Politics of consciousness-raising that represents multiple perspectives

Rights

The core argument for food sovereignty is that people have the right to democratically determine their agrifood systems (Patel, 2009). This is important because the right to food can be co-opted. For instance, Germann (2009) argues that the Voluntary Guidelines issued by the FAO stripped the human right to food of its critical potential and instead reaffirmed the neoliberal project by "recast[ing] as a *policy goal*, instrumentalized in terms of a *policy approach* and proposed as *the economic freedom of the individual*" (p.138, emphasis in original). However, a food sovereignty analysis begins by attending to groups facing rural poverty and malnutrition, whose participation in the policy process is irreducible with neoliberal ideology (Windfuhr and Jonsén, 2005). Thus, having fair processes is a prerequisite for achieving fair outcomes.

Participation in the policy creation process, though, is necessary but insufficient. Marginalized agrarian communities maintain a right to demand what ought to be done, based on notions of sovereignty that transcend the nation state and access to mechanisms for involvement in the processes and politics responsible for securing rights. Moreover, food sovereignty requires the freedom to organize for collective action. This "[s]elf-determination needs to be reconceptualized with an eye toward specifying the hundreds and thousands of strategies that local communities use to assert control over the organization of economic activities and their inevitable (but not necessarily harmful) anthropogenic effects" (Peña, 2005, p. 145). Activists may also carve out space for alternative ways of knowing that reflect local knowledge and iterations of autonomy, while working to integrate collective concerns that challenge an individualistic understanding of rights (Peña, 2005).

Self-determination and self-governance

Food sovereignty activists emphasize control of local, regional, and national agrifood systems as a distinct alternative to the infiltration of food aid and extraction of natural resources (Menezes, 2001). Many traditional knowledge systems – women's knowledge systems in particular – contain agricultural sustainability models that simultaneously maintain biodiversity and cultural diversity (Shiva, 2000). Delinking from the corporate food regime and industrialized modes of production often happens concurrently with relocalization efforts premised on community control. Moreover, the movement's polycentrism allows it to address and politicize both local and translocal agrarian issues in specific times and spaces (McMichael, 2008c).

While autonomy and self-determination are important, social equity must also be present (Menezes, 2001). One approach is to develop a shared class perspective across various scales among dispossessed peasants (Walker, 2008). While this provides a shared point of departure in much of the Global South, there is also differentiation tied to gender and nationality that must be recognized and respected. For example, in working out alliances between female activists from

World March of Women (WMW) and La Vía Campesina in Mali in 2007, there was agreement that capitalism produces poverty and food insecurity, but disagreements over caregiving. Urban women from WMW prioritized sharing roles with men, while rural women in La Vía Campesina viewed working in the kitchen as a cultural expression and place where their knowledge can combat processed foods. Through such debate, WMW came to respect the needs of rural women for self-determination. They also helped them in campaigns to resist GMOs and privatization, while improving their economic power (Nobre, 2013), using events such as International Women's Day to link land rights in the Philippines, discrimination against women in Paraguay, and native seed recovery in Europe and Africa (La Vía Campesina, 2014).

Social equity and egalitarianism

Paul Nicholson, one of the founding members of La Vía Campesina notes, "We have certainly defended the concept of 'patrimony of humanity' when it comes to seeds, water, and for land ... We have a common good, and it must be protected and defended" (Wittman, 2009a, p. 679). In this sense, individual rights are embedded in collective rights. By striving for a 'radical egalitarianism,' food sovereignty seeks to alter power imbalances in the global agrifood system by dismantling oppression in all its forms, particularly in terms of race, ethnicity, class, and gender (Patel, 2009). Moreover, food sovereignty is premised on active solidarity to achieve a fundamental philosophical and cultural shift in our use of food, away from individual wealth accumulation and toward building equitable ties, including trade, between diverse groups of people (Jackson and Mitchell, 2009).

Decentralized, non-hierarchical forms of organization provide a foundation for developing such egalitarian rural livelihoods, particularly between men and women (Desmarais, 2007). For instance, women farmers experience disadvantages in terms of resource access as well as sexism in their daily lives. Therefore, women created the Women's Commission of La Vía Campesina, which furthered a feminist analysis that resists neoliberalism, and led to the spread of egalitarian practices both within this transnational organization and in the rural communities where these women reside. Strengthening community ties also helps to better resist the divide and conquer tactics of large corporations such as Monsanto.

Healthy agro-ecosystems

Biodiversity and ecological health is vital to maintain in and around any local or regional agrifood system (Shiva, 2008). While there are varying alternative models of food production, biodynamic polycultures and permaculture farming (i.e. agroecological) practices are more widely practiced and promoted. These models protect ecosystem health with integrated practices that respect the soil, plants and animals, minimize external inputs, and where possible mimic natural processes. A commitment to systems like agroecology is important for bringing people together to engage in the organizing necessary to ensure food security (Altieri and Nicholls, 2008).

Such approaches are viewed as fundamental to achieving food sovereignty and slowing climate change (McMichael, 2008b). Small-scale sustainable farms emit between one-half and two-thirds less carbon dioxide for every acre of production (IPPC, 2007). These place-based experiences compel farmers to make context specific changes that include developing new social relations (Patel, 2009) alongside new biological methods of pest control (Altieri *et al.*, 1999), especially in the tropics where there is no seasonal cold.

Counter-hegemonic narratives

The commodification of land and industrialization of agriculture uses food as a means to profit, which furthers the separation between humans and the ecosystems they occupy. Peasants, particularly women, are often viewed as disposable within the corporate food regime, whose labor is used as a 'natural resource' to be exploited by male capitalists in the Global North (Salleh, 2010). Or, for those experiencing hunger, this is a technological problem in need of patronizing scientific expertise. In both cases, many transnational development entities assume that poor countries need to adopt industrialized agriculture and neoliberal structural adjustments to avoid hunger and develop their economies to favor export based practices (Holt-Giménez *et al.*, 2009).

The food sovereignty movement works to overcome these social and ecological divides (Wittman, 2009b). Specifically, food sovereignty provides a set of narratives that push back against development models that drive a wedge between groups and instead emphasize the importance of empowering education that builds unity in the midst of diversity. Examples of this include La Vía Campesina Women's Commission that organizes regional, national, and international conferences to bring women together to increase their participation in global summits and meetings. The narrative of equity at the heart of food sovereignty helps retake the language of development from male dominated Northern governments and multilateral trade and development institutions without sacrificing democracy in the name of sustainability (Karriem, 2009). By centering agriculture as the foundation for social and ecological healing, the food sovereignty movement is poised to spread their message, and build strong social networks that take care of peasants and offer solutions to GCC (McMichael, 2008b).

New subjectivities

Imposed trade regimes create severe ecological and social consequences for peasant populations, which currently frame how agrarian communities understand and respond to these conditions. Food sovereignty, though, offers unique cultural lenses to understand these relational complexities from the perspective of marginalized social groups. For example, the food sovereignty movement represents a shift away from social movements that seek concessions from the state or access to the market. Instead, it focuses on alternative political organization, diverse rationalities, and non-capitalist and decolonial forms of thought (Rojas, 2007). As discussed below, these subjectivities also provide a foundation for understanding how those most impacted develop culturally, economically, and ecologically relevant agrifood systems that simultaneously resist the corporate food regime (Salleh, 2010).

For food sovereignty, let's begin to cool the planet

As should be clear by now, the institutions supporting such an ecologically and socially destructive global agrifood system make transformation difficult. Nonetheless, the movement has developed a comprehensive set of alternative strategies. Importantly, many food sovereignty activists are directly linking to the climate justice movement. At the annual UN Framework Convention on Climate Change in Cancun in 2010, otherwise known as the Conference of the Parties (COP16), conflict between countries such as the United States and China and India led to non-binding resolutions to expand carbon trading markets, clean development projects, and the use of techno-fixes involving transnational firms, without clear carbon reduction targets deemed imperative by the IPCC. The same year, La Vía Campesina organized members from 29 states in Mexico, and 36 countries, to an international conference occurring simultaneously with COP16. This international food sovereignty alliance is pushing for more radical proposals given the Global South's disproportionate experience of the quadrupling of climate-related disasters between 1980 and 2006 (World Food Program, 2011, n.p.).

In order to begin cooling the planet, they demand the following:

> 1) Resume the principles of the Peoples' Accord in Cochabamba; 2) Establish a binding agreement to reduce by 50 percent greenhouse gas emissions in industrialized countries by 2017; 3) Allocate 6 percent of developed countries' GDP to finance actions against the Climate Crisis in countries of the global south; 4) Total respect for Human Rights, Indigenous Peoples' Rights and Rights of Climate Migrants; 5) The formation of an International Tribunal for Climate Justice; 6) State policies to promote and strengthen sustainable peasant agriculture and food sovereignty.

Subsequently, global days of action such as the 2011 International Food Sovereignty Day to Cool Down the Earth, were organized to coincide with COP17 in Durban, South Africa. During this day of action, activists strategized how to bring about agrarian reform for food sovereignty, develop seed sovereignty strategies and scale agroecological solutions to mitigate GCC, end corporate practices of biopiracy, and fully integrate equal participation for women.

Given that the corporate food regime produces different outcomes at different scales, the tactics and strategies activists adopt reflect local concerns and resist the integrative logic of an international trade regime that brings the labor and resources of these agrarian communities into export-dependent relationships with the North. Such resistance also responds to the expansion of export-only agricultural commodity markets for Northern consumption that increase GHGs. The mitigation of GHGs, then, includes local management strategies that reduce reliance on energy intensive systems that release high levels of carbon, nitrogen, and methane, and simultaneously increase carbon storage. By one estimate, if 10,000 medium sized farms in the US were to convert to organic production, they would store enough carbon in the soil to equivalently take almost 1.2 million cars off the road (Rodale Institute, 2003). Many farmers employ organic and agroecological methods in a flexible manner depending upon the climate within which farming takes place. This includes practices that better manage cropland, improve grazing/pasture land and the management of livestock, restore, retain, and increase organic soils and land, reuse manure as fertilizer, and create bioenergy from waste (Smith *et al.*, 2007).

The food sovereignty movement claims the right of peasants and small-scale and subsistence farmers in shaping food policy. Local or traditional knowledge often directly challenges the conventional wisdom of industrialized agriculture and biotechnology. For example, instead of applying artificial fertilizers to increase yields, many farmers will gather manure from livestock and blend this with household and field waste compost to create a sustainable natural fertilizer. Moreover, activists appeal explicitly for autonomy from globalized systems in order to allow for versatility in developing social and ecological alternatives that respect difference. Strategies to achieve self-determination, then, are different across scale.

Yet, the local expression likely varies based on specific social and ecological requirements. For example, women within the movement fought for female economic and political self-determination after being relegated to subordinate roles and positions in peasant and farm organizations for years (Desmarais, 2007). At the same time, neoliberal development models and patriarchal norms have prevented many peasant women from achieving the same yields as men because they have less access to land, water, seeds, training, and credit. Yet, women's agricultural knowledge is imperative to integrate into strategies that ameliorate the causes of climate change and hunger, particularly in places such as India where women also resist corporations like Monsanto and the GHG-producing biotechnology model they represent. After all, women grow most of the world's food, and in subsistence agricultural communities, oftentimes possess generations of passed knowledge of specific growing conditions (Shiva, 2008).

By contextualizing the problems faced by local communities the movement can flexibly adapt to challenge the homogenizing tendencies of the corporate food regime and come up with a variety of alternatives. However, solving problems at the scale of global capitalism becomes difficult when universalized claims to rights clash with the particularities of local struggles, preventing activists from building solidarity across difference (Harvey, 1996). Food sovereignty activists attempt to overcome these contradictions. As McMichael (2011) argues, to challenge neoliberalism on a global scale requires the practice of "multifunctionality", namely a "method of valuing and farming" premised on ecological, social, and democratic values (pp. 810–811). The movement embodies such an approach, which is needed to bridge difference within the movement, while providing a means to mitigate the causes of GCC.

Because agriculture takes place within particular local ecosystems, the movement preferences agroecology as it is more resilient to climate change, conserves soil, and enhances biodiversity (Altieri, 2009). Moreover, productivity on farms using agroecological farming rivals industrialized agricultural systems. One research project covered 286 sustainable farming projects in fifty-seven poor countries, covering 3 percent of cultivatable land in the Global South. Productivity increased on 12.6 million farms by an average of 79 percent (Pretty *et al.*, 2006). Equally important, agroecology reduces poverty by providing an agricultural practice that increases economic self-sufficiency (De Schutter, 2010b). This is especially urgent given predictions of climate change contributing to increases in food prices from between 50 percent and 90 percent by 2030 (Bailey, 2011).

Structural challenges that restrict the spread of agroecological farming practices remain. Consider land tenure: while experiments by governments in Cuba and Venezuela attempt to ameliorate poverty and the country's ecological footprint by redistributing land to peasants, and social movements such as the Brazilian Landless Workers Movement successfully occupy many large landed estates, this is by no means universal. Development agencies tend to favor policies that privatize agricultural production and increase land use for a few export commodity crops, thus exacerbating food insecurity and increasing levels of GHGs. The food sovereignty movement gives some clear indications as to how to proceed, but requires land to do so (Rosset, 2009). Thus, food sovereignty provides new discursive tools for those within the UN and FAO to push for redistributive land policies.

These counter-hegemonic narratives motivate the food sovereignty movement. Instead of social relations being determined by neoliberal ideology, and token forms of democracy, agriculture is viewed as a means to create more just and sustainable social relations. Media tools are used to organize agrarian communities around a global movement for food sovereignty. Communication technologies are now commonly used to organize food sovereignty activists around shared interests, evidenced by thousands of activists showing up at all the recent COP conferences to link food and climate issues. The impact in global policy circles is beginning to emerge. Take for instance the UN Special Rapporteur on the Right to Food aligning his reports with much of the food sovereignty agenda, and spreading this through mainstream papers such as the *Guardian* (De Schutter, 2010a; De Schutter, 2010b).

The food sovereignty movement represents the most global and diverse challenge to the energy-intensive and GHG-emitting global agrifood system. As a critique, it exposes the exploitative tendencies of neoliberal policies, an overreliance on biotechnology such as GMOs, and the largely unquestioned commitment to industrialization. As an alternative, food sovereignty works to actualize new forms of human rights, self-determination, social equity, and healthy agro-ecosystems based on a set of counter-hegemonic narratives and new subjectivities. Our global inability to adequately mitigate the impacts of the corporate food regime on anthropogenic climate change is clear. Grassroots alternatives, though, that challenge and reorganize agrifood systems while also inviting society to celebrate and work across the intersections of race, ethnicity, class, and gender are contained in the food sovereignty movement.

References

Altieri, M., 2009. Agroecology, small farms, and food sovereignty. *Monthly Review*, 61(3), pp. 102–113.

Altieri, M. A. and Nicholls, C. I., 2008. Scaling up agroecological approaches for food sovereignty in Latin America. *Development*, 51(4), pp. 472–80.

Bailey, R., 2011. Growing a better future: food justice in a resource-constrained world. *Oxfam International*. Great Britain: Grow Campaign.

Declaration of Maputo: V International Conference of La Vía Campesina, 2008. [pdf] Available at: http://viacampesinanorteamerica.org/en/viacampesina/conferencias/V%20conferencia%20Declaration%20of%20Maputo.pdf [Accessed 10 December 2010].

De Schutter, O., 2010a. It's time to tackle climate change and agricultural development in tandem. *The Guardian*, October 16, 2010 [online]. Available at: www.guardian.co.uk/global-development/poverty-matters/2010/oct/16/climate-change-agricultural-development-policymakers [Accessed 18 November 2011].

De Schutter, O., 2010b. Agroecology and the right to food. *Special Rapporteur on the Right to Food, 16th Session of the UN Human Rights Council*.

Desmarais, A.A., 2007. *La Via Campesina: Globalization and the Power Of Peasants*. London: Pluto Press.

Food and Agriculture Organization, 2010. The state of food insecurity in the world 2010. Rome: United Nations [pdf]. Available at: www.fao.org/docrep/013/i1683e/i1683e.pdf [Accessed 3 February 2012].

Germann, J., 2009. The human right to food: 'voluntary guidelines' negotiations. In: Y. Atasoy, (ed) *Hegemonic Transitions, the State and Crisis in Neoliberal Capitalism*. New York: Routledge. pp. 126–143.

Goodland, R., 2009. Livestock and climate change. *World Watch Magazine*. November/December, pp. 10–19.

Grubinger, V., Berlin, L., Berman, E., Fukagawa, N., Kolodinsky, D.N., Parsons, B., Trubek, A., Wallin, K., 2010. *University of Vermont Transdisciplinary Research Initiative Spire of Excellence Proposal: Food Systems*. Burlington, V T: University of Vermont.

Harvey, D., 1996. *Justice, Nature, and the Geography of Difference*. Oxford: Blackwell.

Harvey, D., 2005. *A Brief History of Neoliberalism*. Oxford: Oxford University Press.

Holt-Giménez, E., Patel, R., and Shattuck, A., 2009. *Food Rebellions: Crisis and the Hunger for Justice*. Oxford: Pambazuka Press.

International Panel on Climate Change, 2007. *Climate change 2007: Fourth Assessment Report of the Intergovernmental Panel on Climate Change*. New York: Cambridge University Press.

Jackson, A. and Mitchell, E., 2009. Food sovereignty: time to choose sides. *Soundings: A Journal of Politics and Culture*, (41), pp. 100–106.

Karriem, A., 2009. The Brazilian landless movement: mobilization for transformative politics. In: Y. Atasoy (ed.) *Hegemonic Transitions, the State and Crisis in Neoliberal Capitalism*. New York: Routledge, pp. 262–280.

La Vía Campesina, 2010. What is La Vía Campesina? [online]. Available at: http://viacampesina.org/en/index.php?option=com_content&view=category&layout=blog&id=27&Itemid=44 [Accessed 6 April 2010].

La Vía Campesina, 2014. March 8: Special Report on the International Women s Day [online]. Available at: http://viacampesina.org/en/index.php/main-issues-mainmenu-27/women-mainmenu-39/1605-march-8-special-report-on-the-international-women-s-day [Accessed 1 February 2015].

Magdoff, F., Foster, J.B., and Buttel, F., eds., 2000. *Hungry For Profit: the Agribusiness Threat to Farmers, Food, and the Environment*. New York: Monthly Review Press.

McMichael, P., 2008a. Peasants make their own history, but not just as they please … In: S.M. Borras Jr., M. Edelman, and C. Kay (eds) *Transnational Agrarian Movements Confronting Globalization*. Oxford: Wiley-Blackwell.

McMichael, P., 2008b. The peasant as 'canary'? Not too early warnings of global catastrophe. *Development*, 51(4), pp. 504–511.

McMichael, P., 2008c. Food sovereignty, social reproduction and the agrarian question. In: P. McMichael, A.H. Akram-Lodhi, and C. Kay (eds.) *Peasants and Globalisation. Political Economy, Rural Transformation and the Agrarian Question*. London: Routledge, pp. 288–312.

McMichael, P., 2009. A food regime genealogy. *The Journal of Peasant Studies*, 36(1), pp. 139–169.

McMichael, P., 2011. Food system sustainability: Questions of environmental governance in the new world (dis)order. *Global Environmental Change*, 21(3), pp. 804–812.

Menezes, F., 2001. Food sovereignty: a vital requirement for food security in the context of globalization. *Development*, 44(4), pp. 29–33.

Nobre, M., 2013. Women's autonomy and food sovereignty. In: E. Holt-Giménez (ed.) *Food Movements Unite! Strategies to Transform Our Food System*. Oakland, CA: Food First Books, pp. 293–306.

O'Brien, K.L. and Leichenko, R.M., 2000. Double exposure: assessing the impacts of climate change within the context of economic globalization. *Global Environmental Change*, 10(3), pp. 221–232.

Patel, R., 2009. Food sovereignty. *The Journal of Peasant Studies*, 36(3), pp. 663–706.

Peña, D., 2005. Autonomy, equity, and environmental justice. In: D.N. Pellow and R.J. Brulle (eds) *Power, Justice, and the Environment: A Critical Appraisal of the Environmental Justice Movement*. Cambridge, MA: MIT Press.

Pretty J., Noble A., Bossio D., Dixon J., Hine R.E., Penning de Vries, P. and Morison J.I.L., 2006. Resource conserving agriculture increases yields in developing countries. *Environmental Science and Technology* 40 (4), pp. 1114–1119.

Rodale Institute, 2003. Organic farming combats global warming – big time [online]. Available at: www.rodaleinstitute.org/ob_31 [Accessed 19 February 2012].

Roberts, J.T. and Parks, B.C., 2009. Ecologically unequal exchange, ecological debt, and climate justice: the history and implications of three related ideas for a new social movement. *International Journal of Comparative Sociology*, 50(3–4), pp. 385–409.

Rojas, C., 2007. International political economy/development otherwise. *Globalizations*, 4(4), pp. 573–587.

Rosset, P., 2009. Fixing our global food system: food sovereignty and redistributive land reform. *Monthly Review*, 61(3), pp. 114–128.

Salleh, A., 2010. From metabolic rift to 'metabolic value': reflections on environmental sociology and the alternative globalization movement. *Organization and Environment*, 23(2), pp. 205–219.

Srinivasan, U.T., Carey, S.P., Hallstein, E., Higgins, P.A.T, Kerr, A.C., Koteen, L.E., Smith, A.B., Watson, R., Harte, J., and Norgaard, R.B., 2007. The debt of nations and the distribution of ecological impacts from human activities. *Proceedings of the National Academy of Sciences of the United States*, 105(5), pp. 1768–1773.

Shiva, V., 2000. *Stolen Harvest: the Hijacking of The Global Food Supply*. Cambridge, MA: South End Press.

Shiva, V., 2005. *Earth Democracy: Justice, Sustainability, and Peace*. Cambridge, MA: South End Press.

Shiva, V., 2008. *Soil Not Oil: Environmental Justice in an Age of Climate Crisis*. Cambridge, MA: South End Press.

Smith, P., Martino, D., Cai, Z., Gwary, D., Janzen, H., Kumar, P., McCarl, B., Ogle, S., O'Mara, F., Rice, C., Scholes, B., and Sirotenko, O., 2007. Agriculture. In: B. Metz, O.R. Davidson, P.R. Bosch, R. Dave, L.A. Meyer (eds) *Climate Change 2007: Mitigation* Contribution of Working Group III to the Fourth Assessment Report of the Intergovernmental Panel on Climate Change, Cambridge: Cambridge University Press, pp. 497–540.

Stehfest, E., Bouwman, L., Vurren, D., Elzen, M., Eicjhout, B., and Kabat, P., 2009. Climate benefits of a changing diet. *Climate Change*, 95, pp. 83–102.

United Nations Millennium Ecosystem Assessment, 2005. *Ecosystems and Human Well-being: Biodiversity Synthesis*. Washington D.C.: World Resources Institute.

Wallerstein, I., 2004. *World-systems Analysis: an Introduction*. Durham, NC: Duke University Press.

Walker, K., 2008. From covert to overt: everyday peasant politics in China and the implications for transnational agrarian movements. *Journal of Agrarian Change*, 8(2–3), pp. 462–488.

Windfuhr, M., and Jonsén, J., 2005. *Food sovereignty: towards democracy in localized food systems*. Warwickshire: ITDG Publishing.

Wittman, H., 2009a. Interview: Paul Nicholson, La Vía Campesina. *The Journal of Peasant Studies*, 36(3), pp. 676–682.

Wittman, H., 2009b. Reworking the metabolic rift: La Via Campesina, agrarian citizenship, and food sovereignty. *Journal of Peasant Studies*, 36 (4), pp. 805–826.

World Food Program, 2011. 7 facts about climate change and hunger [online]. Available at: www.wfp.org/stories/7-facts-about-climate-change-and-hunger [Accessed 7 December 2011].

45 Children in a changing climate

How child-centered approaches can build resilience and overcome multiple barriers to adaptation

Paul Mitchell

There is near universal agreement that children have rights, and agreement on the scope and content of those rights. Children's rights, enshrined and safeguarded in the Convention on the Rights of the Child (CRC) (UN, 1989), have been ratified by every United Nations (UN) member state, except Somalia and the United States (US). While the US has not ratified the CRC, it is obliged to respect the object and purpose of the convention as it is broadly considered customary international law due to its overwhelming ratification (UN, 1969, art. 18). Nevertheless, many nation states have not implemented policies to secure and protect children's rights.

Global climate change (GCC) is a key issue for children's rights. The impacts of global warming will further hinder children's ability to claim their rights by placing more obstacles in the way of states' (already mediocre) provisions of protection. Children are susceptible to nearly all the conceivable impacts of GCC, including direct physical impacts, health impacts, and the more gradual impacts of under-nutrition and malnourishment. The ways in which communities and governments plan for and respond to the impacts of GCC through policy processes and practical actions will, in large part, shape the ways in which children experience a climate changed future – and may compound the existing layers of discrimination many children already face due to the complex intersections of age, race, class and gender, as well as where they are geographically located. However, as GCC impacts intensify, children's rights will become even more difficult to achieve and sustain. The inevitable result will be more lives unnecessarily cut short as governments fail to safeguard the rights of children in a changing climate.

The world is heading towards a future with a much harsher climate. While there is a global commitment to holding temperature increases to no more than two degrees centigrade above pre-industrial levels (UNFCCC, 2010, p. 3), this goal is unlikely to be met. Temperature increases may well be double the goal by century's end (International Energy Agency, 2011;) largely due to global capitalism's insatiable need to continually expand production, the Global North's addiction to high-consumption lifestyles, and the Global South's understandable desire to achieve similar levels of development.

The impacts on children of a world that is 4°C (39°F) hotter will be detrimental to the achievement of the rights enshrined in the CRC. Children's rights have contributed to advances in children's welfare (the under-five mortality rate, for example, has halved since 1990 (UN, 2014, p. 24)), but progress is already stalling in the face of climate-linked diseases, which are the largest killer of young children (UN, 2014, p. 24), and are set to increase in severity due to GCC. In this context, 'development as usual' approaches, even those adopting a child-centered approach, are likely to prove inadequate to meeting children's needs.

Children are one of the most vulnerable groups to the impacts of GCC (Save the Children, 2009; Seballos *et al.*, 2011; UNICEF, 2011). They are physiologically and psychologically less able to cope with its impacts and their exclusion from decision-making processes exacerbates these inherent vulnerabilities. Combined, these issues intensify the existing multiple layers of discrimination many children experience on a daily basis. Furthermore, it is the most vulnerable children who will be most deprived of their right to a healthy environment and a sustainable future.

This chapter focuses on the impacts of GCC on children in the Global South, highlighting the general vulnerabilities of children as a group, but also examining the multiple layers of discrimination that many children suffer (including poverty, gender, class, religion, race and disability) and how GCC will exacerbate these challenges. In so doing, I argue that treating children as a generic 'vulnerable group' when devising strategies and actions to manage climate change risks masking the significant differences within and between groups of children. While children are more exposed than adults to most GCC impacts, not all children are equally vulnerable to all impacts, nor are all children more vulnerable than all adults to all impacts.

I argue that taking a child-centered (contextualized and rights-based) approach to reducing climate risks and building resilience will make a significant contribution to helping children claim their rights and reach their aspirations in a climate changed world. Finally, I conclude by arguing that integrating this risk reduction and resilience building approach into broader sector-based programs and activities can help address the multiple drivers of children's vulnerability to GCC.

Linking children's rights and mortality

Even without considering the additional burdens of GCC, the fulfilment of child rights remains a problem globally. A key indicator of progress in implementing children's rights is the under-five mortality rate, which measures the number of children who die before their fifth birthday per 1,000 live births. In 2013, 6.3 million children died before reaching age five (UNICEF, 2014a, p. 29). The majority of under-five deaths globally have climate-linked causes, including diarrheal diseases, malaria and malnutrition (UNICEF, 2012, p. 25). In the Global South, child survival remains a challenge, with the Least Developed Countries having an average under-five mortality rate of 80 (per 1,000 live births in 2013) (UNICEF, 2014a, p. 41), in stark contrast to industrialized countries' average of 6 (in 2010) (UNICEF, 2012, p. 129).

While some of these failures are related to economic development, they also reflect the level of priority being accorded to child rights. This is demonstrated by the fact that increasing a country's economic resources does not necessarily result in an increase in access to rights. For example, the US ranks worst in under five mortality (tied with the Slovak Republic) of the 31 high-income members of the Organization for Economic Cooperation and Development. The US also ranks eight places below Cuba in terms of child survival, while Cuba's gross domestic product per capita is a fifth of that of the US. To make a difference, children's rights must be more substantive than rhetorical.

Vulnerability or vulnerabilities?

In annual *State of the World's Children* reports, UNICEF highlights that it is the most vulnerable children (those facing multiple levels of discrimination, particularly due to their race, class and/or gender, on top of age-based disadvantage and geographic location that experience the most pervasive violation of their rights (UNICEF, 2012, p. 13; UNICEF, 2014b, p. 3-4)). While there are many vulnerable populations in relation to many different climate-related challenges, children "constitute an extremely large percentage of those who are most vulnerable, and the implications, especially for the youngest children, can be long term" (Bartlett, 2008, p. 514).

But vulnerability is a loaded term

While children are more exposed than adults to a variety of GCC impacts, due to their developmental stage, not all children are equally vulnerable, nor are all children more vulnerable than all adults. Given the complex socio-economic and cultural interactions inherent in the climate crisis, the field of Intersectionality Studies (Cho, Crenshaw and McCall, 2013) provides a foundation for examining how issues of race, class and gender as well as age impact on how GCC is (or isn't) addressed at local, national and international levels. The CRC defines children as people up to the age of 18. Thus, it is necessary to take into account the different capacities, capabilities, and vulnerabilities of different age groups. A one-year old has vastly different needs and capacities compared to a 17-year old. Equally important are the differences within age groups that a homogenous approach masks. A middle class 10-year old boy living with two parents in a well-resourced environment in the Global North is likely to be significantly less vulnerable to GCC than a 10 year old girl in a female-headed household in a slum in the Global South.

Children are generally lumped into a broad 'vulnerable' group in GCC policy and strategy documents from international to local levels, actively eroding their rights by systemic exclusion (Tanner, 2010). The constituents of this group tend to include women, indigenous peoples, children and people with disabilities – essentially everyone but white males. This results in a generalized failure to meet the needs of any one sub-group and serves to erase differences within groups. Three examples should suffice to highlight this issue. The Cancun Agreements (a key guiding document for the post-2015 GCC framework) mentions children once when referring to the potential consequences of actions to curb emissions on vulnerable groups, "in particular women and children" (UNFCCC, 2010, p. 15). Given that women and children make up the majority of the world's population, one can only hope that all GCC action would be cognizant of the potential impacts it has on them. Things get slightly better at the national level in some countries. The Bangladesh National Climate Change Strategy and Action Plan, for example, mentions children 15 times, but only ever in the context of a homogenous status of vulnerability, along with women (Government of Bangladesh, 2008). There is no mention of increasing children's agency or voice. Children are also largely absent even where national governments have sought to localize their adaptation strategies. As an example, the process adopted by the Government of Nepal to take their National Adaptation Program of Action and create a Local Adaptation Program of Action renders children completely invisible by subsuming them under the catch-all 'community' (Government of Nepal, n.d. and 2010).

Vested political and socio-economic interests are a crucial factor constraining action to curb emissions as governments and corporations struggle to maintain the status quo where the environment is treated as an externality and short-term profits and political expediencies outweigh long-term sustainability. Similarly, existing power dynamics—including those relating to race, class and gender and their intersections—within and between countries are a significant factor in determining which adaptation strategies are adopted and to what extent they are implemented. As Pelling (2011) argues:

> The power held by an actor in a social system, translated into a stake for upholding the status quo, also plays a great role in shaping an actor's support or resistance towards adaptation or the building of adaptive capacity when this has implications for change in social, economic, cultural or political relations, or in the way natural assets are viewed and used.
>
> (p. 5)

Although scientific understanding of GCC is increasing, policy is not keeping abreast of science. If we are to avoid the worst consequences for the most vulnerable, a significant shift towards evidence-based policy needs to occur. If it has been too difficult for States to realize child rights in the largely stable climatic conditions that have characterized the Holocene, what will happen to children as the impacts of climate change intensify in the Anthropocene?

The implications of GCC are far-reaching and potentially catastrophic, but governments and communities delay action. These delays will inevitably cost lives and livelihoods. Children are inherently vulnerable to the impacts of GCC and, in many ways, have the most to lose in a changing climate. The following section provides an overview of children's vulnerability, setting the scene for a shift to a child-centered approach to action on GCC.

Children's vulnerability to climatic change

There are several reasons why children, in general, are particularly vulnerable to the impacts of GCC: physiology, psychology, and exclusion; which, combined, exacerbate the existing multiple layers of discrimination that many children experience on a daily basis. Discrimination based on poverty (relative or absolute), gender, class, religion, race and disability is also just as common for children as adults. But for children, the fact that age-based discrimination is pervasive and institutionalized (Child Rights Information Network, 2009) makes these other challenges more complex.

Physiological impacts

The most evident and obvious reason children are more vulnerable than (most) adults to GCC is physiological as "they are physiologically and metabolically less able than adults at adapting to heat and other climate-related exposure" (UNICEF, 2011, p. 1). Children are affected by nearly all the conceivable impacts of GCC – including immediate direct physical impacts (from extreme weather events like typhoons and floods); health impacts (from increased diarrheal and respiratory disease and changing ranges of vector-borne diseases); and the more gradual impacts of under-nutrition and malnourishment (Sheffield and Landrigan, 2001).

Children are more likely than adults to be injured or killed in a disaster (UNICEF, 2007). A 2007 estimate found that up to 175 million additional children will be affected by natural disasters annually by 2015 (Save the Children, 2007). Taking a broader view, the Global Humanitarian Forum (2009, p. 9) estimated that "325 million people are seriously affected by climate change every year," leading to an estimate that around 300,000 deaths are caused by GCC annually. With children making up a significant proportion of the population in the Global South, the number of children killed by the impacts of GCC is already unacceptably high and the potential future impacts on the world's children are catastrophic.

Sheffield and Landrigan (2001, p. 292) estimate the number of deaths directly attributable to an increased disease burden due to GCC in the year 2000 at more than 150,000. Eighty-eight per cent of these deaths were children under five. With estimates that an additional 310 million people will suffer ill health attributable to GCC by 2030 (Global Humanitarian Forum, 2009), the number of children suffering will increase exponentially. It is the most marginalized children (those suffering multiple layers of discrimination) who will be the first and worst affected as GCC impacts intensify.

Psychological impacts

Children are also among those most susceptible to the psychological impacts of GCC (Frumkin 2008; Farrant, Armstrong and Albrecht 2012). Doherty and Clayton (2011, p. 265) distinguish three levels of psychological impacts of GCC: acute and direct impacts, such as mental health injuries associated with more frequent and powerful weather events; indirect and vicarious impacts, for example, intense emotions associated with observation of climate change effects at broader scales; and psychosocial impacts that include heat-related violence, resource conflicts, migration and dislocation, post-disaster adjustment and chronic environmental stress. Children are susceptible to all of these impacts, either directly (by witnessing and experiencing the death and destruction associated with extreme weather events) or indirectly (with disaster events or chronic climate-related malnutrition impacting on the cognitive development of fetuses and infants).

The physical and psychological impacts will, in many cases, reinforce each other. For example, the physical impacts of climate-linked food insecurity include malnourishment and stunting, and inadequate nutrition has been linked to developmental and behavioral problems in children (Wachs, 2000; Tanner and Finn-Stevenson, 2002). Psychological issues in children have also been linked to their experience of disasters (the majority of which are weather-related). A study of children impacted by Hurricane Katrina found that they were four times more likely than before the storm to be depressed or anxious and twice as likely to have behavioral problems. The most affected children were those who were more socio-economically marginalized prior to the disaster (Abramson, Garfield and Redlener, 2007).

Exclusion

Article 12 of the CRC provides:

> States Parties shall assure to the child who is capable of forming his or her own views the right to express those views freely in all matters affecting the child, the views of the child being given due weight in accordance with the age and maturity of the child".
>
> (UN 1989, art. 12)

While Article 12 should not be interpreted as giving children complete autonomy in decision-making processes, "it does introduce a radical and profound challenge to traditional attitudes, which assume that children should be seen and not heard" (Lansdown, 2001, p. 2). Sadly, the general assumption seems to be that children have little to offer in the way of solutions to local challenges and that they should, essentially, be quiet and do what they are told. Seballos *et al.* (2011, p. 39) found that adults' perceptions of children's agency is a crucial factor in whether children are able to effectively engage in disaster risk reduction

(DRR) activities. This finding is highly likely to be echoed in children's engagement in community-based adaptation actions given the significant overlaps between community-level responses to disasters and GCC.

While education systems can reinforce existing social inequalities (Lynch and Baker, 2005; Haveman and Smeeding, 2006), it may also be true that the rote learning methodology still widely used in the Global South reinforces the perception that children are 'empty vessels' to be filled with knowledge. Anecdotal evidence from child-centered DRR and GCC projects that the international NGOs Plan International and Save the Children have supported in the Global South shows that a child-centered approach to teaching and learning creates an atmosphere where children feel more confident to provide input and create innovative solutions to issues their communities struggle to solve (Mitchell and Borchard, 2014). This is particularly relevant to the impacts of natural disasters, climate change and environmental degradation, which are all too often pushed aside. These projects generally run outside of school hours through 'clubs' but children involved have often expressed a desire to have their regular classes conducted in a more child-centered way. Curriculum development, however, is the purview of government agencies and, in many countries, making changes to the way classes are run is extremely difficult. Even when the government is largely supportive and the topics are uncontroversial, introducing new ideas and topics is a slow process. As long as most children remain voiceless at home and in their communities, and have their agency disregarded at school, their ability to influence decision-making processes and outcomes will remain minimal.

The child-centered approach in the context of climate change

The child-centered approach to development is based on child rights and places children at the heart of efforts to secure their rights and fulfill their development aspirations. It works to directly target children – particularly the most vulnerable, excluded and marginalized – and works to overcome the disadvantages children face by helping them (and their caregivers) understand and combat multiple layers of discrimination. The approach is also about engaging with support structures and institutions at all levels (households, communities, local and national governments and international organizations) to secure children's rights. And it can be an effective way for development and civil society organizations to access and support communities to manage the impacts of GCC, especially as members may be (understandably) more concerned with more immediate needs.

While community-based responses to GCC are rapidly growing, the participation of children in these actions is minimal, despite research showing that children have much to contribute to building community resilience to GCC (Tanner *et al.*, 2009). Putting children at the center of responses to GCC can have unexpected and innovative results. For example, a program that Save the Children supported in Kenya worked with school students across the country to help them better understand the implications of GCC for them and their communities and asked them to find local solutions to reduce risk and build resilience. Participating children developed innovative solutions including: enhancing rain water harvesting (using plastic bottles to create low-cost gutter systems to channel water to tanks); more sustainable fuel sources (using coal ash and waste paper to make cleaner burning cooking briquettes); waste reduction/recycling (weaving discarded plastic bags into re-useable shopping bags to sell in the local market); and food security (creating kitchen gardens utilizing drought-tolerant crop varieties and organic principles) (Interclimate Network, 2012).

Save the Children is not alone in reaching the conclusion that child-centered approaches can reap significant benefits. A study by UNICEF and Plan International (2011) found:

> the benefits of child-focused approaches to adaptation are likely to be high – because children are numerous and experience the impacts of climate change more acutely than other groups and over a longer period, the avoided losses associated with adaptation to both sudden disasters and systemic climate change are significant".
>
> (p. 24)

To be effective, child-centered approaches cannot just be 'business-as-usual' with token child representation: children from all demographics must be involved in decision-making processes from the very start of a program.

In the context of continuing uncertainty of timing and magnitude, but increasing certainty that things will get worse, how can we most effectively reduce all children's vulnerability, while balancing the need to recognize the high variability in children's needs and capacities with the pragmatism of treating them as a homogenous group? One means of doing so is to shift from building adaptive capacity to fostering resilience.

From adaptive capacity to resilience

One of the most effective means of reducing children's vulnerability to the impacts of GCC is to build their capacity to adapt to the range of changes it may bring in their lifetime. The Intergovernmental Panel on Climate Change (2007, p. 21) defines adaptive capacity as "the ability of a system to adjust to climate change … to moderate potential damages, to take advantage of opportunities, or to cope with the consequences." Building children's adaptive capacity is likely to prove one of the most effective strategies to enhance the resilience of whole communities over time. After all, it is today's children who will bear the brunt of GCC.

While including children's specific needs and capacities in adaptation planning and actions may be perceived as a burden on governments, communities and organizations, Bartlett (2008) argues:

> there are strong synergies between what children need and the adaptations required to reduce or respond to more general risks … It has generally been found that neighbourhoods and cities that work better for children tend to work better for everyone, and this principle undoubtedly applies also to the adaptations…called for by climate change.
>
> (p. 514)

While taking a child-centered approach to building adaptive capacity can pay dividends for whole communities, there is a risk that simply looking to build children's adaptive capacity will result in one-size-fits-all approaches and, consequently, fail to address the range of underlying structural inequalities that many children face. This pitfall can be avoided by integrating climate risk and resilience building into existing child-centered community development projects working within specific sectors. Moving from an 'adaptive capacity' approach to a 'resilience' approach helps to broaden understandings of what children need and the multiple challenges they face in achieving their rights. A healthier, wealthier, better educated individual will have a higher adaptive capacity and, therefore, will be likely to have a greater level of resilience to the impacts of GCC (Dodman, Ayers and Huq, 2009).

Whereas adaptation can be seen as a series of actions designed to avoid or minimize the negative impacts of GCC, resilience should not be understood as a fixed state or an outcome. Rather, it should be seen as a process that helps communities cope with a variety of climatic and non-climatic shocks, and helps address the broader range of challenges that face communities in the Global South (Dodman, Ayers and Huq, 2009).

Resilience is "context specific, and will change over time as children, communities and institutions evolve" (Save the Children, 2013, pp. 3-4). While the approach can be an effective tool to address myriad challenges, the concept can be so broad that it becomes impractical. To avoid this, it is helpful to view resilience as a product of testing and creativity rather than something that can be developed in the same way across place and time. We need to let communities and individuals experiment with different tools and approaches to building their own resilience – and this will necessarily involve "safe failures" (Wilkinson, 2011, p. 162).

A key means of achieving these outcomes is to include resilience building options and opportunities within existing processes and programs. When compared to standalone adaptation projects, integrated actions are more likely to: be more sustainable (building on existing structures rather than working in parallel); promote local ownership and trust; help safeguard the outcomes of other development activities by introducing 'climate-smart' strategies; and ensure that adaptive actions are tailored to the specific local context and anchored in a sector already identified as significant to the community. Risk reduction and resilience building approaches also have the benefit of being flexible enough to target the specific vulnerabilities and inequalities of a specific group of children (or even a particular child), thereby avoiding the one-size-fits-all approach to adaptation (Mitchell and Bouchard, 2014).

Conclusion

It is clear that GCC presents an unparalleled challenge to the sustainability of human development, and that children, as a group, are particularly vulnerable to its impacts. Children are almost completely absent from the development and implementation of GCC policy and action at all levels. Yet, having a different range of vulnerabilities (often) distinct to other 'vulnerable groups,' their specific needs and capacities should be distinctly addressed within policy documents and development processes. Indeed, how children's vulnerabilities to GCC are shaped by layers of discrimination faced on a daily basis is also a key element of an effective, child-centered response. Adopting a child-centered approach to reducing climate risk and building resilience at local, national and international levels would, necessarily, increase the voice and agency of children in the decision-making processes that directly impact their levels of vulnerability to GCC. Initial implementation of this approach is likely to result in a highlighting of children as a homogeneously vulnerable group, thereby masking the important differences within and between age groups and erasing the impacts of other identifiers. However, as a first step, getting children on the agenda in the formation and implementation of climate policy is more important than highlighting the degrees of difference of vulnerability and capacity between children. Adults are, after all, the caretakers of the planet for today's children and future generations. It is our responsibility to ensure our institutions respect and protect children's rights – especially in the context of a changing climate.

References

Abramson, D., Garfield, R. and Redlener, I., 2007. *The Recovery Divide: Poverty and the Widening Gap Among Mississippi Children and Families Affected by Hurricane Katrina.* New York: Columbia University Mailman School of Public Health.

Bartlett, S., 2008. Climate change and urban children: impacts and implications for adaptation in low- and middle-income countries. *Environment and Urbanization*, 20, pp. 501–519.

Cho, S., Crenshaw, K. and McCall, M., 2013. Toward a field of intersectionality studies: theory, applications, and praxis. *Signs*, 38(4), pp. 785–810.

Child Rights Information Network, 2009. *Global Report on Laws Protecting Children from Age Discrimination.* London: CRIN.

Dodman, D., Ayers, J, and Huq, S., 2009. Building resilience. In World Watch Institute, ed, *State of the World 2009: Into a Warming World.* Washington DC.: World Watch Institute.

Doherty, T. and Clayton, S., 2011. The psychological impacts of global climate change. *American Psychologist*, 66(4), pp. 265–276.

Farrant, B., Armstrong, F. and Albrecht, G., 2012. Future under threat: climate change and children's health. *The Conversation* [online]. Available at: http://theconversation.com/future-under-threat-climate-change-and-childrens-health-9750 [Accessed 7 May 2013].

Frumkin, H., Hess, J., Luber, G., Malilay, J. and McGeehin, M., 2008. Climate change: the public health response. *American Journal of Public Health*, 98(30), pp. 435–445.

Global Humanitarian Forum, 2009. *Human Impact Report: Climate Change – The Anatomy of a Silent Crisis*. Geneva: Global Humanitarian Forum.

Government of Nepal, 2010. *National Adaptation Programme of Action to Climate Change* [pdf]. Available at: http://unfccc.int/resource/docs/napa/npl01.pdf [Accessed 10 September 2012].

Government of Nepal, (n.d.) *Adaptation to Climate Change: NAPA to LAPA* [pdf]. Available at: www.moenv.gov.np/cdkn/knowledge%20products/NAPA%20TO%20 LAPA.pdf [Accessed 10 September 2012].

Government of Bangladesh, 2008. *Bangladesh National Climate Change Strategy and Action Plan*. Dhaka: Ministry of Environment and Forests.

Haveman, R. and Smeeding, T., 2006. The role of higher education in social mobility. *The Future of Children*, 16(2), pp. 125–150.

InterClimate Network, 2012. *Kenya Impact Report 2008–2011: International Climate Challenge* [online]. Available at: www.blurb.co.uk/books/3228343-icc-impact-report-kenya-2008-2011?redirect=true [Accessed 12 August 2013].

International Energy Agency, 2011. *World Energy Outlook 2011*. Paris: IEA.

International Monetary Fund, 2012. *World Economic Outlook October 2012: Coping with High Debt and Sluggish Growth*. Washington, DC.: IMF.

Intergovernmental Panel on Climate Change, 2007. *Climate Change 2007: Synthesis Report*. Geneva: IPCC.

Lansdown, G., 2001. *Promoting Children's Participation in Democratic Decision-making*. Florence: UNICEF.

Lynch, K. and Baker, J., 2005. Equality in education: an equality of condition perspective. *Theory and Research in Education*, 3(2), pp. 131–164.

Mitchell, P. and Bourchard, C., 2014. Mainstreaming children's vulnerabilities and capacities into community-based adaptation to enhance impact. *Climate and Development*, 6(4), pp. 372–381.

Pelling, M., 2011. *Adaptation to Climate Change: From Resilience to Transformation*. London: Routledge.

Save the Children, 2007. *Legacy of Disasters: The Impact of Climate Change on Children*. London: Save the Children.

Save the Children, 2009. *Feeling the Heat: Child Survival in a Changing Climate*. London: Save the Children.

Save the Children, 2013. *Reducing Risk, Enhancing Resilience*. London: Save the Children.

Seballos, F., Tanner, T., Tarazona, M. and Gallegos, J., 2011. *Children and Disasters: Understanding Impact and Enabling Agency*. Brighton: IDS.

Sheffield, P. and Landrigan, P., 2011. Global climate change and children's health: threats and strategies for prevention. *Environmental Health Perspectives*, 119(3), pp. 291–298.

Tanner, E. and Finn-Stevenson, M., 2002. Nutrition and brain development: social policy implications. *American Journal of Orthopsychiatry*, 72, pp. 182–193.

Tanner, T., Garcia, M., Lazcano, J., Molina, F., Molina, G., Rodriguez, G., Tribunalo B. and Seballos, F., 2009. Children's participation in community-based disaster risk reduction and adaptation to climate change. *Participatory Learning and Action*, 60, pp. 54–64.

Tanner, T., 2010. Shifting the narrative: child-led responses to climate change and disasters in El Salvador and the Philippines' *Children & Society*, 24, pp. 339–351.

UNICEF and Plan International, (n.d.) *The Benefits of a Child-centred Approach to Climate Change Adaptation* [pdf]. Available at: www.childreninachangingclimate.org/database/plan/Publications/The-Benefits-of-a-child-centred-appraoch-to-climate-change-adaption.pdf [Accessed 14 August 2012].

UNICEF, 2007. *Climate Change and Children*. New York: UNICEF.

UNICEF, 2011. *Children's Vulnerability to Climate Change and Disaster Impacts in East Asia and the Pacific*. Bangkok: UNICEF.

UNICEF, 2012. *The State of the World's Children 2012: Children in an Urban World*. New York: UNICEF.

UNICEF, 2014a. Statistical tables. New York: UNICEF.

UNICEF, 2014b. *The State of the World's Children 2015: Executive Summary*. New York: UNICEF.

UNICEF, (n.d.) *Child-centered Development: the Basis for Sustainable Human Development*. New York: UNICEF.

UNICEF and Plan International, (2011) *The Benefits of a Child-centred Approach to Climate Change Adaptation* [pdf]. Available at: www.childreninachangingclimate.org/database/plan/Publications/The-Benefits-of-a-child-centred-appraoch-to-climate-change-adaption.pdf [Accessed 14 August 2012].

United Nations, 1969. *Vienna Convention on the Law of Treaties* [pdf]. Available at: https://treaties.un.org/doc/Publication/UNTS/Volume%201155/volume-1155-I-18232-English.pdf [Accessed 20 August 2014].

United Nations, 1989. *Convention on the Rights of the Child* [online]. Available at: www2.ohchr.org/english/law/crc.htm [Accessed 14 August 2012].

United Nations, 2014. *The Millennium Development Goals Report 2014* [pdf]. Available at: www.un.org/millenniumgoals/2014%20MDG%20report/MDG%202014%20English%20web.pdf [Accessed 20 August 2014].

UN Framework Convention on Climate Change, 2010. *Decision 1/CP.16 The Cancun Agreements: Outcome of the Work of the Ad Hoc Working Group on Long-term Cooperative Action under the Convention* [pdf]. Available at: http://unfccc.int/resource/docs/2010/cop16/eng/07a01.pdf#page=2 [Accessed 14 August 2012].

Wachs, T, 2000. Nutritional deficits and behavioral development. *International Journal of Behavioral Development*, 24, pp. 435–441.

Wilkinson, C., 2011. Social-ecological resilience: Insights and issues for planning theory. *Planning Theory*, 11(2), pp. 148–169.

World Bank, 2010. *World Development Report 2010: Development and Climate Change*, Washington D.C.: World Bank.

46 Banter from a Repressed Heart

James Elias Hamue Torres

I sit here at the keys of my message board
Having heartfelt 'motions and dreams of galactic hoodies
Calling teachers moms because that's how I feel.
Spent time after school because I'm worried about my education
See, kids in this nation aren't worried about grades;
The needs of my population aren't being portrayed.
There's ignorance and bliss and so many words
There's blindness and sadness with a pinch of suicidal depression
There's goals, there's needs, but people still don't succeed.

I still haven't learned to appreciate
Creating rifts between people who've stayed with me
Getting hugs because I want to see some affection
Because that's all I need.
Wonderin' how this is all we do, sleep eat hate repeat.
Been a preacher of hate
Still can't accept my fate
But after all this nonsense I still hide my heart.

I need a recorder because these conversations in my head keep goin' on repeat
So many variations
So many words I want to say
So many ideas pouring right out of my brain
From my heart.
Straight onto the page.

Little exposure to this predicament
Eyes still unable to predict this.
Yet I, the lonely individual
Have seen what little media has to offer
Sickness leads to heat for destruction
This planet of ours is no different
Global warming is the fever, we're the virus

Covering our eyes while we make this home sick
And this is the sad tale that describes our world changin'
The host kills the virus, or the virus kills the host.

All I ever wanted to do was eat Gummi Bears and string my fingers right through her hair
Cuddle up in that blanket
Close my eyes and stop thinking
Hold my heart and stop thanking
Tie my hands and stop resisting
But I'm punished.
Part of a generation that disregards notions
Of the impending change
The change that involves the fever, the inhabitants of this home
The change that regards a virus, endangering this place
The change that kills a host, something this world will face.

Figure 46.1 Jose Gonzalez, "Voz"

Index

Printed and bound by CPI Group (UK) Ltd, Croydon, CR0 4YY
17/10/2024
01775686-0017